The Development of Catalysis

The Development of Catalysis

A History of Key Processes and Personas in Catalytic Science and Technology

Adriano Zecchina
Salvatore Califano

For general information on our other products and services or for technical support, please contact our Customer Care Department within the United States at (800) 762-2974, outside the United States at (317) 572-3993 or fax (317) 572-4002.

Wiley also publishes its books in a variety of electronic formats. Some content that appears in print may not be available in electronic formats. For more information about Wiley products, visit our web site at www.wiley.com.

Library of Congress Cataloging-in-Publication Data:

Names: Zecchina, Adriano, 1936- | Califano, S.
Title: The development of catalysis : a history of key processes and personas in catalytic science and technology / Adriano Zecchina, Salvatore Califano.
Description: Hoboken, New Jersey : John Wiley & Sons, Inc., [2017] | Includes bibliographical references and index.
Identifiers: LCCN 2016039869| ISBN 9781119181262 (cloth) | ISBN 9781119181309 (pdf) | ISBN 9781119181293 (epub)
Subjects: LCSH: Catalysis–History. | Photocatalysis–History. | Biocatalysis–History. | Chemistry, Physical and theoretical–History.
Classification: LCC QD505 .Z43 2017 | DDC 541/.395–dc23 LC record available at https://lccn.loc.gov/2016039869

Typeset in 10/12pt, Warnock by SPi Global, Chennai, India

Printed in the United States of America

10   9   8   7   6   5   4   3   2   1

Contents

Preface

Alchemists and chemists have always known how to increase reaction rates by raising the temperature. Only much later, chemists realized that the addition to the reaction of a third chemical substance, the catalyst, could give rise to the same effect. Catalysis, whose discovery date is difficult to establish, goes back to the period of transition from alchemy to chemical science. Actually it is a type of science with a prominent interdisciplinary character. Several branches of scientific disciplines have contributed to its development, including chemical kinetics; inorganic, organic, and protein chemistry; material and surface science; advanced physical methods, and computer modeling. The early development of catalysis science received a great impulse from the contribution of many scientists active in the nineteenth century and in the first decade of twentieth century. To mention only a few, Berzelius coined the word "catalysis," van't Hoff and Arrhenius are remembered for their discoveries in the field of chemical kinetics, and then Davy, Dobereiner, Thenard, and Philips, for having discovered the catalytic properties of platinum. This group could be considered the fathers of heterogeneous catalysis. They were great scientists in a period of great discoveries and advancements, and their work has formed the basis of catalytic science. However, it is only with Sabatier and Ostwald in the first decade of the twentieth century that catalysis in its heterogeneous version became worthy of Nobel Prize recognition. Homogeneous catalysis came later with the seminal contributions of Roelen (use of cobalt carbonyls as catalysts in hydroformylation reaction) and of Wilkinson and Osborn (hydrogenation reactions using organometallic complexes as catalysts), and they are only a few who can be considered the initiators. The start of enzymatic catalysis could coincide with the isolation and crystallization of the first enzymes by Sumner (urease) and Northrop (pepsin) and the structural determination of lysozyme by Chilton Phillips. However, it is worth recalling that already in eighteenth century, Payen correctly recognized the catalytic action of a particular biological substance, which he termed "diastase," and then at the end of nineteenth century, Fischer had the remarkable intuition to use the so-called key and lock model to explain the selectivity of enzymes.

After these starts, the number of publications involving explicitly the word "catalysis" in all versions (heterogeneous, homogeneous, photo, and enzymatic) gradually increasing from few tens in the 1920s to several hundred in the 1950s and to more than ten thousands in the 2000s, which is a consistent fraction of all the documents having chemistry as a key word. This exponential growth is continuing today. In a parallel way the number of Nobel laureates directly or indirectly involved in catalysis greatly increased in the twentieth century, approaching the total number of about 40 in 2014. This number is only a small fraction of all valuable researchers who have contributed to the development of catalysis science and technology. In this book we document the importance and richness of many of these contributions. We describe the main branches of catalytic science (heterogeneous, homogeneous, and enzymatic) and how they evolved separately with little reciprocal contamination. Today the need to design selective and super-selective catalysts is being inspired by the research on enzymes and by the environmental issues that are gaining increasing consideration and contributing to fruitful contamination. As the development of catalysis and its industrial applications in the twentieth century are definitely a complex matter, the scope of this small book is more to give a picture of the major trends than a detailed illustration of all discoveries and contributions. We only underline catalysis science and technology are still evolving and can be a source of inspiration for the present and future generation of scientists.

15 June 2016

Adriano Zecchina
Torino

Salvatore Califano
Firenze

1

From the Onset to the First Large-Scale Industrial Processes

1.1 Origin of the Catalytic Era

Chemists have always known, even before becoming scientists in the modern term (i.e., during the long alchemist era), how to increase reaction rates by raising the temperature. Only much later on, they realized that the addition to the reaction of a third chemical substance, the catalyst, could give rise to the same effect.

Formerly the word "affinity" was used in chemical language to indicate the driving force for a reaction, but this concept had no direct connection with the understanding of reaction rates at a molecular level.

The first known processes involving reactions in solution accelerated by the addition of small amounts of acids are normally defined today as homogeneous catalysis. Experimental evidence for such processes dates back to the sixteenth century, when the German physician and botanist Valerius Cordus published posthumously in 1549 his lecture notes with the title *Annotations on Dioscorides.*

Valerius Cordus (1515–1544), born in Erfurt, Germany, organized the first official pharmacopoeia (φαρμακοποιΐα) in Germany. He wrote a booklet that described names and properties of medicaments, completing and improving the famous pharmacopoeia written by the Roman natural philosopher Pliny the Elder and listing all known drugs and medicaments. In 1527, he enrolled at the University of Leipzig where he obtained his bachelor's degree in 1531. During these years, he was strongly influenced by his father Euricius, author in 1534 of a systematic treatise on botany (*Botanologicon*). Valerius Cordus, after completing his training in the pharmacy of his uncle at Leipzig, moved in 1539 to Wittenberg University. As a young man, he also made several trips to Europe, the last one to Italy where he visited several Italian towns, including Venice, Padua, Bologna, and Rome. There he died in 1544 at the age of only 29 and was buried in the church of Santa Maria dell'Anima.

His role in pharmacy was based on the *Dispensatorium*, a text he prepared in 1546 that, using a limited selection of prescriptions, tried to create order in

The Development of Catalysis: A History of Key Processes and Personas in Catalytic Science and Technology,
First Edition. Adriano Zecchina and Salvatore Califano.
© 2017 John Wiley & Sons, Inc. Published 2017 by John Wiley & Sons, Inc.

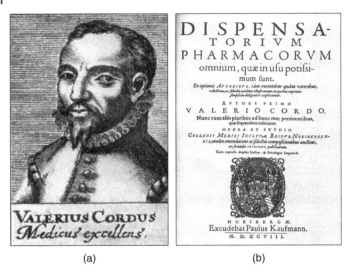

(a)   (b)

Figure 1.1 (a) Valerius Cordus, discoverer of ethyl ether formation from ethyl alcohol in the presence of an acid (oil of vitriol). (b) Cover page of *Dispensatorium Pharmacorum*. Images in the public domain.

the unsystematic corpus of medicaments existing at that time. Soon his dispensatory became obligatory for the complete German territory. In 1540 Cordus discovered ether and described the first method of preparing this special solvent in the *De artificiosis extractionibus liber*. Following a recipe imported to Europe from the Middle East by Portuguese travelers, he discovered how to synthesize ethyl ether by reacting oil of vitriol, *"oleum dulci vitrioli"* (sweet oil of vitriol), with ethyl alcohol (Califano, 2012, Chapter 2, p. 40). The synthesis was published in 1548 (Cordus, 1548) after his death and again later in the *De artificiosis extractionibus liber* (Cordus, 1561) (Figure 1.1).

He, of course, did not grasp the fact that the presence of an acid in the solution had a catalytic effect on the reaction. Only at the end of the eighteenth century did chemists realize that a few drops of acid or even of a base added to a solution could speed up reactions in solutions, giving rise to the era of homogeneous catalysis.

The chemical importance of these processes became evident only several years later, when the French agronomist and nutritionist Antoine-Augustin Parmentier (1737–1813) realized in 1781 that the addition of acetic acid accelerated the transformation of potato flour into a sweet substance. Parmentier was known for his campaign in which he promoted potatoes as an important source of food for humans not only in France but also throughout Europe (Block, 2008) (Figure 1.2).

(a)   (b)

Figure 1.2 (a) Augustin Parmentier (1737–1813) and (b) Anselme Payen (1795–1871) (images in the public domain). Parmentier discovered the accelerating action of acetic acid in the transformation of potato flour into a sweet substance. Payen attributed the starch transformation induced by few drops of sulfuric acid previously discovered by Constantin Kirchhoff to the concomitant action of a particular biological substance named *diastase*. Thus, we can consider him a true precursor of the modern *enzyme* science (vide infra: Chapter 8).

During the Seven Years' War, while performing an inspection at the first front lines, Parmentier, captured by a Prussian patrol, was sent on probation to the shop of a German pharmacist Johann Meyer, a person who became his friend and had a great influence on his scientific formation. After his return to Paris in the year 1763, he pursued his research in nutrition chemistry. His prison experience came back to his mind in 1772 when he proposed, in a contest sponsored by the Academy of Besançon, to use the potato as a convenient food for dysenteric patients, a suggestion that he soon extended to the whole French population. This suggestion, complemented in 1794 by the book *La Cuisinière Républicaine* written by Madame Mérigot, definitely promoted the use of potatoes as food for the common people first in France and subsequently over the entire continent. In 1772, he won a prize from the Academy of Besançon with memoirs in which he further emphasized the praise of the potato as a source of nutrients (Parmentier, 1773, 1774).

An additional early example of catalytic processes was found by the Russian chemist Gottlieb Sigismund Constantin Kirchhoff (1764–1833) born in Teterow in the district of Rostock, in Mecklenburg-Western Pomerania (Germany), who was working in St. Petersburg as an assistant in a chemist's shop. In 1811, he became the first person who succeeded in converting starch into sugar (corn syrup), discovering that the hydrolysis of starch in glucose was made faster by heating a solution to which he had added only a few drops of sulfuric acid (Kirchhoff, 1811a, b). This gluey juice was a kind of sugar, eventually named glucose. Kirchhoff showed at a meeting of the Imperial Academy of Sciences in St. Petersburg three versions of his experiments. He apparently discussed the problem with Berzelius who then told the Royal Institute in London about Kirchhoff's experiments, remarking upon the treatment with sulfuric acid.

At the suggestion of Sir Humphry Davy, members of the Royal Institution in London repeated his experiment and produced similar results. It was, however, only in 1814 that the Swiss chemist Nicolas-Théodore de Saussure showed that the syrup contained dextrose.

1.2 Berzelius and the Affinity Theory of Catalysis

The first who coined the name *catalysis* was Berzelius, one of the founders of modern chemistry, in 1836. Born in 1779 at Väversunda in Östergötland, Sweden, although in continual financial difficulties and suffering many privations, he was able to study at the Linköping secondary school and then enroll at Uppsala University to study medicine during the period between 1796 and 1801, thanks to the moral support of Jacob Lindblom, Bishop of Linköping. At Uppsala, he studied medicine and chemistry under the supervision of Anders Gustaf Ekeberg, the discoverer of tantalum and supporter of the interest in the chemical nomenclature of Lavoisier.

He worked then, as a medical doctor near Stockholm, until Wilhelm Hisinger, proprietor of a foundry, discovered his analytical abilities and decided to provide him with a laboratory where he could work on his research on looking for new elements.

In 1807, the Karolinska Institute appointed Berzelius as professor in chemistry and pharmacy. In 1808, he was elected as a member of the Royal Swedish Academy of Sciences and, in 1818, became secretary of the Academy, a position that he held until 1848. During his tenure, he revitalized the Academy, bringing it into a significant golden era (Figure 1.3).

In 1822, the American Academy of Arts and Sciences nominated him as Foreign Honorary Member, and in 1837, he became a member of the Swedish Academy. Between 1808 and 1836, Berzelius worked with Anna Sundström, who acted as his assistant (Leicester, 1970–1980).

Berzelius developed a modern system of chemical formula notation in which the Latin name of an element was abbreviated to one or two letters and superscripts (in place of the subscripts currently used today) to designate the number of atoms of each element present in the atom or molecule.

Figure 1.3 Jöns Jacob Berzelius (1779–1848) (image in the public domain), one of the founders of modern chemistry. He coined the word *catalysis*.

Berzelius discovered several new elements, including cerium and thorium. He developed isomerism and catalysis that owe their names to him. He concluded that a new force operates in chemical reactions, the *catalytic force* (Califano, 2012, Chapter 2, p. 42).

A first attempt to interpret the mechanism of catalysis was made by Berzelius who, in a report to the Swedish Academy of Sciences of 1835 published in 1836 (Berzelius, 1836a), had collected a large number of results on both homogeneous and heterogeneous catalytic reactions that he reviewed, proposing the existence of a "new catalytic force," acting on the matter. In 1836, he wrote in the *Edinburgh New Philosophical Journal* (Berzelius, 1836a):

> *The substances that cause the decomposition of H_2O_2 do not achieve this goal by being incorporated into the new compounds (H_2O and O_2); in each case they remain unchanged and hence act by means of an inherent force whose nature is still unknown… So long as the nature of the new force remains hidden, it will help our researches and discussions about it if we have a special name for it. I hence will name it the catalytic force of the substances, and I will name decomposition by this force catalysis. The catalytic force is reflected in the capacity that some substances have, by their mere presence and not by their own reactivity, to awaken activities that are slumbering in molecules at a given temperature.*
>
> *Berzelius J. J. Quoted by Arno Behr and Peter Neuber, Applied Homogeneous Catalysis, Wiley-VCH Verlag GMbH and &Co KGaA, 2012*

He coined also the word catalysis, combining the Greek words κατά (down) and λύσις (solution, loosening). According to Berzelius a catalyst was a substance able to start a reaction without taking part in it and thus without being consumed. In his famous paper of 1836 (Berzelius, 1836b), he wrote:

> *the catalytic power seems actually to consist in the fact that substances are able to awake affinities, which are asleep at a particular temperature, by their mere presence and not by their own affinity.*
> Berzelius J. J., the *Edinburgh New Philosophical Journal* XXI, 223, 1836c

In 1839, Justus von Liebig, one of the most important organic chemists of his time, tried an interpretation of catalysis based on the concept that a third body, the catalyst, added to the reactants, although not taking part in the reaction, was able to speed up the process (Liebig, 1839). After a few years, the German physicist and physician Julius Robert von Mayer, in the framework of his studies of photosynthetic processes, developed in 1845 a different interpretation of the catalytic mechanism. Mayer put forward the idea that the catalyst was able to release large amounts of *sleeping energy* that could allow the reaction to break out. Christian Friedrich Schönbein (1799–1868), discoverer of ozone, further developed the idea that the catalyst, without interacting with the reagents, could speed up the reaction producing intermediate products able to open new and faster paths to the reacting molecules. He asserted that a reaction is not a single process but occurs as the consequence of a time-ordered series of intermediate events (Schönbein, 1848). After some years, the German Friedrich Karl Adolf Stohmann (1832–1897) proposed the possibility that a catalyst could release energy to facilitate the reaction. He pointed out that catalysis is a process in which the energy released by the catalyst transforms into the motions of the atoms of the reacting molecules. These in turn reorganize themselves, giving rise to a more stable system by emission of energy (Stohmann, 1894).

The research of Ludwig Wilhelmy (1812–1864) complemented Berzelius's idea of the existence of substances activating the ability of chemical compounds to react. He found that the addition of inorganic acids made the inversion process of cane sugar easier. Augustus George Vernon Harcourt, who discovered the importance of acid catalysis in clock reactions (Shorter, 1980), reached the same conclusion (1834–1919). In 1896, the Scottish mathematician William Esson (1838–1916) interpreted Harcourt's data in terms of a differential equation.

1.3 Discovery of the Occurrence of Catalytic Processes in Living Systems in the Nineteenth Century

In the same period, it became evident that catalytic effects also occur in living systems. Actually, the fact that living organisms contain substances able to facilitate or even trigger chemical reactions was known for a long time, but was never considered the consequence of catalytic processes occurring in the body. As documented in the chapter 8 devoted to enzymes, the use of yeasts for the production of wine was, for instance, a very old technique, already known to the Bronze Age Minoan and Mycenaean civilizations. The ability of the acid juice contained in the stomach of animals to digest meat and even bones was demonstrated by the French scientist René Antoine Ferchault de Réaumur (1683–1757), and later by the Italian biologist Lazzaro Spallanzani (1729–1799) and by the Scottish physician Edward Stevens (1755–1834).

The occurrence of different mechanisms involving living substances and contributing to orient the course of a reaction was proved by a series of fundamental research at the beginning of the nineteenth century. In 1833, Anselme Payen (1795–1871) and Jean-François Persoz (1805–1868) attributed the starch transformation, discovered by Kirchhoff, to the action of a particular biological substance. They called it *diastase* and further proved that at 100°C it loosed its catalytic activity (Payen and Persoz, 1833).

Anselme Payen studied chemistry at the École Polytechnique under the supervision of the chemists Louis Nicolas Vauquelin and Michel Eugène Chevreul. Besides the discovery in 1833 of the first enzyme (diastase), the synthesis of borax from soda and boric acid and a process for refining sugar can be attributed to him. He also isolated and named the carbohydrate cellulose (Payen, 1838). In 1835 Payen became a professor at the École Centrale and later at the Conservatoire National des Arts et Métiers in Paris. His friend Jean-François Persoz was *préparateur* of Louis Thénard at the Collège de France, before becoming professor of chemistry at the University of Strasbourg. In 1830 he had become director of the École de Farmacie and in 1850 succeeded Jean-Baptiste Dumas at the Sorbonne. Persoz studied the solubility of chemical compounds and their molecular volumes. His collaboration with Payen led to the discovery of diastase and of its presence in human saliva.

In 1835, in collaboration with Jean-Baptiste Biot, he showed how to follow experimentally the inversion of cane sugar, simply observing with a polarimeter the variation of its rotatory power after acidification.

The German Johann Wolfgang Döbereiner (1780–1849), who became famous for his discovery of similar triads of elements that paved the route to Mendeleev's organization of the elements in the famous periodic table, also investigated starch fermentation. In 1822, he was one of the first to observe the fermentative conversion of starch paste into sugar and gave a correct explanation of alcoholic fermentation, finding that starch transforms into alcohol, only after conversion to sugar (Döbereiner, 1822). Döbereiner, son of a coachman, had in his youth a poor education. Despite very poor schooling, he succeeded in attending the University of Jena where he eventually reached the position of professor.

In the field of catalysis, he worked on the use of platinum as catalyst. For his discovery of the action of solid catalysts, he is considered as one of the initiators of heterogeneous catalysis.

In 1877 the German physiologist Wilhelm Kühne (1837–1900), a pupil of several outstanding chemists and physiologists of the time, including Claude Bernard in Paris, isolated trypsin from gastric juice (Kühne, 1877) and coined the word ἐνζυμων (enzumon), enzyme, from the Greek, ἐν in and ζυμων ferment, to describe cellular fermentation.

1.4 Kinetic Interpretation of Catalytic Processes in Solutions: The Birth of Homogeneous Catalysis

The first interpretation of catalytic events at the beginning of the nineteenth century dealt mainly with reactions occurring in solution. At that time the affinity concept dominated the interpretation of chemical processes. This idea, inherited from the alchemist's vision of the interaction between chemical elements or compounds, formally corresponded to the attraction between human beings. This metaphoric explanation of catalysis, however, did not satisfy the members of the new branch of chemical physics, educated by their training in mechanics and thermodynamics to a mechanistic approach toward the interpretation of chemical reactions. The route to this new comprehension of chemical reaction was paved by the pioneering work of Ludwig Ferdinand Wilhelmy (1812–1864), usually credited for publishing the first quantitative study of chemical kinetics.

Wilhelmy, born in 1812 at Stargard in Pomerania, studied pharmacy in Berlin. After a period spent in an apothecary shop, he studied chemistry and physics at Berlin, Giessen, and Heidelberg, where he obtained his PhD in 1846.

In his life Wilhelmy was always an amateur, not bound to the university system. He conducted the largest part of his research in his private house, a villa that he reorganized as his private laboratory. Nevertheless, he was highly respected in the German physical society that he had founded with Heinrich Gustav Magnus (1802–1870). Together with Paul du Bois-Reymond, Rudolf Clausius, Hermann von Helmholtz, and Carl Wilhelm Siemens, he was considered by all his friends as a leader of the young European chemical physics. After traveling chiefly in Italy and Paris, he returned to Heidelberg and became a *Privatdozent* in 1849. He remained at the university for only five years.

In 1850 Wilhelmy, in the framework of a series of polarimetric research, studied the inversion of cane sugar catalyzed by inorganic acids and proved experimentally that this reaction leads to the conversion of a sucrose solution into a 1:1 mixture of fructose and glucose (Figure 1.5). Under the assumption that the initial velocity of the reaction is proportional to the concentration of both the cane sugar and the acid and counting the time from the moment in which the sugar solution is brought in contact with the acid, Wilhelmy succeeded in describing the time evolution of the process in terms of the differential equation

$$dZ/dt = MZS$$

where Z and S are the amount (concentration) of sugar and acid at time t, respectively, whereas M is a "velocity coefficient," which is constant over a large time interval (Califano, 2012, Chapter 2).

Wilhelmy wrote in his paper that the

> *process is certainly only one member of a greater series of phenomena which all follow general laws of nature*

and that these laws can be expressed mathematically. The time, however, was not yet mature to appreciate the importance of his work, and it remained practically ignored for a long time, until Wilhelm Ostwald, the true father of modern chemical physics, realized its importance and even developed a quantitative analytical method to measure the strength of the acids from their ability to catalyze the sugar inversion.

The English chemist George Vernon Harcourt (1834–1919) complemented Wilhelmy's research (Figure 1.5). In the early 1860s Harcourt embarked on a research project on the rates of chemical reactions. He settled on two reactions for which, during definite time intervals, the amount of chemical change could be accurately measured. In close partnership with William Esson, mathematical fellow and tutor of Merton College, he studied the acid-catalyzed clock reaction of iodide and hydrogen peroxide (Harcourt and Esson, 1866a) as well as the oxidation of oxalic acid with potassium permanganate (Harcourt and Esson, 1866b), showing that the reaction rate was proportional to the concentration of reactants present. Their work for the first time gave detailed treatments of the kinetics of different types of reactions, anticipating several later formulations of equilibrium reactions. In 1912, when they both were well in their seventies, they again collaborated on the effect of temperature on the rates of chemical reactions (Harcourt and Esson, 1913). An interesting outcome of this work is that they predicted a "kinetic absolute zero" at which all reaction ceases; their value of $-272.6°C$ is in remarkable agreement with the modern value of $-273.15°C$.

The results obtained by Wilhelmy and Harcourt were later formalized by the chemist Peter Waage and his brother-in-law Cato Maximilian Guldberg as the law of mass action (Figure 1.4).

The Norwegian mathematician and chemist Cato Maximilian Guldberg (1838–1902) entered the University of Christiania in 1854. He worked

Figure 1.4 From left to right: Cato Maximilian Guldberg (1838–1902) and Peter Waage (1833–1900) (image in the public domain). They jointly formulated the famous law of mass action concerning the variation of equilibrium in chemical reactions that is the milestone of chemical kinetics.

independently on advanced mathematical problems, and his first published scientific article won the Crown Prince's Gold Medal in 1859. In 1862, he became professor of applied mechanics and professor at the Royal Military College the following year. In 1869, he developed the concept of "corresponding temperatures" and derived an equation of state valid for all liquids of certain types.

His friend Peter Waage (1833–1900) was also born in Norway. He attended the University of Christiania and passed his matriculation examination in 1854, the same year as Guldberg. After graduation in 1859, in 1861 the University of Christiania appointed him as lecturer in chemistry and promoted him professor in 1866. He became Guldberg's brother-in-law, marrying in 1870 one of Guldberg's sisters.

Cato Guldberg and Peter Waage's names normally occur together, not because of their family relations but for their joint discovery in 1864 of the famous law of mass action, concerning the variation of equilibrium in chemical reactions that is the milestone of chemical kinetics.

For a generic reaction

$$\alpha A + \beta B \rightleftarrows \gamma C + \delta D \tag{1.1}$$

the velocity of the direct reaction is equal to that of the inverse reaction, and both are proportional to the concentrations of the reagents, according to the equations

$$v_{dir} = k_{dir}[A]^{\alpha}[B]^{\beta} \text{ and } v_{inv} = k_{inv}[C]^{\gamma}[D]^{\delta} \tag{1.2}$$

where square brackets indicate concentrations. By equalizing the two velocities $v_{dir} = v_{inv}$, they obtained the relationship

$$\frac{[A^{\alpha}][B^{\beta}]}{[C^{\gamma}][D^{\delta}]} = K \tag{1.3}$$

well known to all first-year chemistry students (Guldberg and Waage, 1864).

An important step toward the understanding of the external factors influencing the reaction rates was realized by the publication of papers by Marcellin Berthelot and his student Léon Armand Pean Saint-Gilles (Berthelot and Saint-Gilles, 1862) concerning the kinetics of the esterification reactions of the type

$$RCOOH + R'OH \rightleftarrows RCOOR'$$

for which they demonstrated that the direct reaction rate is proportional to the product of the concentration of the two reactants (Califano, 2012, Chapter 2).

The Parisian chemist, science historian, and politician Pierre Eugène Marcellin Berthelot (1827–1907), author in 1854 of a PhD thesis *Sur les combinaisons de la glycérine avec les acides*, in 1859 became professor of organic chemistry at the École Supérieure de Pharmacie and in 1865 at the Collège de France (Figure 1.5). He was also involved in social and political activities: he was general inspector of higher education in 1876, life senator in 1881, minister of public instruction in 1886, and minister of foreign affairs in 1895–1896.

Berthelot was an adversary of the *vis vitalis* theory supported by Berzelius, the most important chemist of the time, who maintained that the formation of organic substances was controlled by interactions different from those occurring in the inorganic world (Berthelot, 1860). Berthelot proved instead with the synthesis of several hydrocarbons, natural fats, and sugars that organic compounds obey the same laws that control the formation of the inorganic compounds. His opposition to the vitalistic approach and his belief that the organic world was controlled by the same mechanical laws that operate in the universe gave rise to his interests in thermochemistry and calorimetry, producing an enormous number of experiments and two books on these arguments, *Essai de mécanique chimique fondée sur la thermochimie* (Berthelot, 1879) and *Thermochimie* (Berthelot, 1897).

Figure 1.5 (a) Ludwig F. Wilhelmy (1812–1864), (b) George Vernon Harcourt (1834–1919), (c) Marcellin P. Berthelot (1827–1907), and (d) Leopold Pfaundler von Hadermur (1839–1920). They are the main protagonists of the onset of chemical kinetics. *Source*: (a) The Wilhelmy image reuse is free because it is of uncertain source and can be considered as an orphan work. (b) The Harcourt image is in the public domain. (c) Author of Berthelot image (public domain): Magnus Manske. (d) Hardemur image is by courtesy of Oesper Collections in the History of Chemistry, University of Cincinnati.

Berthelot was also convinced that reactions producing heat (exothermic reactions) are spontaneous, whereas those absorbing heat (endothermic) are not. Berthelot's idea, even if plausible, was not correct, since there are spontaneous reactions that are not exothermic as well as reactions proceeding spontaneously and absorbing heat from the external world.

His research on the heat of reaction led him to study the theory of explosives (Berthelot, 1872). In the important book *La chimie au moyen âge* (1893), he wrote the following general reflection about chemistry:

> *La chimie est née d'hier: il y a cent ans à peine qu'elle a pris la forme d'une science moderne. Cependant les progrès rapides qu'elle a faits depuis ont concouru, plus peut-être que ceux d'aucune autre science, à transformer l'industrie et la civilisation matérielle, et à donner à la race humaine sa puissance chaque jour croissante sur la nature. C'est assez dire quel intérêt présente l'histoire des commencements de la chimie. Or ceux-ci ont un caractère tout spécial: la chimie n'est pas une science primitive, comme la géométrie ou l'astronomie; elle s'est constituée sur les débris d'une formation scientifique antérieure; formation demi-chimérique et demi-positive, fondée elle-même sur le trésor lentement amassé des découvertes pratiques de la métallurgie, de la médecine, de l'industrie et de l'économie domestique.*

In a period in which the legacy of the mechanistic philosophers was overwhelming all other theoretical approaches, it was natural to describe kinetic processes in terms of collisions between particles. Already in 1620, at the beginning of the seventeenth century, Bacon (1620; Rossi, 1978) had used the concept of intestine motion to explain chemical processes.

Similarly Franciscus de la Boë, better known as Franciscus Sylvius (1614–1672), professor of medicine in Leyden in 1658, had also suggested to his friend René Descartes that heat might correspond to some kind of "intestine" motion of either the molecules or the underlying ether, a view later supported in 1798 by Count Rumford and in 1799 by Humphry Davy. The discussion of Bacon's vitalistic idea of the "intestine motion" played an important role in the chemistry of that time up to the beginning of the twentieth century.

The association of heat with the translational motions of molecules was, however, due essentially to the development of the kinetic theory of gases (Herzfeld, 1925).

The true origin of the kinetic theory dates back to the eighteenth century to the *Hydrodynamica* published by Daniel Bernoulli in 1738 (Bernoulli, 1738) in which he maintained that gases consist of great numbers of molecules moving in all directions and that their impact on a surface causes the gas pressure.

Despite the attempts made by both John Herapath (1821–1847) (Herapath, 1821) and John Waterston in 1846 (Waterston, 1893), this theory did not attract general agreement until the 1850s. In the second part of the century, it was revived by Krönig (1856) and Clausius (1857) in Germany as well as by Joule (1851) and Maxwell (1860) in England.

A reaction mechanism based on molecular collisions was actually proposed in 1867 by a young Austrian physicist, Leopold Pfaundler von Hadermur (1839–1920), professor at the University of Innsbruck from 1867 to 1881 and then at Graz, where he succeeded Boltzmann (Califano, 2012, Chapter 2, p. 29) (Figure 1.5). He applied Boltzmann's kinetic theory of gases to equilibrium reactions, assuming that the rate of the direct and of the inverse process was the same (Pfaundler, 1867).

At the age of 28, he published his seminal paper on the application of the kinetic theory of matter and heat to chemical reactions. In 1887, he became a full member of the Vienna Academy of Sciences.

Pfaundler was the first to rationalize the law of mass action in terms of the number and frequency of molecular collisions and anticipated significant aspects of both the collision and the transition-state theories of chemical kinetics via his concepts of critical threshold energies and collision complexes. According to him, not all molecules had the same amount of internal and translational energy. Therefore only a limited number of molecular collisions were effective in determining the reaction, either by forming or by dissociating molecules.

Pfaundler was also the first to interpret chemical affinity in pure mechanistic terms, relying primarily on the work of the French chemist Henri Sainte-Claire Deville and his coworkers on the experimental behavior of equilibrium systems of the type

$$Heat + CaCO_3(s) \rightleftarrows CaO(s) + CO_2(g)$$

involving either solid or gaseous dissociation and concerning the general phenomenon of reversible reactions (Sainte-Claire Deville, 1856, 1865, 1869). On the basis of the analogy of the behavior of equilibrium resulting from the thermal dissociation of solids or from the evaporation of pure liquids, he concluded that, just as with vapor pressure, the dissociation increased with increasing temperature and decreased with decreasing temperature (Sainte-Claire Deville, 1866, 1867).

Pfaundler's work remained practically unknown, until it aroused the attention of the German thermochemist Alexander Nicolaus Franz Naumann, professor at the University of Giessen, who extensively quoted it in a review on dissociation phenomena (Naumann, 1867). In 1868 August Horstmann (1842–1929) attempted to quantify Pfaundler's qualitative arguments by using a probability distribution to calculate the change in the degree of dissociation of various vapors as a function of temperature (Horstmann, 1868) and tried to develop a quantitative theory of dissociation using the kinetic theory of gases (Horstmann, 1873).

Pfaundler's views consequently were criticized after 1870 by several scientists including Horstmann, Pattison Muir (1848–1931) (Pattison Muir, 1885), Bancroft (1861–1953) (Bancroft, 1897), and Pierre Duhem in 1898 (Duhem, 1898), all upholders of a purely thermodynamic approach based on either the maximization of the entropy function or on minimization of the Gibbs free energy. Pfaundler published a rejoinder to Horstmann's critics (Pfaundler, 1876), not fully convincing for the scientific community, including the same Naumann in 1882 who, after a first enthusiastic agreement to Pfaundler's ideas, became disillusioned by 1873 with the kinetic approach, in large part because he felt that it failed to explain why pure solids did not exert a mass action effect.

Finally, the Dutch chemist Jacobus Henricus van't Hoff (1852–1911) mentioned Pfaundler's name in the introduction to the first edition of his *Études de dynamique chimique* of 1884.

The most important representative of the history of science among the critics of Pfaundler was undoubtedly the French philosopher and physicist Pierre Maurice Marie Duhem (1861–1916), an important member of the French cultural milieu of the end of the nineteenth century. In his youth, Duhem was profoundly influenced by his teacher Jules Moutier, an ingenious theorist who had published a number of texts, including *La thermodynamique et ses principales applications*.

In 1882 Duhem enrolled at the École Normale Supérieure where he received a *license* in mathematics and another in physics at the end of the academic year 1883–1884. During the academic year 1884–1885, Duhem presented a doctoral thesis in physics entitled *Le potentiel thermodynamique et ses applications à la mécanique chimique et à l'étude des phénomènes électriques*, which was rejected by the doctoral committee, probably a political decision. The prestigious French scientific publisher, Hermann, published, however, a version of the thesis the following year. At a time when French scientists were predominantly liberal and anticlerical, Duhem was instead openly conservative and deeply religious.

Duhem was one of the first to appreciate the work of Josiah Willard Gibbs, writing the earliest critical examination of Gibbs's *On the Equilibrium of Heterogeneous Substances* in 1887. In the mid-1890s, Duhem published his first essays on the history of science that led him in 1904 to a new understanding of the history of science, considered as a continuity between medieval and early modern science. This path culminated in important historical works such as the *Études sur Léonard de Vinci* and the *Le système du monde*.

The true father of the theoretical approach to catalysis in terms of the kinetic theory developed by Maxwell and Celsius was, however, Jacobus van't Hoff, the first chemist able to assign to chemical physics the structure of the theoretical core of modern chemistry.

Jacobus van't Hoff (1852–1911), after graduation at the Polytechnic of Delft in 1871, studied mathematics at Delft and then attended the laboratory of Kekulé at Bonn and that of Wurtz in Paris. In 1874, he obtained a PhD at Utrecht. In 1876, he was an assistant at the Veterinary College of Utrecht and the following year at the University of Amsterdam. Only in 1878, when already known all over Europe for his theory of the stereochemistry of the carbon atom, he was promoted to the position of professor of chemistry, mineralogy, and geology, a position that he maintained for 18 years until he moved to Berlin as honorary professor and member of the Real Academy of Prussia. A romantic dreamer, lover of music and poetry, van't Hoff was a convinced supporter of the importance of fantasy in scientific research. In the inaugural lecture *Verbeeldingskracht in de Wetenschap* (the power of imagination in science) that he gave at the University of Amsterdam, he defended the role of imagination in scientific investigation.

Trained as an organic chemist, he was one of the first to become interested in chemical physics, thanks to his excellent preparation in mathematics and physics.

Van't Hoff deservedly joined the chemical physics community with the book *Études de dynamique chimique* (1884), in which he faced the problem of identifying the conditions that control the equilibrium of reversible reactions (van't Hoff, 1884a, b) (Figure 1.6).

(a)   (b)

Figure 1.6 (a) Jacobus H. van't Hoff (1852–1911), awarded the 1901 Nobel Prize in Chemistry (image in the public domain: author Nicola Persheid), and (b) Svante Arrhenius (1859–1927), awarded the 1903 Nobel Prize in Chemistry (image in the public domain). They formulated the fundamental laws of chemical kinetics.

In 1886 van't Hoff published a new text in French entitled *L'équilibre chimique dans l'etat dilué gazeux ou dissous* (van't Hoff, 1886) that presented his own ideas on the chemical physics of diluted solutions (van't Hoff, 1887).

The equation

$$k = Ae^{-E_a/RT} \tag{1.4}$$

proposed by van't Hoff in 1884 (Califano, 2012, Chapter 2, p. 33) is universally known as the Arrhenius equation, since Svante Arrhenius was the first to offer in 1889 its physical interpretation (Arrhenius, 1889) (Figure 1.6).

Arrhenius suggested that, in order for a reaction to take place, the reacting molecules have to possess energy greater than a limiting value, what he called *activation energy* E_a. At the temperature T, the fraction of molecules possessing a kinetic energy larger than E_a is defined by the statistical distribution law of Boltzmann and is proportional to the factor $e^{-Ea/RT}$. In the Arrhenius equation, the fraction of free energy available to give rise to the reaction is thus only the one superior to the value E_a.

Arrhenius also supplied a convenient graphical method to evaluate k_a. This procedure is normally used to evaluate the activation energy in homogeneous and heterogeneous processes.

Van't Hoff also studied the effect of temperature on the equilibrium constant of reversible reactions and formulated the famous van't Hoff isochore. From this equation, the conclusion can be derived that in a reversible reaction a shift of the equilibrium tends always to compensate the temperature variation. At lower temperature, the equilibrium shifts in the direction that produces heat, whereas a temperature increase produces the opposite effect (van't Hoff, 1898). This conclusion is in reality a particular case of the more general principle formulated in 1885 by the French chemist Le Châtelier (1850–1936) that states that *each system tends to counteract any change imposed from the exterior by minimizing its effect* (Le Châtelier, 1884).

In 1884, van't Hoff laid down the mathematical basis of chemical kinetics, starting from the idea that each reaction is the sum of a series of elementary events in which the molecules collide and give rise to a reaction (van't Hoff, 1884a, b). The probability that the collision gives origin to a reaction is greater than the concentration of the reacting species. In the case of a monomolecular reaction in which the reaction rate depends on the concentration c of a single species, he formulated the differential equation

$$v = -\frac{dc}{dt} = kc \tag{1.5}$$

where k is a constant that he named *rate constant* because it represents the concentration decrease per unit time for a unitary concentration. The analogy between diluted solutions (van't Hoff, 1894) and ideal gases allowed him to extend the second principle of thermodynamics to solutions and to develop the fundamental relationships controlling the displacement from equilibrium of a reaction with temperature in the form

$$\frac{d \ln K}{dT} = \frac{\Delta H}{RT^2} \tag{1.6}$$

Integration of this equation allowed van't Hoff to obtain the exponential dependence of the reaction rate constant k (Arrhenius equation) from the inverse temperature.

The dependence of the rate constant on temperature had actually been described even before van't Hoff and Arrhenius by John J. Hood on the basis of his experiments on the oxidation of ferrous sulfate by potassium dichromate (Hood, 1878), but he gave no valid explanation of this effect. As van't Hoff said in the 1884 paper (van't Hoff, 1884a, b),

> the equilibrium is to be regarded as a result of two changes taking place with the same velocity in opposite directions.

In 1850 the English chemist Alexander William Williamson (1824–1904) advanced a similar idea, when commenting on the formation of ethers by the reaction of sulfuric acid and alcohols (Williamson, 1850).

Reversible reactions had already been studied by Berthollet in the framework of his research on affinity, reaching the conclusion that chemical reactions do not always proceed up to the end but often reach an equilibrium situation that depends on the amount of reactants involved. Almost at the same time, the Italian chemist Faustino Malaguti (1802–1878) (Malaguti, 1853) also stated that equilibrium is reached when the rates of the two opposite reactions equalize. Malaguti, born near Bologna, joined the University of Bologna where he graduated in pharmacy. Banished for political reasons from Italy, he immigrated to Paris where he became an assistant at the Gay-Lussac laboratory of the École Polytechnique. Malaguti is better known for his research in chemical physics, concerning affinity and chemical equilibria. With this research he anticipated the law of mass action of Guldberg and Waage, who enthusiastically quoted his work.

1.5 Onset of Heterogeneous Catalysis

If van't Hoff played a fundamental role in the development of the theory of reaction rates in solution, no less important was the research of the second man of genius of nineteenth-century chemistry, Wilhelm Ostwald (1853–1932). Ostwald was the leading figure in the study of chemical affinity, who faced the central theoretical question in the chemistry of his time, establishing the theoretical basis of heterogeneous catalysis and even encouraging its industrial applications. Ostwald received the Nobel Prize award in 1909.

At the beginning of the nineteenth century, enough experimental evidence had been already accumulated on the fact that some solid compounds, in particular platinum, could affect the reaction rate without being consumed, that is, they could be true heterogeneous catalysts able to facilitate or even trigger chemical reactions.

Earlier, in 1815, Humphry Davy, during his research that led to the miners' safety lamp, discovered that a combustible gas could be oxidized by atmospheric oxygen on the surface of a platinum wire without the production of a flame and yet with the emission of enough heat to keep the platinum incandescent (Figure 1.7). In 1817, he prepared a platinum sponge able to absorb large quantities of gas and realized that, in the presence of finely divided platinum, alcohol vapors transformed in acetic acid (Davy, 1820). In 1818, the French chemist Louis Jacques Thénard discovered that finely divided Pt favors the decomposition of H_2O_2 into water and oxygen (Thénard, 1818) (Figure 1.7). Furthermore, in 1823, Johann Wolfgang Döbereiner, the German chemist later famous for the theory of triads that gave origin to the periodic system of the elements, noticed that mere contact with Pt powder causes the formation of water from hydrogen and oxygen already at room temperature without alteration of the metal (Figure 1.7).

Döbereiner's experiment (Döbereiner, 1823) attracted the attention of the Italian physicist Ambrogio Fusinieri (1775–1853) who, in the period from 1823 to 1826, researched into the catalytic activity of platinum and suggested that a solid layer of gas adsorbed on the platinum surface was continuously rebuilt as the gas was consumed in the combustion (Fusinieri, 1824). According to him, the platinum catalyst acted like a candlewick with the laminae burning like candle wax. As an explanation of the burning of the concrete laminae, he proposed the concept of "native caloric" (Robertson, 1975). Even Michael Faraday was involved in the study of the catalytic activity of platinum, showing its ability to recombine hydrogen and oxygen obtained from the electrolysis of water (Faraday, 1834). In his monumental paper of 1834, he proposed the idea (increasingly supported since the research of Langmuir) of simultaneous adsorption of both reactants on a platinum surface.

In 1834, he proposed that the reactants have to adsorb simultaneously at the surface but did not offer a reasonable explanation for the process. Later Ostwald gave a more reasonable explanation, suggesting that a catalyst does not influence the thermodynamic equilibrium of reactants and products but only affects the rate of the chemical reactions.

Platinum in the form of wires in sponges became in a few years the most important solid catalyst of the time. In 1824, William Henry, author of the law describing the dissolution of gas in liquids, discovered the inverse process, the one in which ethylene would stop the catalytic action of platinum on the hydrogen–oxygen mixture (Henry, 1824). In 1831, Peregrine Phillips, a vinegar manufacturer in Bristol, England, patented a new process (British

(a) (b)

(c)

Figure 1.7 (a) Humphry Davy (1778–1829) (image in the public domain), (b) Louis Jacques Thénard (1777–1857) (image in public domain), and (c) Johann Wolfgang Döbereiner (1780–1927) (image in the public domain: authors Carl A. Schwerdgeburth, engraver, and Fritz Ries, painter). Together with Peregrine Phillips, this group of contemporary scientists discovered in the period 1818–1831 the catalytic properties of platinum in gas oxidation (Davy), hydrogen peroxide decomposition (Thénard), hydrogen–oxygen reaction (Döbereiner), and SO_2 to SO_3 oxidation reaction (Phillips).

Patent No. 6096), concerning the instantaneous union of sulfur dioxide with atmospheric oxygen when passing the mixture over platinum heated to a strong yellow heat. Sulfur trioxide formed rapidly when in contact with water to produce concentrated sulfuric acid. In 1838, the French chemist Charles Frédéric Kuhlmann (1803–1881), professor at the University of Lille and owner since 1829 of a chemical company, later merged into the Pechiney–Ugine–Kuhlmann group, producing sulfuric acid, succeeded in transforming nitric oxide into ammonia by using a Pt sponge in the presence of hydrogen. Kuhlmann (1803–1881), born in Colmar, started his scientific career at the University of Strasbourg. Upon arriving in Lille, he gave lectures on chemistry and soon founded a society for the production of sulfuric acid. In a little time, he also started on the production of hydrochloric acid, sodium sulfate, nitric acid, and chlorine in new plants.

In 1863, Henry Debus showed likewise that methylamine is produced when a mixture of hydrogen and hydrocyanic acid is passed over platinum black heated to 110°C (Debus, 1863). A similar reaction, involving the synthesis of benzyl amine, was later suggested by Mendius (1862), normally referred to as the Mendius reaction. Only few years later, in 1874, Prosper De Wilde (1835–1916), professor of chemistry at the University of Brussels, discovered that acetylene could be hydrogenated to ethylene and then to ethane over a platinum catalyst (De Wilde, 1874).

Significant support to the theory of catalysis based on the idea that the catalytic process was due to an adsorption mechanism came from the research of the French chemist Jacques Duclaux (1877–1978). The same idea was held by the French chemist Henri Moissan, who won the 1906 Nobel Prize in Chemistry for his discovery of fluorine, studying in collaboration with Charles Moureu (1863–1929) the interaction of acetylene with finely divided nickel, iron, and cobalt metals (Moissan and Moureu, 1896). He was convinced that acetylene was adsorbed in the metal pores where, by the effect of heat, a pyrolytic reaction took place, creating a mixture of carbon, benzene, and hydrogen.

The following year the French chemist Paul Sabatier (1854–1941), professor at the University of Toulouse, in collaboration with the abbot Jean Baptiste Senderens (1856–1937) proved that Moissan's conclusions were wrong (Califano, 2012, Chapter 2, p. 44), since ethylene in contact with the metal produces ethane and not hydrogen. From this result they deduced that the action of nickel was to catalyze the breaking of the ethylene double bond, in other words to give origin to a true chemical reaction (Sabatier and Senderens, 1897). This result drove Sabatier and Senderens to reach the conclusion that catalysis was not, as maintained by Ostwald, a purely physical process but that finely divided metals were able to absorb large amounts of gas and that this property was very specific, confirming a selective action of a pure chemical nature (Figure 1.8).

(a) (b)

(c)

Figure 1.8 (a) Jean Baptiste Senderens (1856–1937), (b) Paul Sabatier (1854–1941), and (c) Wilhelm Ostwald (1853–1932) (images in the public domain). They are considered as the originators of catalysis science. Paul Sabatier obtained the 1909 Nobel Prize *for his method of hydrogenating organic compounds in the presence of finely disintegrated metals.* Wilhelm Ostwald received the 1912 Nobel Prize *in recognition of his work on catalysis and on the fundamental principles governing chemical equilibria and rates of reactions.*

Paul Sabatier (1854–1941), born at Carcassonne, entered in 1874 the École Normale Supérieure where he graduated three years later. He then moved to the Collège de France in 1878 as assistant to Berthelot and received the degree of doctor of science in 1880.

He became professor of chemistry in 1884 at the University of Toulouse, a post that he retained until his retirement in 1930. He was ever faithful to Toulouse and turned down many offers of attractive positions elsewhere, notably the succession to Moissan at the Sorbonne in 1908.

Sabatier's earliest research concerned the thermochemistry of sulfur and metallic sulfates, the subject for his thesis leading to his doctorate, and in Toulouse, he extended his physicochemical investigations to sulfides, chlorides, chromates, and copper compounds. He also studied the oxides of nitrogen and nitrosodisulfonic acid and its salts and carried out fundamental research on partition coefficients and absorption spectra.

In 1834, Michael Faraday had proposed the idea that catalytic processes occur on a platinum surface thanks to the simultaneous adsorption process of the reactants (Faraday, 1834). As soon as Sabatier started his investigations into the phenomenon of catalysis, he pointed out anomalies in Faraday's physical theory of catalysis and formulated his alternative chemical theory, postulating the formation of unstable intermediaries.

His subsequent accurate research concerning the use of metal catalysts under very finely divided forms gave rise to modern oil hydrogenation and synthetic methanol industries. He also proved the selectivity of catalytic processes and the sensitivity of catalysts to poisons, introducing the use of supports to enhance the catalytic activity.

Sabatier's work is accurately recorded in the publications of several learned societies. His most important book, *La catalyse en chimie organique* (catalysis in organic chemistry), was first published in 1913, with a second edition in 1920, of which an English translation was published in 1923.

For his method of hydrogenating organic compounds in the presence of finely divided metals, Sabatier was awarded the Nobel Prize in Chemistry for 1912, sharing the prize with Victor Grignard, who received it on account of his discovery of the so-called Grignard reagent. In an interesting paper devoted to the biography of Sabatier and his scientific contributions, Michel Che asked why the Nobel committee did not use the word catalysis in the nomination concerning the Sabatier Nobel Prize in 1912 (Che, 2013).

Paul Sabatier gave with this experiment and with following research (Sabatier and Senderens, 1897) a fundamental contribution to heterogeneous catalysis developing the catalytic hydrogenation technique by which ethylene was transformed in ethane using finely divided nickel as catalyst. Undoubtedly, Sabatier was one of the basic founders of heterogeneous catalysis.

In 1902, Sabatier and Senderens, intrigued by the discovery in 1890 of the reaction between nickel and carbon monoxide made by the German Jewish chemist Ludwig Mond (1839–1909), also demonstrated that by flowing a mixture of CO and H_2 in the ratio 1:3 in volumes on finely divided nickel at 250°C, there is a 100% yield of methane and water. At higher temperatures, the same kind of reaction occurs with a mixture of carbonic anhydride and hydrogen in the ratio 1:4. This type of reaction became known as the "Sabatier process" (Sabatier and Senderens, 1902):

$$CO + 3H_2 \rightleftarrows CH_4 + H_2O \quad CO_2 + 4H_2 \rightleftarrows CH_4 + 2\,H_2O$$

By 1911 Sabatier reported at length on many hydrogenation and dehydrogenation reactions that could be carried out in the laboratory, and became the leading authority on catalysis. The catalytic hydrogenation method of Sabatier–Senderens gave rise in the period 1906–1919 (Sabatier and Mailhe, 1904; Sabatier and Murat, 1912) to a series of important papers, which were published mostly in the *Comptes Rendus de l'Académie des Sciences* and in the *Bulletin de la Société Chimique de France* (Califano, 2012, Chapter 2, p. 44).

Subsequent research on heterogeneous catalysis extended to the ability of several other elements to catalyze even very complex reactions.

The metals classified by Sabatier as active in hydrogenation (or hydrogen activation) catalysis included not only Pt (the ubiquitous catalyst) but also other finely divided transition metals like Pd, Ni, Fe, and Co (Sabatier, 1913). This observation led to the discovery of the Haber–Bosch and Fischer–Tropsch (FT) processes where activation of hydrogen on Fe and Co catalysts plays a central role. Since Sabatier's time, hydrogenation catalysts with industrial applications (heterogeneous and homogeneous) have been based on these metals, particularly Pt, Pd, and Ni. The official award statement the Nobel Prize, where it mentions that the hydrogenation reactions are carried out on *finely disintegrated metals*, is interesting because the role of surface phenomena in determining the catalytic events is implicit in this expression.

In the book *La catalyse en chimie organique*, Sabatier enounced his famous principle stating that the interactions between the catalyst and the substrate should be neither too strong nor too weak but *just right*. On the one hand, if the interaction is too weak, the substrate will fail to bind to the catalyst and no reaction will take place. On the other hand, if the interaction is too strong, the catalyst is blocked by the substrate or by products that fail to dissociate. This intuition was fully confirmed by the subsequent observations in the twentieth century about the catalytic activity of the metallic elements of the periodic table. In fact, when the bond between the surface atoms and hydrogen is sufficiently weak (as it occurs, for instance, in Pt, Pd, and Ni), hydrogenation of adsorbed olefins occurs at room temperature. By contrast, when the bond between the surface metal atoms and H is too strong, as happens with W, Mo, Cr, Ti, and so on (which form stable hydrides), the catalytic activity is weakened or even absent.

Sabatier was a very reserved man, tied to Toulouse, and indifferent to success to the point that he never accepted a move to Paris to the chair left vacant in 1907 by the death of Moissan. Nevertheless, when Sabatier became famous after the Nobel award, he attracted, without even understanding why, the displeasure of the abbey Senderens, who believed himself to have been driven out from the official Nobel recognition. In consequence of their misunderstanding, their relationship, which had lasted for more than thirty years, eventually deteriorated.

To end the list of results on the catalytic properties of metals, it is appropriate to recall that in 1884 William Ramsay tried, without success, to react directly hydrogen and nitrogen, using Pt as catalyst. Only 30 years later did Fritz Haber successfully perform this reaction.

The contribution to catalytic science of Wilhelm Ostwald (1853–1932), by far the most important representative of the modern chemical physics, is outstanding and comparable to that of Sabatier (Figure 1.8). Ostwald, born in Riga, graduated from the University of Tartu, Estonia, in 1875, where he received his PhD in 1878 under the guidance of Carl Schmidt. He taught at the Riga Polytechnicum from 1881 to 1887. In 1887, he moved to Leipzig as professor of physical chemistry where he worked for the rest of his academic life with only a short interruption for one term, as exchange professor at Harvard University in 1904–1905.

He started his contribution to catalysis by criticizing severely the Stohmann's ideas on the mechanism of catalysis. He dedicated a large portion of a review of Stohmann's paper published in the *Zeitschrift für physikalische Chemie* (Ostwald, 1894) to his own view of the problem, asserting that the catalyst did not alter the reaction mechanism but simply accelerated its kinetics, lowering the energy barrier necessary to prime the reaction. He maintained that a catalyst does nothing but speed up the reaction, which in any case would have occurred without its presence, although at a much slower rate.

In particular, for gas-phase reactions, Ostwald suggested that the catalytic effect of a metal was due to pure physical processes of adsorption in which the gases entered the cavities of the porous metal where their close contact, combined with local heating processes, favored the reaction. This suggestion does not introduce a clear concept of chemical interaction between the catalyst surface and the substrate. Accordingly, Sabatier and Senderens later developed the fundamental concept of the interaction between the substrate and the catalyst surface.

In 1894, Ostwald synthetically formulated his own definition of catalysis as (Califano, 2012, Chapter 2, p. 43):

jeder Stoff, der den Vorgang nicht erst hervorruft, sondern nur einen vorhandenen Vorgang beschleunig[1]

---

1 Any substance that does not originate from the process but is only accelerating a previous process.

and in 1901 gave a more precise definition (Ostwald, 1902):

> *Ein Katalysator ist jeder Stoff, der, ohne im Endprodukt einer chemischen Reaktion zu erscheinen, ihre Geschwindigkeit verändert.*[2]

This definition is still valid today. In a short time it gained an enormous influence on the development of the chemical industry. The theory of catalysis was the principal reason for of the Nobel award to Ostwald in 1909 (Califano, 2012, p. 43). The Nobel lecture, delivered by Ostwald at the Royal Academy of Sweden (Ostwald, 1910), effectively confirmed catalysis as one of the fundamental branches of chemical physics. In the Ostwald lecture presentation, Hans Hildebrand, president of the Royal Academy of Sciences, summarized the Ostwald catalysis concept in the following way:

> *catalytic action consists in the modification, by the acting substance, the catalyst, of the rate at which a chemical reaction occurs, without that substance itself being part of the end-products formed.*

In connection with Ostwald's statement, a definition of heterogeneous reaction was emerging referring to the situation where gaseous or liquid reactants passed over the surface of a solid acting as catalysts. As a solid contains an infinite number of atoms, this differentiated the heterogeneous catalysts from the homogeneous parents, where only a discrete number of atoms are usually involved in the catalyst center. This difference cannot be taken too rigidly, since the borderline between the particles of finely dispersed solids and some homogeneous catalysts based on metal clusters is often elusive.

1.6 First Large-Scale Industrial Processes Based on Heterogeneous Catalysts

1.6.1 Sulfuric Acid Synthesis

In term of tonnage, sulfuric acid production, whose importance grew strongly when the production of explosives increased during the First World War, is the first true industrial process. Even today, sulfuric acid is a very important chemical commodity, and indeed, a nation's sulfuric acid production is a good indicator of its industrial strength.

In the nineteenth century, sulfuric acid was produced using lead condensing chambers, invented in 1746 by the English inventor John Roebuck. Together

---

2 A catalyst is a substance that alters a chemical reaction rate without being part of the final product.

with the Birmingham merchant Samuel Garbett, Roebuck built in 1749 a factory in Scotland, which for several years had the absolute monopoly of the production of this essential component of the industrial revolution in England.

The leaden condensing chambers were then gradually adopted worldwide and became the dominant method for the production of sulfuric acid.

The Anglo-German chemist Rudolph Messel (1848–1920), born at Darmstadt and educated at the universities of Zurich, Heidelberg, and Tubingen, later collaborator of J. C. Calvert and of Sir Henry Roscoe at Manchester, devised, in 1870, a contact process for producing concentrated fuming sulfuric acid (oleum) by passing the vapor of ordinary sulfuric acid over finely divided platinum at red heat (Figure 1.9). A patent was filled by Messel's partner Squire (1875), and the process was made operative, an account being given to the Chemical Society (Messel and Squire, 1876). In 1878, Messel became the managing director of the company known as *Squire, Chapman & Messel Ltd*, holding the position until his retirement in 1915. At that time, some thousands of tons a week were already being produced worldwide by the process he had developed. Messel's multistep process was very complicated because it involved heating sulfuric acid (from the chamber method) in a platinum vessel to decompose it into water, sulfur dioxide, and oxygen, followed by a water-condensing

(a)   (b)

Figure 1.9 (a) Rudolph Messel (1848–1920) (image by courtesy of the Society of Chemical Industry) and (b) Clemens Winkler (1838–1904) (image in the public domain). They greatly contributed to the catalytic synthesis of sulfuric acid. Messel developed the contact process for producing concentrated sulfuric acid (oleum) with finely divided platinum. Winkler, discoverer of germanium, proposed the use of platinized asbestos in the production process. *Source*: (a) Reproduced with permission of Anne Pinheiro.

step and a third step where the mixed gases were passed over finely divided platinum to form sulfur trioxide. The final step was the dissolution of sulfur trioxide in sulfuric acid to give oleum.

An innovation came in 1886 when Clemens Alexander Winkler (1838–1904), professor of chemistry and director of the Freiberg School of Mines, proposed using platinized asbestos in a production process (Winkler, 1875) (Figure 1.9). Winkler was a German chemist who discovered the element germanium in 1886, solidifying Dmitri Mendeleev's theory of periodicity.

Winkler, born in Freiberg, entered Freiberg University of Mining and Technology in 1857. Sixteen years later, Winkler was appointed professor of chemical technology and analytical chemistry at the same university.

In 1886 Winkler obtained a new mineral from the Himmelsfürst mine, near Freiberg. The mineral *argyrodite* that was Winkler's start toward finding germanium is now known to be a double sulfide with formula $GeS_2 \cdot 4Ag_2S$.

Mendeleev, in order to insert germanium into the periodic table, suggested that it might be the element *ekacadmium*, the existence of which he had predicted earlier. In contrast, Lothar Meyer favored the identification of germanium with *ekasilicon*, a different predicted element.

Winkler's production process was similar to that previously proposed by Peregrine (1832) employing sulfur dioxide and oxygen in stoichiometric proportions (Cook, 1926). This process represented a significant advance on the ones previously used, since until then experts had attempted to use finely divided metals without the asbestos support. Before 1920 the industrial catalyst for concentrated sulfuric acid production was Pt supported on kieselgur or silica. The important step of the Pt-based contact process is the oxidation of sulfur dioxide to sulfur trioxide following the scheme

$$2SO_2(g) + O_2(g) \rightleftarrows 2SO_3(g)$$

catalyzed by finely divided platinum, here acting as in the Ostwald process of ammonia oxidation as an oxidation catalyst. In other words, Pt is able to activate not only hydrogen (as largely investigated by Sabatier) but also oxygen. The state of oxygen on the Pt surface has since been debated. With the advent, after 1970, of the modern physical and computational methods, the state of oxygen (molecular or atomic) on the surface of platinum or other noble metal surfaces has been clarified. The coexistence of molecular (Gustafsson and Anderson, 2004) and dissociated species on the surface of platinum has been only recently demonstrated (Bocquet, Cerda, and Sautet, 1999). A detailed discussion concerning the experimental and theoretical results can be found in the recent W. F. Schneider and coworkers contribution (McEwen *et al.*, 2012).

Only after 1920 the expensive Pt catalyst was substituted by vanadium pentoxide (Thomas, 1970; Farrauto and Bartholomew, 1997; Lloyd, 2011), whose catalytic properties had been discovered several years before (1899) by Charles

R. Meyers. An effective vanadium catalyst for the contact process was developed in 1921 and employed in Germany by the Badische Company (Slama and Wolf, 1921). Today the catalyst consists of vanadium pentoxide supported on a SiO_2-based substrate containing promoters. The surface chemistry of vanadium pentoxide has been the subject of many studies in the twentieth century, and only with the advent of surface science methods was his composition and structure correctly understood.

We will return to the surface structure of active sites in a later chapter, when partial oxidation reactions will be discussed (Weckhuysen and Keller, 2003). This final formulation is the result of continuous trial-and-error investigations and accumulated experience based on the studies of a great number of researchers working in industrial laboratories. The contact process (still relying on supported vanadium pentoxide catalyst) was significantly improved only in 1963 when the Bayer AC announced the first double contact adsorption process that reduced the unreacted SO_2 emissions in the atmosphere.

1.6.2 Ammonia Problem

At the beginning of the twentieth century, sodium nitrate extracted by English companies from existing sediments of Chilean guano was practically the only chemical fertilizer utilized in agriculture. The realization that guano could be exhausted in a few years, coupled with the need for European countries to obtain other sources of nitrogen rich material, had, for some time, prompted applied research groups to produce synthetic ammonia. Ammonia, whose name derived from the ancient Egyptian deity Amun, known to the Greeks as Ammon, is the basic compound for the synthesis of nitric acid and of nitrates of enormous importance not only in agriculture but also in the production of explosives. In the latter years of the nineteenth century, the scientific community was strongly involved in the discussion on what later became known as "the nitrogen problem," earnestly addressed by Sir William Crookes in his speech at the meeting of the British Association for the Advancement of Science in Bristol in September 1898 (Crookes, 1898; Topham, 1955):

> ... *all civilized nations stand in deadly peril of not having enough to eat.* ... *the fixation of atmospheric nitrogen is one of the great discoveries awaiting the ingenuity of chemists.*

Crookes was very concerned to show that, owing to the fast rate of increase of the population, the world's supplies of wheat would soon prove insufficient and the land would not continue to produce the same yield year after year, unless adequate quantities of nitrogenous manure were used to fertilize the soil. In a number of European countries, afraid that in the event of war, the Chile nitrates might well prove to be inaccessible, the need to obtain chemical compounds such as ammonia in large amounts was complemented with a similar demand for supplies of nitric acid in the manufacture of explosives.

The problem of succeeding in the fixation of atmospheric nitrogen became a central one in Europe. Nitrogen fixation is a general term to describe the conversion of atmospheric nitrogen N_2 into a form used by plants. One method to fix nitrogen mimicked the natural fixation of nitrogen by lightning. The process involved subjecting air to a high-voltage electric arc to produce nitric oxide:

$$N_2 + O_2 \rightarrow 2NO(g)$$

The nitric oxide was then converted into nitric acid and combined with limestone to produce calcium nitrate. The problem with this process was that the procedure required large amounts of energy and was economically feasible only when there was a cheap supply of electricity.

In 1828 Michel Desfossés, apothecary at Besançon, conceived the possibility of fixing atmospheric nitrogen through chemical reactions. He observed that mixtures of alkali metal oxides and carbon react at high temperatures with nitrogen. With the use of barium carbonate as starting material, the first commercially used process, developed by Louis Joseph Frederic Margueritte and Alfred Lalouel Sourdeval, became available in the 1860s. The resulting barium cyanide could be reacted with steam-yielding ammonia. In 1898 Adolph Frank and Nikodem Caro decoupled the process producing calcium carbide and then reacted it with nitrogen to obtain calcium cyanamide. They founded the Cyanid-Gesellschaft mbH, which would later become the Bayerischen StickstoffWerke AG (BStW) in Trostberg in southeast Bavaria, where today is located in the Trostberg Chemical Park, home for the three major companies AlzChem, Badische Anilin und Soda Fabrik (BASF), and Evonik–Degussa. The Frank–Caro process dominated the industrial fixation of nitrogen until the discovery of the Haber process in 1909.

Another way to fix nitrogen is the synthesis of ammonia. The conversion of hydrogen and nitrogen into ammonia had been studied since the mid-1800s, but serious work on the subject did not occur until the turn of the century.

In 1901 Henry Le Châtelier made the first attempt to produce ammonia by subjecting a mixture of hydrogen and nitrogen to a pressure of 200 atmospheres and a temperature of 600°C in a heavy steel bomb (Le Châtelier, 1936) (Figure 1.10). Contamination with oxygen led to a violent explosion, and Le Châtelier abandoned his attempts to synthesize ammonia. Toward the end of his life, Le Châtelier said (van Klooster, 1938):

> *I let the discovery of the ammonia synthesis slip through my hands. It was the greatest blunter of my scientific career.*

At the beginning of the twentieth century, the fear of shortage of Chile's guano convinced many European scientists to help in finding solutions for

(a)

(b)

(c)

(d)

Figure 1.10 (a) Henry Le Châtelier (1850–1936), (b) Fritz Haber (1868–1934), (c) Carl Bosch (1874–1940), and (d) Paul Alwin Mittasch (1869–1953) (images in the public domain). They are the protagonists of the ammonia synthesis under high pressure. The first attempt to perform the synthesis under high pressure is attributed to Le Châtelier. Haber (1918 Nobel Prize) was the first to succeed in performing the reaction. Bosch (1931 Nobel Prize) and Mittasch developed the catalysts of the industrial Haber–Bosch process.

their countries' long-term needs. One such man was botanist Wilhelm Pfeffer (1845–1920), professor at the University of Bonn who in 1901 expressed his concern about the need for supplies of fixed nitrogen (Pfeffer, 1887) to his friend Ostwald who at this time was investigating the effects of catalysts on chemical reactions. Ostwald's response was immediate; it was obviously his duty as a chemist to play his part in making his country independent of Chile saltpeter and in obtaining nitric acid from other sources.

1.6.3 Ammonia Oxidation Process

Ostwald, in the initial years of the twentieth century, decided to give his attention to the oxidation of ammonia to nitric acid since he was aware of the fact that Charles Frédéric Kuhlmann had already discovered such a process by passing a mixture of ammonia and air over platinum sponge heated to about 300°C in a glass tube. In 1838 Kuhlmann had also filled a patent application for this invention. In the course of a paper presented to the Société des Sciences of Lille in the same year, Kuhlmann made these prophetic remarks:

> *If in fact the transformation of ammonia to nitric acid in the presence of platinum and air is not economical, the time may come when this process will constitute a profitable industry and it may be said with assurance that the facts presented here should serve to allay completely any fears felt by the government on the difficulty of obtaining saltpetre in sufficient quantities in the event of war.*

It was, however, clear to Ostwald that the theoretical basis of the ammonia oxidation reaction had to be well clarified before its industrial organization could be planned on a large scale. He thus decided to convince his private assistant and later his son-in-law, Eberhard Brauer, to work on this problem. In the first experiments, ammonia and air were passed over the catalyst (platinized asbestos) in known quantities and with known velocities, and it was at once clear that the conversion to nitric acid was practicable. A new apparatus and systematic variations of reaction conditions (gas velocities, ammonia–air ratio, and catalyst temperature) finally gave a conversion of 85%. Thanks to this systematic investigation, the foundations of the technical process for producing nitric acid from ammonia were laid on a solid basis in 1902, since then known as the Ostwald–Brauer process. The translation from idea to practice presented many problems before the project became economically convenient. The main problem of the original Ostwald–Brauer process was the large amount of platinum necessary to carry out the reaction. Very soon, however, Karl Kaiser at the Technische Hochschule in Charlottenburg patented the utilization of platinum gauzes that gave much better yields. Furthermore addition of small quantities of rhodium to platinum improved the catalyst activity. A small factory was available to Ostwald and Brauer, and here a pilot chemical plant became operative. Ostwald filed patents for his procedure in 1902, although his German patent was disallowed because of Kuhlmann's earlier disclosures of more than 60 years before (Califano, 2012, Chapter 2, p. 43).

The later date of 1908 is often reported for Ostwald's patent starting time, probably because of the unavoidable bureaucratic delays to make it operative or perhaps because only by then was the Fritz Haber ammonia synthesis (Haber and Le Rossignol, 1910) really effectual and the ammonia price became acceptable.

1.6.4 Ammonia Synthesis

The problem of ammonia synthesis from atmospheric nitrogen found a practical solution in 1908 when the German chemist Fritz Haber, professor at the University of Karlsruhe, discovered the catalytic procedure for the synthesis of ammonia from hydrogen and atmospheric nitrogen (Figure 1.10).

Before describing the ammonia synthesis from H_2 and N_2 discovered by Haber following the reaction

$$N_2 + 3H_2 \rightarrow 2NH_3$$

it is appropriate to summarize the state of the knowledge concerning the industrial production of H_2 whose purity is essential for the ammonia synthesis process. In the first years of the twentieth century, hydrogen was obtainable from the coal gasification when steam is put in contact with incandescent coke (at about 1000°C) following the reaction

$$C + H_2O \rightarrow CO + H_2$$

As CO is a poison for the catalysts used in the ammonia synthesis, liquefaction or copper liquor scrubbing was used to remove it from the $(CO + H_2)$ gas mixture. Both methods are, however, not applicable to large-scale production. A catalytic process was consequently needed to remove the CO from the mixtures. In 1914 Bosch and Wild (Bosch and Wild, 1914) discovered that oxides of iron and chromium could convert a mixture of steam with CO into CO_2 at 400–500°C, according to the reaction in the gas phase (water gas shift reaction)

$$CO + H_2O \rightarrow CO_2 + H_2$$

a process generating additional hydrogen for the Haber process. The water gas produced from the carbonaceous source by steam reforming was passed over the iron–chromium catalyst to convert the CO to CO_2 by the water gas shift reaction. Iron-based catalysts are still used today industrially to transform CO to CO_2 by the water gas shift reaction even if other catalytic systems are also employed.

In modern times the H_2 production is based nearly exclusively on methane, following the Ni-catalyzed process

$$CH_4 + H_2O \rightarrow CO + 3H_2$$

Fritz Haber (1868–1934), born in one of the oldest Jewish families of Breslau, studied chemistry in the period 1886–1891 first at the University of Heidelberg in Bunsen's laboratory, then at the University of Berlin under A. W. Hoffmann, and finally at the Königlichen Technischen Hochschule in Berlin, Charlottenburg (today the technical University of Berlin) under Carl Liebermann. There he obtained his PhD (Haber, 1891). After completing his PhD thesis, he decided to start a university career in chemistry under the supervision of Ludwig Knorr at Jena. In 1894, he obtained a position as an assistant professor of Hans Bunte at the Technische Hochschule of Karlsruhe. In 1898, he published his treatise on electrochemistry *Grundriss der technischen Elektrochemie auf theoretischer Grundlage* and started the study of reduction processes at the cathode.

At the beginning of the twentieth century, Haber was already well known for his papers on the combustion of hydrocarbons (Haber and van Oordt, 1904) and for his research in electrochemistry and thermodynamics. For these he obtained in 1906 the position of professor of chemical physics and electrochemistry at the University of Karlsruhe. In 1911, he succeeded as director of the Institute for Physical and Electrochemistry at Berlin-Dahlem, where he remained until 1933.

The biography of Haber is significant owing to its connection with the tragic German history between the two world wars. Haber was of Jewish origin and a patriotic German proud to serve the German homeland during the World War I. As a chemist, he was assigned to the section for the preparation of poisonous gases where he became leader of the teams developing chlorine gas and other deadly gases, despite the proscription by the Hague Convention of 1907, to which Germany was a signatory. In particular, he was indirectly involved in the creation of Zyklon (a compound used in Auschwitz–Birkenau and Majdanek extermination camps during the Nazi regime). He also participated in the production of the grünkreuz gas, a variation of "mustard gas," invented in France in 1854 by the pharmacist Alfred Riche (1829–1908) and better developed by another Nobel Prize winner, the French chemist Victor Grignard of the University of Nancy. Due to the involvement of Haber in the warfare program and in the development of poisonous gases, it is thought that his wife, Clara Immerwahr, a brilliant chemist and the first woman to receive a doctorate from Breslau University, was so deeply hurt that she committed suicide in May 1915. On the same day the German patriot Fritz Haber, although seriously heartbroken by his wife's tragedy, traveled to the Eastern Front. In the new Germany of the Weimar Republic, Haber continued to strive patriotically, but he was forced to witness the growth of vicious anti-Semitism, spreading around him and, by the early 1930s, realized that his patriotism had no more place in Germany. In 1933, the Nazi race laws compelled nearly all his staff to resign.

Rather than agreeing to this, Haber himself resigned. Sir William Pope then invited him to go to Cambridge, England, and there he remained for a while. He suffered, however, for some time from heart disease and, fearing the English winter, moved to Switzerland, where he died on January 29, 1934, at Basel. The personal history of Haber is relatively common among scientists. In fact, in the period between the two world wars and even beyond, many scientists (chemists and physicists) faced the contradiction between patriotic and humanitarian behavior.

In 1908, he discovered the direct synthesis from N_2 and H_2, a chemical process that has virtually changed the economy of the world, making it possible for the production of millions of tons of fertilizers. Haber succeeded in reacting hydrogen with nitrogen, one of the less reactive existing gases, using a catalyst based on finely dispersed osmium and uranium at a temperature of about 500°C and a pressure of about 150–200 atmospheres. The collaboration of Haber with the British chemist Robert Le Rossignol (1884–1976) from University College London was essential for the realization of the ammonia synthesis. Le Rossignol worked in Germany in the period 1908–1909 with Haber and was instrumental in the development of the high-pressure devices used in the Haber process, eventually producing a tabletop apparatus that worked at 200 atmospheres pressure (Haber and Le Rossignol 1907, 1908a, b). Le Rossignol was interned in Germany in 1914 at the outbreak of the World War I, but was released to work for the industrial firm Auergesellschaft during the war. Le Rossignol returned to the United Kingdom after the war.

Haber was awarded the 1918 Nobel Prize in Chemistry (awarded in 1919) for his discovery that virtually "made bread from the air" and recognized the assistance received from Le Rossignol, whose name appears on Haber's patents for the process.

In 1910, BASF purchased the Haber initial process and charged one of his best chemists, Carl Bosch and his assistant Paul Alwin Mittasch (1869–1953), to transform it into a large-scale industrial process (Figure 1.10). Haber's catalysts were, however, too expensive and not sufficiently stable to be utilized at an industrial level. Bosch and Mittasch therefore conducted a systematic research of possible catalysts at BASF, covering practically the full periodic system and in two years reached the conclusion that finely dispersed iron with the addition of few percent of alumina and small quantities of potassium and calcium supplied an excellent catalyst, well reproducible, and with a long-lasting lifetime (Haber, 1922). The BASF chemists also solved the problem of a reactor able to work without risks at high temperature and pressure and developed the necessary methods of purification of hydrogen and nitrogen. The intensive collaboration between Haber and Bosch made possible the large-scale commercial production of ammonia. As the war between Germany and the other big world powers began, several European countries were excluded from the possibility of importing guano from South America, since England had complete control of the oceanic routes as well as of the Chilean and Peruvian sediments. The Haber process became therefore of fundamental importance, and Germany would not have been able to afford to continue the war without the massive production of ammonium nitrate that was realized thanks to the BASF plants built with the advice of Fritz Haber and the direction of Carl Bosch.

The first "Stickstoffwerke" became operative at Oppau, in Germany in 1913, with a production of 30 tons of ammonia per day, and a larger plant at Leuna, in Saxony, East Germany, soon followed. Leuna is today the center of the great German synthetic chemical industry. Bosch became managing director of BASF in 1919 and in 1925 was nominated director of the IG Farbenindustrie Aktiengesellschaft and in 1935 general director of the IG Farbenindustrie.

Another German industrial chemist who played a basic role in the development of high-pressure devices for the ammonia synthesis was Friedrich Bergius (1884–1949). He studied first at the University of Breslau in 1903 and then at the University of Leipzig where he received a PhD in chemistry in 1907, under the supervision of Arthur Rudolf Hantzsch. In 1909, Bergius worked with Fritz Haber and Carl Bosch at the University of Karlsruhe in the development of the Haber–Bosch process. The collaboration with Haber and Bosch was of extreme importance for his development as a real expert of high-pressure catalytic processes, to the point that by 1913 he succeeded in developing a new and original method of production of liquid hydrocarbons for use as synthetic fuel by hydrogenation of high-volatile bituminous coal lignite at high temperature and pressure. Lignite has a high content of volatile matter, which makes it easy to convert into gas and liquid petroleum products. Together with Carl Bosch, he won the Nobel Prize in Chemistry in 1931 in recognition of their contributions to the invention and development of chemical high-pressure methods.

Both Bosch and Mittasch continued their research of catalytic synthesis at the industrial level. Strangely enough the Nobel committee ignored Paul Mittasch, who had given fundamental contributions to the high-pressure catalytic synthesis. Carl Bosch was critical about Nazi policies and for this reason was gradually relieved from his high position, falling into despair and alcoholism. He died in Heidelberg.

Beside the Haber–Bosch process, other processes based on the same catalysts were developed in Italy and France in the period 1920–1930 (Ammonia Casale, Fauser–Montecatini, and Claude processes).

Luigi Casale (1882–1927), an Italian chemist and industrialist, who developed a new process for the synthesis of ammonia, founded the Ammonia Casale Company.

Casale graduated at the University of Turin where he spent the initial years of his career as assistant professor in the Institute of Chemistry. Then he visited for a year the laboratory directed by Nernst at the University of Berlin, where he studied the engineering aspects of the ammonia synthesis. Back in Italy in 1919, he constructed the first plant for the synthesis of ammonia based on an innovative design. This plant, located at Rumianca site in Domodossola, operating at pressure different with respect to that of BASF, produced 200 kg/day of NH_3. In 1921 he founded the Società Ammonia Casale, the founding company of the Casale Group, which is one of the oldest companies active in the field of synthetic ammonia production, established in Lugano (Switzerland). The Casale process was in direct competition with the Haber–Bosch process, developed at BASF. Contrary to the BASF process, Casale's was the only one offered on the market, and this secured a substantial success at the time, which continues to this day.

In the same period the Fauser–Montecatini company developed another process, carrying the name of Giacomo Fauser (1892–1971) (Figure 1.11). This industrial researcher of Swiss origin, after graduation at the Politecnico of Milano, had the chance to visit the BASF plant in 1919. When he returned to Italy, he proposed a new process characterized by lower operation pressures and temperatures with respect to the Haber–Bosch one. He constructed a pilot plant in Novara producing 90 kg/day NH_3. This achievement attracted the attention of Guido Donegani, president of the Società Generale Montecatini, who promoted the construction of a new larger plant. For this reason the process was patented as Fauser–Montecatini. Both the Casale and Fauser–Montecatini processes are now adopted in numerous plants, worldwide.

For his contribution to ammonia synthesis, Guglielmo Marconi made Giacomo Fauser in 1935 a member of the Italian *Consiglio Nazionale delle Ricerche*. In 1957 he became also a member of the *Accademia Nazionale dei Lincei*.

In France, Georges Claude (1870–1960) started working on an ammonia process in 1917 (Figure 1.11). Besides ammonia synthesis, Georges Claude is known for his early work on the industrial liquefaction of air in the invention and commercialization of neon lighting. Following Linde, Claude had been expert in separating air from its components by cooling to low temperatures (liquefaction of air) since 1896. Claude established Air Liquide in 1902 and built a number of oxygen plants in France and other countries in the following years. In 1919 he constructed the first ammonia plant. Air Liquide and Saint-Gobain established the Société de la Grande Paroisse Azote et Produits Chimiques for the development and exploitation of the Claude ammonia process. Considered by some to be the *Edison of France*, he was an active collaborator of the German occupiers of France during World War II: for this reason he was imprisoned in 1945 and stripped of his honors. To end the chapter devoted to ammonia synthesis, it can be added that after the war, the original patents from BASF, Casale, Fauser–Montecatini, and Claude expired,

(a)   (b)

(c)

Figure 1.11 (a) Luigi Casale (1882–1927), (b) Giacomo Fauser (1892–1971), and (c) Georges Claude (1870–1960) (images in the public domain). They developed the first commercialized ammonia synthesis processes other than the Haber–Bosch process. *Source*: (a) Courtesy of Casale SA.

and new competitors came on the market. However, the catalyst formulation did not change substantially.

The problems of the ammonia synthesis mechanism and of the atomic or molecular structure of adsorbed nitrogen have been at the origin of stimulating scientific debate for several decades. The debate has seen the confrontation of two hypotheses. The first hypothesis on the mechanism of the ammonia synthesis requires as a starting point the dissociation of the N_2 molecules into adsorbed N atoms that can be successively hydrogenated into the final products. The rate of the N_2 dissociation is determined by the height of the potential barrier. In addition the interaction of the N atoms on the catalyst surface should be sufficiently weak (following the Sabatier intuition) to allow the creation of new empty sites on the surface where fresh N_2 molecules can dissociate. Hydrogen also needs to dissociate, but this reaction is fast and hydrogen binds more weakly to the surface than nitrogen.

An alternative mechanism was based on the hypothesis that N_2 is adsorbed in molecular form and is progressively hydrogenated with the formation of hydrazine-type intermediates followed only at the end by the formation of ammonia. The hypothesis of a role of a molecular form received a substantiated credit by the isolation in the second half of the twentieth century of molecular complexes where N_2 is end bonded or bridged to metal centers (Allen and Senoff, 1965; Fryzuk and Johnson, 2000).

Due to this controversy, the challenging problem of the state of N_2 on the catalyst did not escape the attention of the surface scientists and theoretical chemists operating in the last quarter of the twentieth century when new physical and computational methods became available. While the contribution of classical surface science methods will be thoroughly described when the G. Ertl (2007 Nobel Prize) results are illustrated, we anticipate that the computer modeling results, in particular those from Christensen and Nørskov (2008), have confirmed Ertl's model.

As for the catalyst is concerned, it can be underlined that only about 80 years later the industrial production of ammonia partially switched to the utilization of ruthenium instead of iron-based catalysts (the Kellogg Advanced Ammonia Process (KAAP)) because ruthenium is a more active catalyst and allows milder operating pressures.

The mechanism of the synthesis on Ru surfaces has also been studied with computational methods (Christensen and Nørskov, 2008). It has been confirmed that N_2 dissociates on the surface and that the most active sites are those where the two N atoms originated from the dissociation of a N_2 molecule are stabilized by bonding to five atoms on the surface. One N atom is bound to three Ru atoms, whereas the other N atom is bound to two. None of the Ru atoms within the active site are bound to more than one N atom at a time. This gives stronger Ru–N bonds, resulting in a more stabilized transition state. Successive molecular beam experiments on the dissociation probability of N_2 on Ru surface confirmed these conclusions (Murphy *et al.*, 1999). Bjørk Hammer, professor of physics and astronomy at the Aarhus University in Denmark, and his group predicted this mechanism by density functional theoretical calculations (Mortensen *et al.*, 1997) in the framework of a research project on the factors determining the reactivity of simple molecules adsorbed on metallic catalysts (Kratzer, Hammer, and Nørskov, 1996a, b; Kinnersley *et al.*, 1996).

The rate of ammonia synthesis over a nanoparticle of ruthenium catalyst was computed recently by a quantum chemical treatment using the density functional theory theoretical approach (Honkala *et al.*, 2005).

Ammonium nitrate, the fertilizer generated from ammonia produced by the Haber process, is estimated to be responsible for sustaining one-third of the Earth's population. Ammonium nitrate is, however, a very dangerous material that requires careful handling and can explode via auto-combustion or in mechanical accidents. For this reason, his large-scale production has caused several serious disasters. Among them we recall the September 1921 explosion in the BASF plant at Oppau with 561 casualties, the April 1942 explosion in the town of Tessenderlo in Belgium with 189 casualties, and the explosion on April 1947 of a French cargo in the harbor of Texas City with 570 casualties. In addition the nearby Monsanto chemical storage facility exploded, killing 234 workers. More recently, on September 2001, an explosion of ammonium nitrate occurred in Toulouse, France, with 31 deaths and 2442 injured. The danger connected to the handling of ammonium nitrate has therefore prompted all governments to establish severe controls and to impose particularly restrictive conditions on its storage and utilization in agriculture in the hope of eliminating for the future the causes, not always purely accidental, of its explosions.

1.7 Fischer–Tropsch Catalytic Process

Another important catalytic process, which was found during the first quarter of the twentieth century, is the FT process discovered by Franz Fischer (1877–1947) and Hans Tropsch (1889–1935) at the Kaiser Wilhelm Institute for Coal Research in the Ruhr (Fischer and Tropsch, 1926a, b) (Figure 1.12). To illustrate the continuity of the German school of catalysis, it is interesting to recall that before becoming director of the Kaiser Wilhelm Institute in 1913, Franz Fischer worked with Wilhelm Ostwald and Emil Fischer. From 1920 to 1928, he collaborated with Hans Tropsch, a Bohemian industrial chemist, who graduated at Prague, where he received his PhD under the supervision of Hans Meyer. In 1928, Hans Tropsch became professor at the new Institute for Coal Research in Prague. He later accepted a position in the United States

(a)	(b)

Figure 1.12 (a) Franz Fischer (1877–1947) and (b) Hans Tropsch (1879–1935). *Source*: Courtesy of Max Planck Institut für Kohlenforschung. They discovered the process for converting carbon monoxide and hydrogen into liquid hydrocarbons.

at the laboratories of Universal Oil Products and at the Armour Institute of Technology in Chicago in 1931. Hans Tropsch returned to Germany in 1935, where he died shortly after arrival.

The FT process converts a mixture of carbon monoxide and hydrogen to liquid hydrocarbons using finely dispersed iron as catalyst according to the scheme

$$nCO + (2n + 1)H_2 \rightarrow C_nH_{2n+2} + nH_2O$$

As already mentioned, the mixture of CO and H_2 gas (syngas) was obtained from coal, via the reaction $C + H_2O \rightarrow H_2 + CO$ (coal gasification).

As H_2O can further react with CO following the scheme $CO + H_2O \rightarrow CO_2 + H_2$ (water gas shift reaction), a small amount of CO_2 is always present in the mixture.

Depending on the catalysts and on the operative conditions, the Fischer–Tropsch synthesis (FTS) made feasible not only the production of practically contaminant-free transportation fuels (e.g., diesel) but also the synthesis of other valuable chemicals (e.g., short-chain alkenes) from coal and other feedstock alternatives to oil (for instance, natural gas and, later, biomass). Furthermore, alcohols are also formed on iron-based catalysts.

Iron-based catalyst precursors consist of nanometer-sized iron oxide crystallites often added with promoters to improve the catalyst performance. A typical catalyst contains promoters like copper (to improve reducibility) and potassium (to improve CO dissociation), along with some silica or zinc oxide to improve the catalyst dispersion. A complex catalytic system indeed!

The catalyst precursor is treated in H_2, CO, or syngas to convert it to its active form, a complex mixture of metallic iron, carbidic iron, and iron oxides.

The iron-based FT catalyst system is one of the oldest and perhaps most studied system known in the field of heterogeneous catalysis. However, even after more than 80 years of research, many important questions remain unanswered. In particular, the exact structural composition of the active site in these catalysts is still not unambiguously identified. Also the reaction mechanism for the formation of hydrocarbons and alcohols is still debated (Schulz, 1999; de Smit and Weckhuysen, 2008). Cobalt- and ruthenium-based catalysts (with appropriate promoters) are also good for FTS, ruthenium being the best for the production of long-chain hydrocarbons (waxes).

Franz Fischer and Hans Tropsch, being aware of the carbon- and carbide-forming tendency of iron catalysts, proposed CO dissociation as the primary step and, respectively, iron carbides as intermediates. Prior to dissociation, carbon monoxide was thought to be adsorbed through the carbon end on one or more metal atoms to form linear and bridged carbonyls. The existence of these precursor species was confirmed about fifty years later by spectroscopic and theoretical methods. The carbidic carbon was thought to be hydrogenated to CH_2 species, and these were thought to polymerize. This "carbide theory" was, however, discarded when no carbide phases were discovered using cobalt and ruthenium as the catalysts (Pichler, 1952). The exact mechanism for the production of hydrocarbons over iron carbide phases is largely unknown. In fact, even though it is observed that carburized iron is an active catalyst, it is still disputed in the literature whether bulk carbide phases themselves play an active role in the synthesis (Niemantsverdriet and van der Kraan, 1981). The problem of which metals are active in FTS has been discussed in a recent paper of a group of researchers at the Van't Hoff Institute of Molecular Sciences, University of Amsterdam. They established a clear correlation between the experimental and calculated heats of adsorption of CO and H_2 on several transition metals with expected FT activity, as can also be inferred from recent DFT calculations (Toulhoat and Raybaud, 2013; Bligaard *et al.*, 2004).

A qualitative explanation of the better performance of Ru in catalyzing the formation of long-chain hydrocarbons is related to the small decrease of the metal–carbon (Me—C) bond strength on passing from Fe to Ru, a fact that favors the formation of C—C bonds and hence the polymerization of adsorbed (CH_2) groups to form hydrocarbon chains. In relation to this point, it must be mentioned that the FTS has also been attempted, without substantial success, by means of homogeneous complexes of Co, Fe, and Ru (Badhuri and Mukesh, 2000).

The presence of oxygen-containing organic compounds in the product ion of FTS on Fe inspired Henry Herman Storch, Norma Golumbic, and Robert B. Anderson in their famous volume (Storch, Golumbic, and Anderson, 1951) to postulate a mechanism where the atomic hydrogen is first added to the adsorbed CO to form an oxymethylene species. This idea received much attention and was a common belief for many years. Famous authors like Emmett (Kummer and Emmett, 1953) tried to substantiate this hypothesis.

One of the problems of the FT reaction is the production of water, which can keep the metal in a partially oxidized form. This implies that a mixture of phases is always present under reaction conditions and that the fraction of oxidized metal decreases on passing from Fe, Co to Ru. This fact suggested that hydrocarbon chains are formed on the bare metal surfaces, while the formation of oxygenated products (like alcohols) requires the presence of an oxide phase (for instance, Fe_2O_3).

The simultaneous presence of multiple phases is the reason why the identification of a fully accepted mechanism remained controversial for several years. For this reason the surface chemistry of all potential catalytically active phases Fe, Fe_2O_3, Co, Ru, and the reactivity toward CO, H_2, and H_2O has been the object of numerous surface science studies. The novel surface science techniques, which became available in the 1970s and 1980s boosted research efforts for identifying the surface structures and the adsorbed species (in particular, CH_3, CH_2, CH, and carbidic groups) with the aim of correlating them to the catalytic performance of the materials (Barteau *et al.*, 1985). Bent (1996) made an extensive review of the numerous investigations concerning the properties of CH_3, CH_2, CH, and carbidic species present on the surface of many metals. With the advent of modern computing facilities at the end of twentieth century and the initial years of the twenty-first century, theoretical methods were extensively applied. The completely new and innovative consequence of this development is that experimental synthesis, characterization, and testing were rapidly becoming interwoven with theoretical calculations (Dellamorte, Barteau, and Lauterbac, 2009; Nørskov and Bligaard, 2012).

Since the FT process opened the possibility to obtain gasoline from coal, several political reasons influenced the application of this process. Historically the interest in this alternative production route of hydrocarbons has paralleled the crises in the oil feedstock supply chain. A clear-cut example is the production of transportation fuels from coal in Germany during the Second World War, which was cut off from oil supplies by allied forces. A few decades later, South Africa invested significant research in the FT process during its 1970s–1980s oil sanctions. During that same decade, the 1973 and 1979 energy crises initiated new worldwide initiatives for transportation fuels and chemicals from alternative feedstock, including FT.

Today the Fe-based catalysts, being especially suited for the production of liquid hydrocarbon products from biomass, are attracting new attention.

1.8 Methanol Synthesis

An important process, strictly connected with the FT process, is methanol synthesis. In fact the industrial synthesis of methanol is obtained from synthesis gas ($CO/H_2/CO_2$ mixture) or methane using an appropriate catalyst (Oliveira, Grande, and Rodrigues, 2010).

The use of methanol in making other chemicals (for instance, formaldehyde, an important intermediate for plastics, paint, etc.) has stimulated the studies of the synthesis of methanol starting from the second decade of the twentieth century. More recently further incentive is derived by the use of methanol as an important liquid fuel that can be supplied to modified internal combustion engines or fuel cells and by the conversion of methanol to hydrocarbons such as paraffins or olefins. The methanol to olefin process will be illustrated in the chapter devoted to zeolitic and microporous materials. The first industrial catalyst was a mixed ZnO/Cr_2O_3 oxide consisting of about 70% ZnO and 30% Cr_2O_3 (high-temperature, high-pressure methanol synthesis catalysts). The discovery, in 1923, was made by the German chemist Matthias Pier (1882–1965) (Pier, 1923). Pier, working for BASF, developed a means to convert synthesis gas (a mixture of CO and H_2 derived from coke and used as a source of hydrogen in synthetic ammonia production) into methanol. The process used a zinc chromate catalyst and required extremely vigorous conditions, that is, pressures ranging from 30–100 MPa (300–1000 atm) and temperatures of about 400°C.

The biography of this important industrial researcher is relevant for the discovery of this process because it shows the importance of the German chemical school in catalysis in the first part of the twentieth century. In fact, after graduation in Heidelberg he moved to Berlin where, from 1906–1910, he served as assistant to Emil Fischer and Walther Nernst. After this period he became chemist at BASF (Leuna site) where he successfully synthesized methanol. The zinc chromate catalyst has high activity and selectivity and is resistant to sulfur poisoning, fundamental properties when syngas produced from coal gasification is used (as was the case in Germany). An even earlier discovery of methanol and aldehydes synthesis process (which, however, never went into production) is ascribed to the French chemist Georges Patart, inspector of the Bureau of the Explosives, who took out a patent in 1921 describing the preparation of methanol by hydrogenation under pressure of CO in the presence of a catalyst (Patart, 1921).

Indeed the problem of the discovery has been debated, and evidence of this debate can be found in the *Industrial and Engineering Chemistry Journal* (Patart, 1925) where the BASF statement on one side and the Georges Patart reply (where he admits that BASF was the first to market synthetic methanol) on the other side are reported. As a matter of fact during the war, and in the years immediately following, BASF was exploring the possibility of obtaining methyl alcohol from carbon monoxide and hydrogen, under pressure. In 1923, the technical production of synthetic methanol had reached 10 tons per day, and the output was rapidly increasing. Up to 1925, large-scale production of this compound by this method was in the hands of the Badische Company alone. In 1930, Giulio Natta and M. Strada also discussed the synthesis of higher alcohols from CO and hydrogen (Natta and Strada, 1930).

However, as early as 1935, it was recognized that copper-based catalysts provided considerable advantages for methanol synthesis, allowing considerably lower pressures and, above all, lower temperatures. These catalysts proved to be extremely sensitive to sulfur components. Consequently only after the development of suitable syngas purification systems, mainly to remove sulfur, was the first low-pressure methanol process brought onto the market by the Imperial Chemical Industries Ltd, Great Britain, in 1966, based on a catalyst, which is a mixture of copper oxide, zinc oxide, and alumina. At that time Lurgi Gesellschaft für Wärme und Chemoteknik from Germany also developed a similar low-pressure methanol process.

Most of the methanol plants built in the last twenty years of the twentieth century operate according to the ICI or Lurgi processes, while numerous high-pressure units were converted to the low-pressure system in the second half of the nineteenth century.

The most active catalysts have high Cu content (optimum of about 60% Cu), the maximum quantity being limited by the need for sufficient refractory oxide to prevent Cu sintering. While it is accepted that Al_2O_3 prevents sintering, the role of ZnO in the catalyst has been debated, because pure ZnO is able to catalyze the synthesis of methanol (Strunk *et al.*, 2009a). It must be underlined that ZnO interacts with Al_2O_3 to form a spinel that provides a robust catalyst support. It is generally assumed that the coordination, chemisorption, and activation of carbon monoxide take place on Cu^0 or Cu^+ and that the splitting of hydrogen is heterolytic and takes place on ZnO. However, different points of view on the location and coordination state of Cu^+ sites still exist.

The catalytic mechanism has been at the center of several discussions and reviews, which animated the last quarter of the twentieth century (Behrens *et al.*, 2012).

A further impulse on methanol synthesis studies is the hydrogenation of carbon dioxide. This topic has attracted worldwide research interest in the past fifteen years because the use of CO_2, the most important greenhouse gas, as an alternative feedstock replacing CO in the methanol production can potentially afford an effective way of CO_2 utilization. A recent study shows that a mixture of a proper proportion of CO_2 and CO can increase the yield of methanol (Liu *et al.*, 2003).

As happened for other industrial processes, the methanol synthesis process has stimulated the application of advanced surface science methods to the study of the surface properties of pure Cr_2O_3, ZnO (Zecchina *et al.*, 1996), and nano-sized Cu particles (Armelao *et al.*, 2006).

Supported gold nanoparticles are also utilized in methanol synthesis. However, their extensive industrial application is still not adopted (Haruta, 1997; Bond, 2002; Strunk *et al.*, 2009b). The catalytic activity of finely divided gold particles was initially considered in contrast with accepted opinion on the inactive character of gold. This is early evidence of the "quantum size effect," which states that the physical properties of nanosized particles are different from those of the bulk solid.

1.9 Acetylene Production and Utilization

Acetylene C_2H_2 was discovered in 1836 by Edmund Davy, who identified it as a "new carbure of hydrogen". The French chemist Marcellin Berthelot redis-covered it in 1860 and coined the name acetylene. From 1900 to 1950 it was prepared by the hydrolysis of calcium carbide, a reaction discovered in 1862 by Friedrich Wöhler (1800–1882), pupil of Gmelin and for a while collabora-tor of Berzelius, who is better known for the synthesis of urea. The well-known reaction is

$$CaC_2 + 2H_2O \rightarrow Ca(OH)_2 + C_2H_2$$

Calcium carbide production requires extremely high temperatures from coke and lime, necessitating the use of an electric arc oven, following a method dis-covered in 1982 by Henri Moissan (1906 Nobel Prize in Chemistry). Today acetylene is mainly produced by the partial combustion of methane or appears as a side product in the ethylene stream from the cracking of hydrocarbons. Approximately 400,000 tonne are produced this way, annually.

Acetylene can react with many molecules to give a wide range of industrially significant chemicals, via several reactions, often collected under the name of "Reppe chemistry" from the name of their discoverer, the German chemist Wal-ter Reppe (1892–1969). The chemistry of acetylene was an extremely important chapter of German chemical industry in the period between the World Wars I and II. The discovery and application of homogeneous Reppe catalysts in the first quarter of the twentieth century gave us the opportunity to underline that, when considered from an historical point of view, homogeneous and hetero-geneous catalysis developed in a parallel way. We will return to discuss the catalysts involved in the Reppe chemistry in the chapter devoted to homoge-neous catalysis.

1.10 Anthraquinone Process for Hydrogen Peroxide Production

Another extremely important process today is the catalytic hydrogenation of anthraquinone, which is one of the two-step reactions for the industrial production of hydrogen peroxide. Anthraquinone is obtained industrially in several ways, including extraction from tar by the Friedel–Crafts reaction of benzene and phthalic anhydride in presence of $AlCl_3$ and by acid-catalyzed dimerization of styrene to give a 1,3-diphenylbutene, which then can be transformed to the anthraquinone.

The whole process of H_2O_2 synthesis is illustrated in the following scheme:

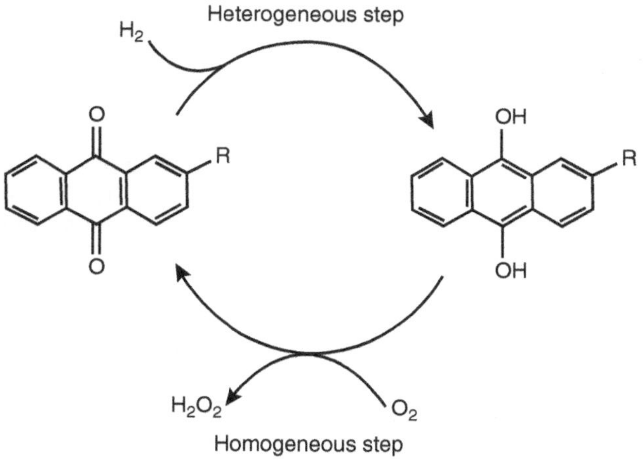

This reaction is playing a fundamental role for the synthesis of the green oxidant H_2O_2, whose current production is increasing at a steady space of 4% and today is estimated to be about 2.2 million ton.

As shown in the scheme reported earlier, the whole process consists of two separate steps, one heterogeneous (hydrogenation) and the other homogeneous (oxidation of reduced molecule to give back anthraquinone). The industrial process was developed in the period 1935–1936 by two researchers of BASF Hans-Joachim Riedl and George Pfleiderer (Riedl and Pfleiderer, 1936). The first commercial production was operated by IG Farbenindustrie in Germany during the World War II. Even today the production of bulk H_2O_2 is based on this process. To illustrate the history of the very innovative ideas that are at the basis of the whole process, in particular, the generation of hydrogen peroxide by reaction of oxygen with hydroquinones under homogeneous

Figure 1.13 Wilhelm Manchot. His fundamental contribution on autoxidation reactions was the basis for the anthraquinone process developed by Hans-Joachim Riedl and George Pfleiderer at BASF in 1936. Source: Photo by Eduard Manchot.

conditions, it is necessary to go back to the early observations in 1901 of the German chemist Wilhelm Manchot (1869–1945) (Manchot, 1901), a scientist of Hungarian origin who in 1895 obtained his PhD at the Technical University of Munich under the supervision of Friedrich Karl Johannes Thiele (Figure 1.13). In 1903 he was nominated professor of inorganic and analytical chemistry at the University of Würzburg where he started to work on problems of autoxidation and activation of catalysts (Manchot and Herzog, 1900, 1901). The fundamental works on autoxidation form the basis of the homogeneous part of the process of H_2O_2 synthesis. The fact that autoxidation could be catalyzed by the presence of certain metallic ions, for instance, copper, iron, manganese, cobalt, and nickel, was studied in detail by James H. Walton and George W. Filson in 1932 (Walton and Filson, 1932). More information can be found in the paper of Goor (1992). As for the heterogeneous step, it is also worthy to underline that the selective reduction of anthraquinone (where only the C=O groups should be attacked by hydrogen leaving the rings unaltered) is not trivial and that the choice of the best hydrogenation (heterogeneous) catalyst has been the subject of many research efforts even in the 1950–2000 period (Campos-Martin, Blanco-Brieva, and Fierro, 2006). Raney nickel catalyst was used initially. However, this catalyst had two serious drawbacks: excessive hydrogenation and rapid deactivation. The second generation of catalysts, based on palladium, was more selective, but due to the excessive activity, hydrogenation of the aromatic rings could not be completely avoided. Thus the need for a highly selective catalyst hydrogenating only the carbonyl group and leaving the aromatic ring intact stimulated the creation (~2000) of a new generation of catalysts based on Ni/B and Ni/Cr/B alloys.

In principle, the obvious alternative to the AO process could be the direct synthesis of H_2O_2 from H_2 and O_2. The first patent on the direct H_2O_2 synthesis using a Pd catalyst was awarded in 1914 to Hugo Henkel and Walther Weber researchers of the Henkel & Cie, in Düsseldorf (Henkel and Weber, 1914). However, little progress was made in the following fifty years because of safety issues (H_2–O_2 mixtures are flammable). Due to the increasing demand for H_2O_2, the direct process attracted renewed interest after 1960, and several industrial and academic laboratories became involved after 1970. Unfortunately, no industrial application has been accomplished to date for the production of bulk H_2O_2, although some successful attempts were realized at DuPont. It is evident that the successful development of a commercialized process would be a major breakthrough in the oxidation process technology. The catalysts used were invariably based on Pd and Pd–Au alloys supported on various oxides, on zeolitic materials, and on resins (Chanchal, 2008). The major problem is the selectivity because the catalysts favoring the synthesis are also very active in the decomposition. For the future of the direct synthesis of hydrogen peroxide, photocatalysis on TiO_2-based catalysts seems to be a good candidate.

References

Allen, A.D. and Senoff, C.V. (1965) Nitrogen-pentammineruthenium(II) complexes. *Journal of the Chemical Society, Chemical Communications*, **24**, 621.

Armelao, L., Barreca, D., Bottaro, G. *et al.* (2006) Recent trends on nanocomposites based on Cu, Ag and Au clusters: a closer look. *Coordination Chemistry Reviews*, **250**, 1294.

Arrhenius, S. (1889) Über die Reaktionsgeschwindigkeit bei der Inversion von Rohzucker durch Säuren. *Zeitschrift für Physikalische Chemie*, **4**, 226–248.

F., Bacon (1620) *Novum Organum Scientiarum*, part II, *prerogative instances*, p. 551.

Badhuri, S. and Mukesh, D. (2000) *Homogeneous Catalysis: Mechanism and Industrial Application*, John Wiley and Sons, N.Y.

Bancroft, W.D. (1897) *The Phase Rule: A Treatise on Qualitative Chemical Equilibrium*. Journal of Physical Chemistry, Ithaca, New York.

Barteau, M.A., Feulner, P., Stengl, R. *et al.* (1985) Formation of methyne intermediates by hydrogenation of surface carbon on Ru(001). *Journal of Catalysis*, **94**, 51.

Behrens, M., Studt, F., Kasatkin, I. *et al.* (2012) The active site of methanol synthesis over Cu/ZnO/Al_2O_3 industrial catalysts. *Science*, **336**, 893–897.

Bent, B.E. (1996) Mimicking aspects of heterogeneous catalysis: generating, isolating, and reacting propose. *Chemical Reviews*, **96**, 1361.

Bernoulli, D. (1738) *Hydrodynamica: sive de viribus et motibus fluidorum commentarii*, Dulsecker, Strasbourg.

Berthelot, M. (1860) Sur la fermentation glucosique du sucre de canne. *Comptes rendus*, **50**, 980–984.

Berthelot, M. (1872) *La Chimie au moyen age*, Imprimerie nationale, Paris.

Berthelot, M. (1879) *Essai de Mécanique Chimique fondée sur la Thermochimie*, Dunod, Paris.

Berthelot, M. (1897) *Thermochimie: donnees et lois numeriques*, vol. 1: *Les lois numeriques*, vol. 2: *Les donnees experimentales*, Gauthier-Villars, Paris.

Berthelot, M. and et Saint Gilles, P. (1862) De la formation et de la décomposition des éthers. *Annales de Physique et de Chimie*, **65**, 385.

Berzelius, J.J. (1836a) *Quelques* Idées sur une nouvelle Force agissant dans les CorpsOrganiques. *Annales de chimie et de physique*, **61**, 146.

Berzelius, J.J. (1836b) Einige Ideen über eine bei der Bildung organischer Verbindungen in der lebenden Natur wirksame, aber bisher nicht bemerkte Kraft. *Jahres-Bericht uber die Fortschritte der Chemie*, **15**, 237–245.

Bligaard, T., Norskov, J.K., Dahl, S. *et al.* (2004) The Brønsted–Evans–Polanyi relation and the volcano curve in heterogeneous catalysis. *Journal of Catalysis*, **224**, 206–217.

Block, B.P. (2008) Antoine-Augustin Parmentier: pharmacist extraordinaire. *Pharmaceutical Historian*, **38**, 6–14.

Bocquet, M.-L., Cerda, J., and Sautet, P. (1999) Transformation of molecular oxygen on a platinum surface: a theoretical calculation of STM images. *Physical Review B*, **59**, 15437.

Bond, G.C. (2002) Gold: a relatively new catalyst. *Catalysis Today*, **72**, 5.

Bosch, C. and Wild, W. (1914) Canadian Patent 153379.

Califano, S. (2012Chapter 2, p. 29, 43,44) *Pathways to Modern Chemical Physics*, Springer Verlag, Berlin Heidelberg.

Campos-Martin, J.M., Blanco-Brieva, G., and Fierro, J.L.G. (2006) Hydrogen peroxide synthesis: an outlook beyond the anthraquinone process. *Angewandte Chemie International Edition*, **45**, 6962.

Chanchal, S. (2008) Direct synthesis of hydrogen peroxide from hydrogen and oxygen: an overview of recent developments in the process. *Applied Catalysis A: General*, **350**, 133–149.

Che, M. (2013) Nobel prize in chemistry 1912 to Sabatier: organic chemistry or catalysis? *Catalysis Today*, **218**, 162–171.

Christensen, C.H. and Nørskov, J.K. (2008) A molecular view of heterogeneous catalysis. *Journal of Chemical Physics*, **128**, 182503.

Clausius, R. (1857) Über die Art der Bewegung welche wir Wärme nennen. *Poggendorf Annalen der Phsiik and Chemistry*, **100**, 353–380.

Cook, E. (1926) Peregrine Phillips, the inventor of the contact process for sulphuric acid. *Nature*, **117**, 419–421.

Cordus, V. (1548) *Pharmacorum Conficiendorum Ratio*. Jacobus Gazellus, Paris.

Cordus, V. (1561) in *De artificiosis extractionibus liber* (ed. C. Gesner). Josias Rihel, Strasbourg.

Crookes, W. (1898) *Rep. of the 68th meeting of the British Association for the Advancement of Science*. Bristol 1898, John Murray, London, p. 3.

Davy, E. (1820) On some combinations of platinum. *Philosophical Transactions of the Royal Society of London*, **110**, 108–125.

Debus, H. (1863) On the conversion of prussic acid into methylamine. *Journal of the Chemical Society*, **16**, 249–260.

Dellamorte, J.C., Barteau, M.A., and Lauterbac, J. (2009) Opportunities for catalyst discovery and development: integrating surface science and theory with high throughput methods. *Surface Science*, Special Issue honoring Prof. G. Ertl, **603**, 1770–1775.

Döbereiner, J.W. (1822) *Zur Gährungs-Chemie und Anleitung zur Darstellung verschiedener Arten künstlicher Weine, Biere u.s.w.* Schmid, Jena.

Döbereiner, J.W. (1823) Propriétés nouvelles et remarquables reconnus au sus-oxide de platine, au sulphure oxidé et â la poussière du même mêtal. *Annales de Chimie et de Physique*, **24**, 91.

Duhem, P. (1898) La loi de phases à propos d'un livre récent *de M. Wilder Bancroft*. *Revue des Questions Scientifiques*, **44**, 54–82.

Faraday, M. (1834) Experimental researches on electricity, Sixth Series. *Philosophical Transactions of the Royal Society of London*, **124**, 55–76.

Farrauto, R.J. and Bartholomew, C.H. (1997) *Fundamentals of Industrial Catalytic Processes*. Chapman & Hall, London, pp. 621–639.

Fischer, F. and Tropsch, H. (1926a) Uber Reduktion und Hydrierung der Kohlenoxids. *Brennstoff-Chemie*, **7**, 97–104.

Fischer, F. and Tropsch, H. (1926b) Process for the production of paraffin-hydrocarbons with more than one carbon atom. US Patent 1746464 A.

Fryzuk, M.D. and Johnson, S.A. (2000) The continuing story of dinitrogen activation. *Coordination Chemistry Reviews*, **379**, 200–202.

Fusinieri, A. (1824) Sulla Causa delle Combustione di Sostanze Gasose per Mezzo delle Superfice di Alcuni Metalli. *Giornale di Fisica*, **7**, 371–376, 443–449.

Goor, G. (1992) in *Hydrogen Peroxide: Manufacture and Industrial Use for Production of Organic Chemicals* (ed. G. Strukul). Kluwer Academic Publishers, Dordrecht, the Netherlands.

Guldberg, C.M. and Waage, P. (1864) Studies concerning affinity. *C. M. Forhandlinger: Videnskabs-Selskabet i Christiana*, **35**, 92–111.

Gustafsson, K. and Anderson, S. (2004) Infrared spectroscopy of physisorbed and chemisorbed O_2 on Pt(111). *Journal of Chemical Physics*, **120**, 7750.

Haber, F. (1891) Promotion in organischer Chemie über das Thema: *Über einige Derivate des Piperonals*, vol. 9, issue 14 of Dissertationen. Berlin Universität, Publisher, G. Schade.

Haber, F. (1922) Über die Darstellung des Ammoniaks aus Stickstoff und Wasserstoff. *Naturwissenschaften*, **10**, 1041.

Haber, F. and Le Rossignol, R. (1907) Mitteilungen Über das Ammoniak-Gleichgewicht. *Chemische Berichte*, **40**, 2144.

Haber, F. and Le Rossignol, R. (1908a) Bestimmung des Ammoniak-Gleichgewichtunter druck. *Zeitschrift fur Elektrochemie*, **14**, 181.

Haber, F. and Le Rossignol, R. (1908b) Die Lage des Ammoniakgleichgewicht. *Zeitschrift fur Elektrochemie*, **14**, 513.

Haber, F. and Le Rossignol, R. (1910) Über die Darstellung des Ammoniaks aus Stickstoff und Wasserstoff. *Zeitschrift fur Elektrochemie*, **16**, 244.

Haber, F. and van Oordt, G. (1904) Über Beryllium verbindungen. *Zeitschrift fur anorganische und allgemeine Chemie*, **43**, 111.

Harcourt, A.V. and Esson, W. (1866a) On the law of connexion between the conditions of a chemical change and its amount. *Philosophical Transactions of the Royal Society of London*, **157**, 117–137.

Harcourt, A.V. and Esson, W. (1866b) The oxidation of oxalic acid with potassium permanganate. *Philosophical Transactions of the Royal Society of London*, **156**, 193.

Harcourt, A.G.V. and Esson, W. (1913) On the variation with temperature of the rate of a chemical change. *Philosophical Transactions of the Royal Society of London*, **212A**, 187–204.

Haruta, M. (1997) Size and support-dependency in the catalysis of gold. *Catalysis Today*, **36**, 153–166.

Henkel, H. and Weber, W. (1914) Manufacture of hydrogen peroxide. US Patent 1108,752, August 22.

Henry, W. (1824) On the action of finely divided platinum on gaseous mixtures and its application to their analysis. *Philosophical Transactions of the Royal Society of London*, **114**, 266–289.

Herapath, J. (1821) A mathematical inquiry into the causes, laws and principal phenomena of heat, gases, gravitation, etc. *Annals of Philosophy*, **1**, 273–293, 340–351, 401–416.

Herzfeld, K.F. (1925) *Kinetische Theorie der Wärme In Müller-Pouillets Lehrbuch der Physik*, vol. **3**. F. Viewig und Sohn, Braunsweig.

Honkala, K., Hellman, A., Remediakis, I.N. *et al.* (2005) Ammonia synthesis from first-principles calculations. *Science*, **307**, 555–558.

Hood, J.J. (1878) On the law of chemical change. *Philosophical Magazine*, **6**, 371–383.

Horstmann, A. (1868) Zur Theorie der Dissociationserscheinungen. *Ber. Deutsch. Chem. Ges.*, **1**, 210–215.

Horstmann, A. (1873) Theorie der Dissociation. *Annalen der Chemie und Pharmacie*, **170**, 192–210.

Joule, J. (1851) Some remarks on heat and the constitution of elastic fluids. *Memoirs of the Literary and Philosophical Society of Manchester*, **9**, 107–114.

Kinnersley, A.D., Darling, G.R., Holloway, S., and Hammer, B. (1996) A comparison of quantum and classical dynamics of H_2 dissociation on Cu(111). *Surface Science*, **364**, 219–234.

Kirchhoff (1811a) Mémoires de l'Académie impériale des science de St. Pétersbourg, **4**, 27–34.

Kirchhoff, S.K. (1811b) Die Entdeckung der leichten Umwandlung der Stärke in Zucker. *Bulletin Neusten Wiss Naturwiss*, **10**, 88–92.

van Klooster, H.S. (1938) Le Chatelier and the synthesis of ammonia. *Journal of Chemical Education*, **15**, 289.

Kratzer, P., Hammer, B., and Nørskov, J.K. (1996a) Geometric and electronic factors determining the difference in reactivity of H_2 on Cu(100) and Cu(111). *Surface Science*, **359**, 45–53.

Kratzer, P., Hammer, B., and Nørskov, J.K. (1996b) A theoretical study of CH_4 dissociation on pure and gold-alloyed Ni(111) surfaces. *Journal of Chemical Physics*, **105**, 5595–5600.

Krönig, A.K. (1856) Grundzüge einer Theorie der Gase. *Poggendorf Annalen der Phsiik and Chemistry*, **99**, 315–322.

Kühne, W. (1877) *Über das Trypsin(Enzym des Pankreas)*, vol. 1, Verhandlungen des naturhistorisch-medicinischen Vereins zu Heidelberg. Neue Folge, Heidelberg, pp. 194–198.

Kummer, J.T. and Emmett, P.H. (1953) Fischer–Tropsch synthesis mechanism studies. The addition of radioactive alcohols to the synthesis gas. The synthesis gas. *Journal of the American Chemical Society*, **75**, 5177–5183.

Le Châtelier, H. (1884) Sur un énoncé général des lois des équilibres chimiques. *Comptes Rendus*, **99**, 786–789.

Le Châtelier, H. (1936) *De la Méthode dans les Sciences Expérimentales*. Dunod, Paris.

Leicester, H. and Berzelius, J.J. (1970–80) *Dictionary of Scientific Biography*, vol. 2. Charles Scribner's Sons, New York, pp. 90–97.

Liebig, J. (1839) Über die Erscheinungen der Gährung, Fäulniss und Verwesung, und ihre Ursachen. *Annales Pharmaceutiques Francaises*, **30**, 250–287.

Liu, X.M., Lu, G.Q., Yan, Z.F., and Beltramini, J. (2003) Recent advances in catalysts for methanol synthesis via hydrogenation of CO and CO_2. *Industrial & Engineering Chemistry Research*, **42**, 6518–6530.

Lloyd, L. (2011) *Handbook of Industrial Catalysts*, Springer Publisher, pp. 35–39.

Malaguti, F. (1853) Exposition de quelques faits relatifs a l'action réciproque, des sels solubles. *Annali di Chimica*, **37**, 198–206.

Manchot, W. (1901) Über Sauerstoffactivirung. *Liebigs Annalen der Chimie*, **314**, 177–199.

Manchot, W. and Herzog, J. (1900) Über das Verhalten des Kobaltocyankaliums und der Chromoverbindungen gegen Sauerstoffgas. *Chemische Berichte*, **33**, 1742–1750.

Manchot, W. and Herzog, J. (1901) Die Autoxydation des Hydrazobenzols. *Justus Liebigs Annalen der Chemie*, **316**, 331–332.

Maxwell, J.C. (1860) Illustrations of the dynamical theory of gases. *Philosophical Magazine*, **19**, 19–32, 20, 21–27.

McEwen, J.S., Bray, J.M., Wu, C., and Schneider, W.F. (2012) How low can you go? Minimum energy pathways for O_2 dissociation on Pt(111). *Physical Chemistry Chemical Physics*, **14**, 16677–16685.

Mendius, O. (1862) Benzylamine synthese. *Ann.*, **121**, 144.

Messel, R. and Squire, W.S. (1876) On the manufacture of sulphuric anhydride. *Chemical News*, **33**, 177.

Moissan, H. and Moureu, C. (1896) Action de l'acétyléne sur le fer, le nickel et le cobalt réduits par l'hydrogène. *Comptes Rendus*, **122**, 1241–1247.

Mortensen, J.J., Morikawa, Y., Hammer, B., and Nørskov, J.K. (1997) Density functional calculations of N_2 adsorption and dissociation on a Ru(0001) surface. *Journal of Catalysis*, **169**, 85–92.

Murphy, M.J., Skelly, J.F., Hodgson, A., and Hammer, B. (1999) Inverted vibrational distributions from N_2 recombination at Ru(001): evidence for a metastable molecular chemisorption well. *Journal of Chemical Physics*, **110**, 6954–6966.

Natta, G. and Strada, M. (1930) La Sintesi dal gas d'acqua di alcoli superiori al metilico. *Giornale di chimica industriale e applicata*, **XII**, 169.

Naumann, A.N.F. (1867) Über specifische Wärme der Gase für gleiche Volume bei constantem Drucke. *Justus Liebigs Annalen der Chemie*, **142**, 265–283.

Niemantsverdriet, J.W. and van der Kraan, A.M. (1981) On the time-dependent behavior of iron catalysts in Fischer-Tropsch synthesis. *Journal of Catalysis*, **72**, 385–388.

Nørskov, J.K. and Bligaard, T. (2012) The catalyst genome. *Angewandte Chemie International Edition*, **52**, 776–777.

Oliveira, E.L.G., Grande, C.A., and Rodrigues, A.E. (2010) Methane steam reforming in large pore catalyst. *Fuel Processing Technology*, **65**, 1539–1550.

Ostwald, W. (1894) Definition der Katalyse. *Zeitschrift für physikalische Chemie*, **15**, 705–706.

Ostwald, W. (1902) Summary of a lecture hold by Ostwald at Amburg in 1901 at the Deutsche Naturforscher Versammlung. Über Katalyse. *Nature*, **65**, 522.

Ostwald, W. (1910) Über Katalyse. *Annalen der Naturphilosophie*, **9**, 1–25.

Parmentier, A.A. (1773) *Examen chimique des pommes de terre, dans lequel on traite des parties constituantes du froment et du riz*. Ed. Didot, Paris.

Parmentier, A.A. (1774) *Méthode facile pour conserver à peu de frais les grains et les farines*. Ed. Barrois l'aîné, Paris.

Patart G. (1921) French Patent, 540, 543.

Patart, G. (1925) Synthetic methanol controversy. *Industrial & Engineering Chemistry*, **17**, 859.

Pattison Muir, M.M. (1885) *The Elements of Thermal Chemistry*. Macmillan, London.

Payen, A. (1838) Mémoire sur la composition du tissu propre des plantes et du ligneux. *Comptes Rendus*, **7**, 1052–1056.

Payen, A. and Persoz, J.-F. (1833) Mémoire sur la diastase, les principaux produits de ses réactions et leurs applications aux arts industriels. *Annales de Chimie et de Physique*, **53**, 73–92.

Peregrine, P. (1832) Patent granted to Peregrine Phillips, Jr. of Bristol, in the county of Somersetshire March 21, 1831. *Journal of the Franklin Institute*, new series, **9**, 180–182.

Pfaundler, L. (1867) Beiträge zur chemischen Statik. *Poggendorf Annalen der Phsiik and Chemistry*, **131**, 55–85.

Pfaundler, L. (1876) Über A. Horstmann's Dissociations theorie und über die Dissociation der fester Körper. *Chemische Berichte*, **9**, 1152–1157.

Pfeffer, W.F.P. (1887) *Osmotische Untersuchungen*, W. Engelmann, Leipzig.

Pichler, H. (1952) in *Advances in Catalysis and Related Subjects*, vol. **IV** (eds. W. Frankenburg, E. Rideal, and V. Komarewsky). Academic Press, New York, p. 271.

Pier M. (1923) Badische Anilin Soda Fabrik. US Patent 1,558,559; 1,569,775.

Riedl, H. and Pfleiderer, G. (1936). US Patent 2,158,525.

Robertson, A.J.B. (1975) The early history of catalysis. *Platinum Metals Review*, **19**, 68–76.

Rossi, P. (1978) *Francis Bacon: From Magic to Science*. Taylor & Francis Publisher.

Sabatier, P. (1913) *La Catalyse en chimie organique*. Ed. Ch. Béranger, Paris.

Sabatier, P. and Mailhe, A. (1904) Synthesis of a new series of tertiary alcohols from cyclohexanol. *Comptes Rendus*, **138**, 245–250.

Sabatier, P. and Murat, M. (1912) Direct reduction of diphenylethanes: preparation of 11 cvclohexylethanes. *Comptes Rendus*, **154**, 1771–1776.

Sabatier, P. and Senderens, J.-B. (1897) Action du nickel sur l'éthylène. Synthèse de l'éthane. *Comptes Rendus de l'Academie des Sciences Paris*, **124**, 1358–1360.

Sabatier, P. and Senderens, J.B. (1902) *Journal of the Chemical Society*, **82**, 333, 1902) New synthesis of methane. *Comptes Rendues*, Action of hydrogen on acetylene in presence of nickel, **134**, 514–516.

Sainte-Claire Deville, H. (1856) Mémoire sur la production des températures très elevées. *Annales de Chimie et de Physique*, **46**, 182–203.

Sainte-Claire Deville, H. (1865) Du phénomène de la dissociation dans les flammes homogènes. *Comptes Rendus*, **60**, 884–889.

Sainte-Claire Deville, H. (1866) *Leçons sur la dissociation*, in *Leçons de chimie*. Société chimique de Paris, Hachette & Cie, Paris, pp. 255–378.

Sainte-Claire Deville, H. (1867) *Leçons sur la dissociation*. Hachette & Cie, Paris, p. 85.

Sainte-Claire Deville, H. (1869) Leçons sur l'affinité 1867, in *Leçons de chimie*, Société chimique de France, Paris, pp. 1–85.

Schönbein, C.F. (1848) On some chemical effects produced by platinum. *Memoirs and Proceedings of the Chemical Society*, **3**, I, 7.

Schulz, H. (1999) Short history and present trends of Fischer–Tropsch synthesis. *Applied Catalysis A: General*, **186**, 3–12.

Shorter, J. (1980) A. G. Vernon Harcourt, a founder of chemical kinetics and a friend of Lewis Carroll. *Journal of Chemical Education*, **57**, 411–416.

Slama, F. and Wolf, H. (1921) German Patent 291792, 1921; US Patent 1371004.

de Smit, E. and Weckhuysen, B.M. (2008) Fischer–Tropsch synthesis: on the renaissance of iron-based multifaceted catalyst deactivation behavior. *Chemical Society Reviews*, **37**, 2758–2781.

Squire, W.S. (1875) Oil of vitriol. British Patent Number: 3278 published 18 September 1875, by London Eyre and Spottiswood at the Great Seal Patent Office c.

Stohmann, F. (1894) Über dem Wärmewerth der Bestandtheile der Nahrungsmittel. *Zeitschrift für Biologie*, **31**, 364–391.

Storch, H.H., Golumbic, N., and Anderson, R.B. (1951) *The Fischer–Tropsch and Related Syntheses*. Wiley, New York.

Strunk, J., Kahler, K., Xia, X., and Muhler, M. (2009a) The surface chemistry of ZnO nanoparticles applied as heterogeneous catalysts in methanol synthesis. *Surface Science*, **603**, 1776–1783.

Strunk, J., Kahler, K., Xia, X. *et al.* (2009b) Au/ZnO as catalyst for methanol synthesis: the role of oxygen vacancies. *Applied Catalysis A: General*, **359**, 121–128.

Thénard, L.J. (1818) Observations sur des nouvelles combinaisons entre l'oxigène et divers acides. *Annales de Chimie et de Physique*, **8**, 306–312.

Thomas, C.L. (1970) *Catalytic Processes and Proven Catalysts*. Academic Press, New York, pp. 182–184.

Topham, S.A. (1955) in *Catalysis, Science and Technology*, vol. 7 (eds J.R. Anderson and M. Boudart), Springer Book, Berlin, p. 1.

Toulhoat, H. and Raybaud, P. (eds) (2013, Chapters 1–3) *Catalysis by Transition Metal Sulphides from Molecular Theory to Industrial Application*. Editions TECHNIP, Paris, XXXI, pp. 1–73.

van't Hoff, J.H. (1884a) in *Etudes de dynamique chimique* (ed. F. Muller), Amsterdam, pp. 114–118.

van't Hoff, J.H. (1884b) *Etudes de dynamique chimique*, vol. **3**, Recueil des Travaux Chimiques, des Pays-Bas, pp. 333–336.

van't Hoff, J.H. (1886) *L'Équilibre chimique dans l'état dilué gazeux ou dissous*, Ed. Norstedt & Söner, Stockholm.

van't Hoff, J.H. (1887) Die Rolle des osmotischen Druckes in der Analogie zwischen Lösungen und gases. *Zeitschrift für physikalische Chemie*, **1**, 481–508.

van't Hoff, J.H. (1894) Wie die Theorie der Lösungen entstand. *Berichte der Deutschen Chemischen Gesellschaft*, **27**, 6–19.

van't Hoff, J.H. (1898) *Vorlesungen über Theoretische und Physikalische Chemie*, Friedrich Viewig und Sohn, Braunschweig.

Walton, J.H. and Filson, G.W. (1932) The direct preparation of hydrogen peroxide in a high concentration. *Journal of American Chemical Society*, **54**, 3228–3229.

Waterston, J.J. (1893) On the physics of media that are composed of free and perfectly elastic molecules in a state of motion. *Philosophical Transactions of the Royal Society of London*, **183A**, 5–79.

Weckhuysen, B.M. and Keller, D.E. (2003) Chemistry, spectroscopy and the role of supported vanadium oxides in heterogeneous catalysis. *Catalysis Today*, **78**, 25–46.

de Wilde, P. (1874) Vermischte Mittheilungen. *Chemische Berichte*, **7**, 352–357.

Wilhelmy, L. (1850) Über das Gesetz, nach welchem die Einwirkung der Säuren auf den Rohrzucker stattfindet. *Annals of Physics*, **81**, 413–428, 499–526.

Williamson, A. (1850) Theory of aetherification. *Philosophical Magazine*, **37**, 350–356.

Winkler, C. (1875) Versuche über die Überfurung der schwefligen Saure in Schwefelsaureanhydrid durch Contactwirkung berufs. Darstellung von rauchender schwefel Saure. *Politechnisches Journal von Dingier, B*, **218**, 128–139.

Zecchina, A., Scarano, D., Bordiga, S. *et al.* (1996) IR studies of CO and NO adsorbed on well characterized oxide single microcrystals. *Catalysis Today*, **27**, 403–435.

2

Historical Development of Theories of Catalysis

2.1 Heterogeneous Catalysis

The development of heterogeneous catalysts by Ostwald, Haber, Fischer, and Tropsch and coworkers, strongly oriented toward industrial applications, had such a deep influence on the economy of the world as to reframe the economic and social structures of all advanced societies. In the meantime, however, the chemical community strongly focused its interests on the problem of understanding the mechanisms underlying the progress of heterogeneous catalysis. This interest fueled the development of surface physical research aimed at clarifying the role of surface electron plasma and its relationship with crystal morphology, defects, and dislocations.

The understanding of the elementary mechanisms of heterogeneous catalysis made a great step forward in 1916 when Irving Langmuir developed a theory of the chemisorption of gases on metallic supports, which rapidly became the starting point for the modern theories of heterogeneous catalysis (Langmuir, 1916). According to Langmuir's model, residual valences favor the adsorption of a gas on a metallic surface. In this way true compounds of variable composition were formed between the gas and metal, and these favored the reaction (Langmuir, 1917, 1918). The chemical character of the surface forces responsible for catalytic activation of molecules, already hypothesized by Sabatier and Senderens, was thus fully recognized.

Irving Langmuir (1881–1957) was an American chemist and physicist (Figure 2.1). Trained as an industrial chemist, he soon showed a high inventive ability and profound technical skill. Very important in his life was a doctoral thesis work at Göttingen under the supervision of Walther Nernst, who introduced him to the subject of catalysis through the study of the formation of nitric oxide from air in the vicinity of a glowing Nernst filament.

After the stimulating experience in Nernst's laboratory, he went back to the United States in 1909, where he joined the General Electric Company and worked there until 1950. Thanks to his technical capability, he produced important innovations such as a new type of gas-filled incandescent lamp

The Development of Catalysis: A History of Key Processes and Personas in Catalytic Science and Technology,
First Edition. Adriano Zecchina and Salvatore Califano.
© 2017 John Wiley & Sons, Inc. Published 2017 by John Wiley & Sons, Inc.

Figure 2.1 Irving Langmuir (1857–1981). Langmuir obtained the Nobel Prize *for his discoveries and investigations in surface chemistry.* Image of Langmuir is in the public domain.

(Langmuir and Orange, 1913) that later led to the discovery of atomic hydrogen used in the atomic hydrogen welding process. In addition he invented the first high-vacuum electron tubes and the first high-emission electron cathode tube. Attracted by Robert Williams Wood's preparation of concentrated atomic hydrogen in an electric discharge tube, he also developed the atomic hydrogen welding torch, in which large amounts of atomic hydrogen are produced by an arc between tungsten electrodes in hydrogen, and the atoms are allowed to recombine on the metal to be heated.

Langmuir's research interests had a sudden change when he read Gilbert N. Lewis's paper on the structure of the atom as well as on chemical bonding theory. Fascinated by atomic structure, he developed his "octet theory" of the atomic structure, in which Bohr's centrally orbiting electrons were replaced by electrons distributed in regions throughout the atom, each electron being stationary in its region or describing a restricted orbit within the region. His formidable capability of explaining even complex problems with fascinating clarity gave rise to his most noted publication, the famous 1919 article *The arrangement of electrons in atoms and molecules* (Langmuir, 1919a, b), which made him known as the true representative of the theory of the chemical bond. In a very short while, he published in the same year another stimulating paper on the same subject (Langmuir, 1919a, b), followed in 1920 by a third paper (Langmuir, 1920) that ensured his reputation as the true American expert of the chemical bond theory. Langmuir's presentation skills were largely responsible for the popularization of the "octet theory," although the credit for the theory itself belongs mostly to Lewis.

In 1932 Langmuir, while still at General Electric, received the 1932 Nobel Prize in Chemistry for his fundamental contributions to the understanding of surface chemistry.

The Langmuir Laboratory for Atmospheric Research near Socorro, New Mexico, was named in his honor as was the *Journal of the American Chemical Society for Surface Science*. He was the first to observe stable adsorbed monatomic films on tungsten and platinum filaments, and was able, after experiments with oil films on water, to formulate a general theory of adsorption. He also studied the catalytic properties of such films.

Langmuir's theory offered a simple mechanism of attack of the gas molecules on the catalyst surface.

To simplify the mathematical treatment, Langmuir assumed that the metallic surface was uniform and formed by an array of sites all energetically identical and non-interacting, which would adsorb just one molecule from the gas phase. He also assumed that the adsorbed molecules did not react together and that they formed at most a single monolayer. The sites involved are normally considered active sites. The Langmuir adsorption isotherm results from this model. With these simplifications, the theory assumed a connection at constant temperature between the monolayer fraction θ of adsorbed gas molecules at pressure P, according to the famous relationship

$$\theta = \frac{\alpha \cdot P}{1 + \alpha \cdot P} \tag{2.1}$$

where α is a constant characteristic of the bond energy between the gas and the substrate, representing the ratio between the equilibrium constants of the direct (adsorption) and inverse (desorption) reactions. The first approximation α is inversely proportional to temperature. The curves of θ as a function of P are known as Langmuir's adsorption isotherms.

Langmuir was already aware that the assumption of identical and non-interacting sites was an approximation that would not hold for real surfaces, when he wrote (Langmuir, 1922):

> *Most finely divided catalysts must have structures of great complexity. In order to simplify our theoretical consideration of reactions at surfaces, let us confine our attention to reactions on plane surfaces. If the principles in this case are well understood, it should then be possible to extend the theory to the case of porous bodies. In general, we should look upon the surface as consisting of a checkerboard.*

In other words, exposed faces of a solid catalyst can contain terraces, ledges, kinks, and vacancies at sites having different coordination numbers.

Similarly, nanoscopic crystals have edges and corners exposing atoms with different coordination numbers (Taylor, 1925; Che and Bennett, 1989), which are associated with different reactivity and catalytic activity. Between 1937 and 1938, the hypothesis of a single gas monolayer adsorbed onto the metal surface was considered too limiting, except at very low pressures. The Hungarian chemist Stephen Brunauer (1903–1986) in collaboration with Paul Hugh

Emmett (1900–1985) and with the future father of the atomic bomb, Edward Teller (1908–2003) at that time at the George Washington University, modified Langmuir's adsorption theory, developing the Brunauer–Emmett–Teller (BET) theory. The theory's name originated from the initials of their names to take into account the possibility of formation of multilayers (Emmett and Brunauer, 1937; Brunauer, Emmett, and Teller, 1938).

Stephen Brunauer emigrated in 1921 to the United States, and in 1929 gained an MS degree from the George Washington University. At the same time, he had already started a long-standing association with Paul Emmett, publishing in collaboration with him a paper on synthetic ammonia catalysts in 1933. In this period, he studied at the Johns Hopkins University, where he met Professor Joseph Christie Whitney Frazer who stimulated his interest in the adsorption of gases and liquids on crystal surfaces. In 1933, he obtained a PhD from Johns Hopkins University with a thesis entitled "Adsorption of nitrogen on iron synthetic ammonia catalyst" (Emmett and Brunauer, 1933). This work led him to develop in 1938 with Paul Emmett and Edward Teller the "point B" method for surface area determination, in the framework of a theory of the adsorption isotherms. The model considered multilayer adsorption processes in contrast with existing monolayer mechanisms (Brunauer, Emmett, and Teller, 1938).

The BET isotherms explain better the physisorption on non-microporous materials in which the molecules are bound to the solid surface by weak van der Waals forces.

In 1937, Paul Emmet, after his PhD thesis at the California Institute of Technology, became professor at the Chemical Engineering Department at Johns Hopkins University, where he collaborated with Brunauer. After a period as participant to the Manhattan project, he went back to Johns Hopkins University in 1955 and retired in 1971.

Edward Teller was a Hungarian-born American theoretical physicist of Jewish origin, known as "the father of the hydrogen bomb." He actively participated in the Manhattan project. His contribution to the adsorption theories, mainly realized when he was at the George Washington University, represents only a small part of his scientific activity.

2.2 Chemical Kinetics and the Mechanisms of Catalysis

The advent of the twentieth century has seen a new and interwoven evolution of chemical kinetics and reaction mechanism theories. These two aspects of science are often present together in the scientific biographies of many scientists. This is the case in the Hungarian polymath Mihály (Michael) Polanyi (1891–1976), a very important and complex personality involved not only in science but also in philosophy and economy (Figure 2.2).

(a)   (b)

(c)

Figure 2.2 (a) Michael Polanyi (1891–1976), (b) Henry Eyring (1901–1981), and (c) John Polanyi (1929–). They can be considered as the main protagonists of the development of twentieth-century chemical kinetics. To Michael Polanyi and Eyring, we owe the formulation of the famous transition-state theory. John Polanyi, son of Michael, received the 1986 Nobel Prize for *contributions concerning the dynamics of chemical elementary processes.*
Source: (a) Reproduced with permission of J. Polanyi. (b) The Eyring image is in the public domain. (c) Reproduced with permission of J. Polanyi.

Although very young, he published in 1913 his first scientific paper (Polanyi, 1913). After serving in the Austrian army during World War I, in 1917 he completed a doctoral thesis in physical chemistry at the University of Budapest (Polanyi, 1917) where he became associated with Leo Szilard and John von Neumann.

Polanyi, although Jewish, was baptized in the Catholic Church. Nevertheless, he left Budapest during the anti-Semitic purges of the Horthy regime in 1919 and visited Karlsruhe where, under the supervision of Georg Bredig, professor of physical chemistry, he completed his thesis work on adsorption and began to work on reaction kinetics (Polanyi, 1920).

In the autumn of 1920, Michael Polanyi was appointed at the Kaiser Wilhelm Gesellschaft in Berlin-Dahlem, where he had the opportunity to meet Fritz Haber, director of the Institute of Physical Chemistry.

In 1921, when working on surface adsorption, Polanyi pursued investigations in different areas (Polanyi, 1921a, b, 1934) of solid materials and on long-range intermolecular interactions. A significant focus of its activity was the adsorption of gases on solid surfaces where he assumed the operation of van der Waals-like forces. For this research he was invited by Haber to give an account of this adsorption theory. In the discussions both Nernst and Einstein pointed out that they did not believe in the existence of the long-distance interactions postulated by Polanyi, interactions impossible in the framework of the Bohr's theory. Despite this discouraging event, Polanyi did not abandon his idea, as proved in his paper in collaboration with Fritz London on the explanation of these forces based on quantum mechanical ideas (London and Polanyi, 1930). Meanwhile, he was also involved, in collaboration with his friend Wigner, in attempts to describe chemical reactions at the molecular level (Polanyi and Wigner, 1925).

In the autumn of 1923, Polanyi was promoted to a position of independent member of the Institute of Physikalische Chemie. The American chemist Henry Eyring came from Wisconsin to study with Polanyi in Berlin in 1929–1930. He arrived just when Polanyi was thinking of using contour maps for representing the potential energy of a hydrogen/hydrogen bromide system as the bromine comes near one hydrogen and the other hydrogen recedes. A ball rolling on a surface would describe the position of all three atoms. This approach to the energy of activation, with the representation of the transition state as a ball rolling over a peak or hump, proved extremely fruitful, both in the short and long run. They used semi-empirical methods, based in quantum mechanics and relying on experimental data for vibrational frequencies and energies of association. Polanyi and Eyring published together two papers during the 1930–1931 period, followed by Polanyi's publication of the book *Atomic Reactions* in 1932 (Polanyi, 1932a, b).

In his career, Michael Polanyi carried on research with some of the most prominent scientists of the age, including Eugene Wigner with whom he shared a lifelong friendship (Polanyi and Wigner, 1925, 1928). In 1926 he became professor of chemistry at the Kaiser Wilhelm Institute in Berlin, where he worked from 1920 to 1933 (Polanyi, 1928, 1932a, b; Polanyi and Schmid, 1929). The new anti-Semitic policy of Hitler convinced him to accept a chair in physical chemistry at the University of Manchester where he continued with the studies of chemical kinetics begun in Berlin, publishing three papers of fundamental importance for the theory of kinetic processes (Evans and Polanyi, 1935, 1937, 1938).

The Manchester period was also a landmark for the development of the theory of chemical kinetics for other reasons. In Manchester Polanyi met Meredith Evans (1904–1952), who arrived from Princeton and who would become a strict Polanyi collaborator, leading in the next few years to the joint publication of a series of papers on the theory of chemical reactivity that has had a profound influence on all subsequent developments in this field.

Some of his students or coworkers came with him from Berlin; among them was Juro Horiuchi, whose activity will be detaily described in the following. The next years at Manchester, Meredith G. Evans was to become one of Polanyi's principal coworkers on reaction kinetics and the theory of the transition state.

Among the products of this collaboration was the formulation of a theory of chemical reactions now well known as "the transition-state theory (TST)". These studies, which included the consideration of the nature of the forces of interaction between atoms and molecules, were extended by Evans and his coworkers to the structure and nature of the forces in liquids and solutions, a field to which in later years he devoted considerable attention (Evans, 1938).

In 1931, Henry Eyring and Michael Polanyi constructed a potential energy surface for the reaction

$$H + H_2 \rightarrow H_2 + H$$

This surface is a three-dimensional diagram, based on quantum mechanical principles as well as experimental data, on vibrational frequencies, and on energies of dissociation.

A year after the Eyring and Polanyi construction, Hans Pelzer and Eugene Wigner made an important contribution by following the progress of a reaction on a potential energy surface (Wigner, 1938; Pelzer and Wigner, 1932). The importance of this work was that it was the first discussion of the concept of a saddle point in the potential energy surface.

Parallel developments had been made by Eyring and his coworkers; as a consequence the synergic collaboration of the Manchester–Princeton school opened up a new era in the study of rate processes (Heyne and Polanyi, 1928).

One of the most important features introduced by Eyring, Polanyi, and Evans was the notion that activated complexes are in quasi-equilibrium with the reactants. The rate is then directly proportional to the concentration of these complexes multiplied by the frequency ($k_B T/h$).

Perhaps the most important Polanyi associate during the Manchester period was the Japanese chemist Juro Horiuti (1901–1979), whom he had already met when working in Germany. Horiuti followed him to Manchester where they collaborated, writing pivotal papers on electrode kinetics and heterogeneous catalysis (Horiuti and Polanyi, 1933–1934, 1933a, b, 1934). The result of this collaboration is the Horiuti–Polanyi mechanism on the hydrogenation of alkenes by heterogeneous catalysts proposed by Horiuti and Polanyi in 1934. This mechanism consists of three steps: (1) alkene adsorption on the surface of the hydrogenated metal catalyst, (2) hydrogen migration to the β-carbon of the alkene with formation of a σ-bond between the metal and α-C, and finally (3) reductive elimination of the free alkane. The validity of this mechanism was fully checked and then confirmed in the successive 80 years by later experimental and theoretical studies.

During and after World War II, Polanyi, although continuing to work in the field of theoretical processes of catalysis (Cremer and Polanyi, 1933), increasingly turned his attention from science to economics and philosophy. In 1948 Polanyi became the chair of social studies at Manchester. He founded in 1940 the Society for Freedom in Science to defend and promote a liberal conception of science as free enquiry against the instrumental view that science should exist primarily to serve the needs of society.

Very interesting and peculiar is the life of the second protagonist of chemical kinetics, Henry Eyring (1901–1981). Eyring cannot be fully understood without taking into account that he was born in a Mormon community and that he was in fact intimately connected all his life to Mormon religious beliefs. This fact strongly influenced his future career (Figure 2.2). After his doctoral degree in chemistry in 1927 at the University of California, Berkeley, he was recruited by the University of Princeton in 1931, where he rapidly made very significant progress, being soon promoted to a position of associate professor and then full professor (1938). At Princeton, his reputation as a researcher became well established. However, in 1946 the University of Utah offered him a position that he accepted, despite the much higher rank offered by the University of Princeton. It is clear that his religious belief definitely influenced this unexpected and peculiar decision. The chemistry building on the University of Utah campus is now named in his honor.

While in Princeton, he published several important papers mostly in collaboration with Michael Polanyi (Polanyi and Eyring, 1930; Eyring and Polanyi, 1931). In these papers, he presented the first potential energy surface for chemical reactions, using a new language based on saddle points or cols, valleys, barriers, passes, and reaction coordinates to describe the energy barrier (E) between reactants to products. Among the most famous of Eyring's contributions, there are two most frequently cited papers (Eyring, 1935, 1936) as well as the standard text *Quantum Chemistry* (Eyring, Walter, and Kimball, 1944).

Figure 2.3 Potential energy profile from reactants to products and transition-state energy following the Eyring theory.

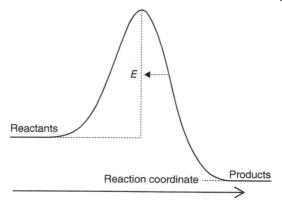

Eyring's formulation in 1935 of the TST (Eyring, 1935) is probably the one, among his many important scientific contributions, that has had the most far-reaching influence and lasting value. His famous equations linking the rate constant with activation energy is of general validity and is one of the milestones of chemical kinetics. A pictorial view of the reaction energy profile on passing from reactants to products now in use in all textbooks of kinetics and catalysis is shown in Figure 2.3. An important factor ensuring the success of this equation is that the TST led to a better understanding of mechanisms of very complicated reactions.

Despite its virtual simplicity and practical applicability, it took a long time for TST to gain wider popularity among chemists. Since it was based on quite new quantum mechanical techniques and statistical ideas, not widely known in the chemist's milieu, it was necessary to introduce the concept of the "activation barrier" to make the potential profile understandable for the nonexpert. It was not before the late 1950s that the concept of potential barrier slowly became part of university curricula all over the world (Figure 2.3).

It is now fully recognized that all chemical transformations pass through unstable transition-state structures with lifetimes near 10^{-13} s, the time of a single-bond vibration.

Linus Pauling in 1946, who was certainly inspired by Polanyi and Eyring's works, advanced an elegant theory linking the TST and catalysis. Following this theory, originally intended to explain the acceleration rate induced by the biological catalysts "enzymes" (Pauling, 1946), the rate acceleration is associated to the capacity of the catalyst to stabilize the transition state and hence to lower the energy barrier. The stabilization is the result of the complementarity between the structure of the catalysts and that of the transition state (selective recognition). It is now fully accepted that this theory is also valid for nonbiological catalysts, like heterogeneous and homogeneous catalysts.

Michael Polanyi's son, John Polanyi, inventor of the infrared luminescence technique, continued the research on reaction dynamics (Figure 2.2). Son of art, John Polanyi, born in Germany but educated in England, emigrated in 1952 to Canada where he first worked at the National Council, moving in 1956 to the University of Toronto where he became professor in 1974. John Polanyi studied in depth the distribution of energy in the excited quantum levels of the reaction products, measuring their infrared emission in simple exothermic reactions (Cashion and Polanyi, 1958). Over the course of a number of years, he developed the technique of infrared emission for several reactive processes and succeeded in building a clear picture of the energy distribution in the vibro-rotational levels of the reaction products (Polanyi, 1963, 1967). Of extreme interest for successive studies in catalysis was the experimental evidence of the existence of transition states that he collected by means of crossed molecular beam experiments (Polanyi and Zewail, 1995). In 1986 John Polanyi obtained the Nobel Prize in Chemistry together with Yuan Lee and Dudley Herschbach, authors of fundamental research in reactive systems by means of the molecular beams technique. These studies have revealed their utility in the surface science studies performed on a single crystal surface (Libuda, 2004).

Two Hungarian brothers, Adalbert and Ladislas Farkas, proposed the first quantum mechanical treatment of the mechanism involved in a heterogeneous catalytic process. Younger than some of their more famous Hungarian compatriots and friends, such as Michael Polanyi, Eugene Wigner, and Leo Szilard, they arrived on the Berlin scene somewhat later. The Farkas brothers contributed to the "Hungarian phenomenon" through their remarkable research on the physical chemistry of molecular hydrogen.

In 1927, they went to the Fritz Haber Institute in Germany where they both worked on the transformation of *ortho-* to *para*-hydrogen catalyzed by tungsten. In 1933, the Farkas brothers were forced to escape Germany owing to the Nazi persecution and immigrated to Israel. Later Adalbert went to the United States where he reached a convenient academic position.

Adalbert (Farkas, 1931) and Ladislas Farkas produced in this period two seminal papers on the *ortho–para* transformation induced by the presence of paramagnetic oxygen molecules (Farkas and Sachsse, 1933). The two Farkas brothers (Farkas, Farkas, and Harteck, 1934), following a previous treatment due to Eugene Paul Wigner (1902–1995) (Pelzer and Wigner, 1932; Wigner, 1933), developed a general theory of *ortho-* to *para*-hydrogen transformation, catalyzed by the inhomogeneous magnetic field in the vicinity of a paramagnetic ion or molecule acting as a perturbation.

The reaction mechanism proposed by the Farkas brothers for the ortho–para transformation process opened the road to intense research activity on the elementary mechanisms of the catalytic processes. After a few years, the transformation was also studied by one of the best English experts of catalysis, the chemist Eric Rideal (1890–1974), pupil of Hugh Stott Taylor (1890–1974) (Figure 2.4). Borrowing the idea from the BET isotherm, he proposed a mechanism bearing his name, which assumes that a *para* H_2 molecule, weakly

(a)   (b)

Figure 2.4 (a) Eric Rideal (1890–1974) and (b) Cyril Hinshelwood (1897–1967). To them we owe the main theories of surface reaction mechanisms. Rideal is well known for the Eley–Rideal mechanism. Hinshelwood, awarded the 1956 Nobel Prize for his research into the mechanisms of chemical reaction, contributed the Langmuir–Hinshelwood reaction mechanism. *Source*: (a) Courtesy of Society for Chemical Industry. (b) Hinshelwood's image is in the public domain.

physically adsorbed in the second layer on the solid surface, reacts with a hydrogen atom strongly chemisorbed in the underlying layer (Rideal, 1939) exchanging an H atom. In 1930, he became professor at Cambridge where he founded a world-renowned laboratory of colloid chemistry.

A more general mechanism not restricted to the *ortho–para*-hydrogen transformations, known as the Eley–Rideal mechanism, was jointly proposed successively by Daniel Douglas Eley and by Rideal himself for reactions between two molecules, one adsorbed on the catalyst and the other coming from the gas phase (Eley and Rideal, 1940, 1941). Eley, who had started his research activity with Michael Polanyi (Eley and Polanyi, 1935), joined Rideal's research group and eventually became editor of the famous series *Advances in Catalysis* published by Academic Press.

An example of a reaction following the Eley–Rideal mechanism is the hydrogenation of CO_2 to give methanol (Ogawa *et al.*, 2002), in which H_2 is the adsorbed species.

An important advancement on heterogeneous catalysis based on a mathematical formula similar to that of the Eley–Rideal theory was proposed in 1954 by Dirk Willem van Krevelen (1914–2001) in collaboration with his colleague Pieter Mars (later a professor at the University of Twente, Holland). van Krevelen and Mars presented their findings with regard to the oxidation of aromatic compounds, such as benzene and toluene, at an international conference on oxidation processes in Amsterdam. Mars and van Krevelen are said to have discovered that oxidation takes place via a two-step mechanism: in the first step, the absorbed reactant is oxidized by the oxygen present in the crystal lattice of the catalyst, followed by desorption of the product; in the second step, the reduced catalyst is oxidized by oxygen from the gas phase (Mars and van Krevelen, 1954).

Van Krevelen (1914–2001), born in Rotterdam, was a prominent industrial chemist who successfully combined a career in the management of research and industry with a scientific career in three different fields. He was among a small number of scientists who introduced the American chemical engineering approach to Europe, and he was one of the first scientists who emphasized that "unit operations" could be better understood in terms of the transfer of mass, momentum, and energy. His research on chemical gas absorption led to an improved understanding of the combined effects of physical transport and chemical reactions. Van Krevelen was the founder of an entirely new branch of chemical engineering: chemical reaction engineering. He was also an important contributor to coal science; he applied the graphical-statistical method of his former supervisor Hein Waterman to coal, obtaining results that were relevant both for geology and for the coal industries. His handbook on coal went through several widely used editions. Van Krevelen published extensively on polymer science as well.

The oxidation of toluene into benzaldehyde, which is still an important process, as it represents the first step toward creating penicillin, was investigated by Mars and van Krevelen using vanadium pentoxide (V_2O_5) as a catalyst. The catalyst splits oxygen molecules into atoms and then acts as a reservoir that can take up these atoms and release them to form molecules. This mechanism is internationally accepted as the Mars–van Krevelen mechanism, and it holds for the oxidation step of SO_2–SO_3 already described in the previous chapter. It provides an outstanding example of how a catalyst can be the entangled combination of a "divorce lawyer" and a "marriage broker."

In the Mars–van Krevelen mechanism, the surface itself is an active part in the reaction: reactants form a chemical bond with the catalytic surface building thin surface layers (for instance, metal oxide, carbide, and sulfide layers). Then the other reactant reacts directly from the gas phase with the atoms from the surface layer. When the reaction product desorbs, a vacancy is left behind in the surface. This vacancy will be filled again by the first reactant. In principle, in the mechanism, as described in 1954 by van Krevelen, the vacancy created by the reaction is filled by a reactant atom from the bulk rather than from the gas phase.

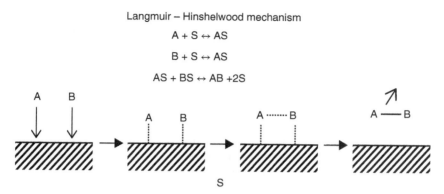

Langmuir – Hinshelwood mechanism

$$A + S \leftrightarrow AS$$

$$B + S \leftrightarrow AS$$

$$AS + BS \leftrightarrow AB + 2S$$

Figure 2.5 The Langmuir–Hinshelwood mechanism. The dashed part is the solid catalyst.

An example of the Mars–van Krevelen reaction mechanism is CO-oxidation under high oxygen pressure on platinum, where the surface forms a labile surface oxide phase with which the CO interacts (Hendriksen and Frenken, 2002). In this particular case an oxygen vacancy is created, which is highly mobile, leading to spontaneous reaction oscillations (Hendriksen *et al.*, 2010).

An efficient and general mechanism, based on Langmuir's adsorption theory and known as the "Langmuir–Hinshelwood mechanism," was developed by Sir Cyril Norman Hinshelwood (1897–1967) (Hinshelwood, 1940), professor of chemistry at the University of Oxford from 1937 (Figure 2.4). The mechanism is based on the idea that two molecules A and B, both adsorbed on the catalyst surface, could give rise to a bimolecular reaction according to the scheme shown in Figure 2.5 (Gadsby, Hinshelwood, and Sykes, 1946).

It is widely recognized that the Langmuir–Hinshelwood mechanism is verified in the majority of reactions catalyzed by metal surfaces. Cyril Hinshelwood also studied in depth the effect of pressure on the kinetics of reaction in the gas phase (Stubbs and Hinshelwood, 1951) (Figure 2.5).

Cyril Hinshelwood was awarded the 1956 Nobel Prize in Chemistry for his research on kinetics (Thompson and Hinshelwood, 1929) and for his fundamental contributions to the theory of chain reactions (Gibson and Hinshelwood, 1928), which is an important chapter of chemical kinetics.

The chain reaction theory, developed by Hinshelwood, is an evolution of the Nernst theory conceived in 1918 to explain the photochemical reaction of chlorine with hydrogen (Nernst, 1918a, b), and it was further developed by his successor at Berlin, Max Ernst August Bodenstein (1871–1942) (Bodenstein, 1913, 1922; Bodenstein and Dux, 1913). Hinshelwood was a key representative of the great scientific tradition of Oxford University. A scientist with a solid classical and philosophical education and passionate reader of Dante (he was fluent in Italian), he created a style of humanistic-scientific writing of great elegance and courtliness. His books, *The Kinetics of Chemical Change* (Clarendon Press, Oxford, 1926), *The Structure of Physical Chemistry* (1951), and *The Chemical Kinetics of the Bacterial Cell*, remain unsurpassed.

At the end of the 1930s, Hinshelwood oriented his research to the study of bacterial growth (Hinshelwood, 1952) and to the ability of bacterial cells to adapt their enzymatic balance to the external environment (Hinshelwood, 1953).

Fundamental to the understanding of chain reactions were the contributions of the Russian chemist Nikolay Semyonov (1896–1986), one of the greatest experts on combustion reactions (Semyonov, 1928a) and of explosive processes. Semyonov, who studied at St. Petersburg, became in 1928 professor at the Polytechnic Institute until in 1931 when he went to direct the Chemical Physics Institute of the Russian Academy of Sciences. In the 1930s Semyonov founded the mathematical theory of auto-ignition reactions (Semyonov, 1928b) that allows the prediction of the auto-ignition temperature of mixtures from knowledge of rate and heat dissipation constants (Semyonov, 1940). In the same period, he developed together with his coworkers the general theory of flame propagation processes in gases. These observations drove him to be interested in chain reactions, for which he developed, in parallel to Hinshelwood, a general theory of reaction kinetics and of branching processes and a detailed analysis of the mechanisms of chain ignition (Semyonov, 1929), as well as of those leading to the end termination of reactive chains. His treatise *Chain Reactions*, written in Russian in 1934 and translated into English the next year (Semyonov, 1934), represented a milestone in the study of chain reactions. For his research activity, he shared with Hinshelwood the Nobel Prize in Chemistry in 1956.

2.3 Electronic Theory of Catalysis: Active Sites

In the period 1930–1960, research in homogeneous and heterogeneous catalysis was focused on the role played by the electronic structure of metals and semiconductors in catalytic processes. In this way, the electronic theory of catalysis attracted increasing attention as a valid combination of solid-state physics and adsorption theories. In the early times, the so-called electronic theory of catalysis was essentially phenomenological and had considerable development in Russia after the pioneer researches of the Ukrainian chemist Lev Vladimirovich Pisarzhevsky (1871–1938), the first to correlate in 1916 the catalytic activity of solids with their electronic structure (Pisarzhevsky, 1955). The Russian school of heterogeneous catalysis continued, as is well documented in the book of Georgii K. Boreskov, *Heterogeneous Catalysis* (Boreskov, 2003), and reached its heights with the establishment in 1958 of the Siberian Institute of Catalysis dedicated today to its founder and first director, Boreskov, a great supporter of the importance of the chemical nature of heterogeneous catalysis (Figure 2.6). Even if unable to explain the intimate mechanisms of catalysis, the electronic theory opened the way to the applications of the concepts of transition state and of collective surface effects later incorporated in the quantum theory of catalysis.

Dowden (1950), and a few years later Hauffe (1955) and Vol'kenshteïn (1963), introduced the concept of surface collective phenomena. They contributed a large amount of experimental research to correlate the catalytic activity to the chemical structure of semiconductors and alloys used in heterogeneous catalysis.

In Germany, a fundamental contribution to the problem of the structure of active sites was given by the Berliner Georg-Maria Schwab (1899–1984), who started his interest in catalysis working at the catalytic decomposition of methane and ammonia (Schwab and Pietsch, 1926) (Figure 2.6). In 1929, in collaboration with Erich Pietsch (1902–1979), Schwab developed the "adlineation" theory, describing the reaction between two surface phases, assigning particular catalytic efficiency to defective structures with high density of adsorption (Schwab and Pietsch, 1929). Schwab discussed in detail the catalytic activity of lattice defects and dislocations at the surface of solid catalysts that expose different crystal faces to adsorption, giving rise to an increase of the density of adsorbed gas molecules on particularly active sites. Later, Pietsch would become one of the principal figures of scientific divulgation in Germany, as editor of the *Gmelin Handbuch der anorganischen Chemie*. In 1929, in collaboration with Erika Cremer (1900–1996), a German pioneer in gas chromatography, he introduced in his kinetic approach to catalysis the concept of compensation effect. He named this effect the "theta rule." In 1930, Schwab and Hermann Schultes also discussed the importance of finely divided metal catalysts dispersed on a support of metal oxides, an argument that Schwab reconsidered in 1950 (Schwab, 1950). He wrote in 1931 his comprehensive treatise *Katalyse vom Standpunkt der Kemischen Kinetik* (Schwab, 1931) translated into English in 1937 as *Catalysis from the Standpoint of Chemical Kinetics,* and was editor of the *Handbuch der Katalyse* (Springer Verlag, Wien) from 1940 to 1957. In 1939, he migrated to Greece as director of the Chemical Physics Department of the Nikolaos Kanellopoulos Institute in the Piraeus. In 1950, he went back to Germany as director of the Institute of Chemical Physics of the University of Munich where he directed a broad research project on catalysis.

In the second half of the century, the previously described contributions formed an ideal background to the understanding of catalyst structure and of applications of heterogeneous catalysis to industrial processes developed by the Belgian professor Michel Boudart (1924–2012) (Figure 2.6).

In 1940, when Hitler's divisions attacked his homeland, Boudart was 16. In order not to be drafted or sent to German factories, Boudart worked as a volunteer stretcher-bearer for the Red Cross. After the end of the war,

Figure 2.6 (a) Georg-Maria Schwab (1899–1984), (b) Georgy Boreskov (1907–1984), (c) Michel Boudart (1924–2012), and (d) Roald Hoffmann (1937–) They realized that the surface structure of catalysts (surface sites) and the surface orbitals play a fundamental role on the catalytic activity. Schwab directed the attention on surface defects and dislocations, while Boreskov dealt with the industrial consequences of catalyst surface composition. Boudart contributed to the spread of the concept of catalysis science. From his research, he derived the reputation of international ambassador of catalysis. Hoffmann (1981 Nobel Prize in Chemistry) was the first to use approximate quantum methods to study metal complexes and surfaces of potential catalytic interest. *Source*: (a) https://de.wikipedia.org/wiki/Datei: GMSchwab1.jpg. Used under CC BY-SA 3.0 https://creativecommons.org/licenses/by-sa/3.0/ deed.de. (b) Courtesy of Boreskov Institute of Catalysis. (c) Reproduced from a drawing of A. Zecchina. (d) The Hoffmann image is in the public domain.

he earned his MS degree in 1947 at the University of Louvain. Then he left Belgium to attend Princeton University, where he took his PhD in chemistry in 1950 under the supervision of Hugh Taylor, one of the most important English experts of catalysis and author with Eric Rideal of the first book on this subject (Rideal and Taylor, 1919).

The ideas of Taylor and Schwab on "active sites" strongly influenced the future activity of Boudart, during his stay at Princeton until 1961, at Berkeley until 1964, and finally at Stanford where he was chair of the Department of Chemical Engineering at Stanford from 1975 to 1978.

During his graduate studies, Michel Boudart recognized the catalytic consequences of the electronic properties of solids and brought Linus Pauling's concepts of the d-character of metals (Pauling, 1947) into the study of catalysis (Boudart, 1950). Later, he combined these concepts with knowledge emerging from theory and surface science to propose a classification of reactions based on their sensitivity to surface structure, measured in practice from the effects of the size and composition of metal particles on the rates of chemical reactions (Boudart *et al.*, 1966). Such pioneering breakthroughs required the development of protocols to counting the numbers of exposed metal atoms using molecular titrants (Benson and Boudart, 1965) and for assessing the structure of such atoms using spectroscopic methods. He pioneered the use of Mössbauer, infrared, and X-ray absorption spectroscopies in heterogeneous catalysis. Michel Boudart was also interested in industrial developments and in clarifying how catalysis can help solve major societal problems (Boudart, Vannice, and Benson, 1969). Michel Boudart transformed chemical engineering into a true science in the United States. He was an international ambassador of catalysis throughout his entire career and played a crucial role in establishing the reputation of Stanford's Chemical Engineering Department (Boudart and Djéga-Mariadassou, 1982). The central theme of his research was the catalytic properties of metals, particularly small metal particles (Boudart, 1985). He was known among friends and colleagues as a "gentleman scientist." In 1974, in the wake of the first oil crisis, Boudart and two associates founded Catalytica in Santa Clara, California, which worked on highly complex catalytic problems for petrochemical, chemical, and pharmaceutical firms as well as government agencies. In 1985, the University of Utah hosted a 5-day symposium on catalysis solely in Boudart's honor. In 2005, the *Journal of Physical Chemistry* dedicated an entire issue to Boudart's legacy. In 2006, the Danish company Haldor Topsøe sponsored the Michel Boudart Award for the advancement of catalysis, which is administered jointly by the North American Catalysis Society and the European Federation of Catalysis Societies.

As underlined by Boudart (Boudart and Djéga-Mariadassou, 1982) the development along the whole 1900–1970 period of catalysts containing finely dispersed metal particles supported on insulating (SiO_2, Al_2O_3) or semi-conducting (TiO_2, ZnO) oxides stimulated the work of scientists on the physical properties (quantum size effect) of these particles and on the so-called metal–support interaction. This fact led to many studies performed by solid-state physicists, summarized in the early Halperin review (Halperin, 1986). It also drew contributions to catalysis and surface science from experts in colloid science (Boutonnet *et al.*, 1982; Xu *et al.*, 1994; Subramanian, Wolf, and Kamat, 2004; Daniel and Astruc, 2004). The cited references are only a small fraction of the numerous contributions published in the 1980–2005 period.

Studies on the optical properties of metal particles (either under colloidal form or supported) further fueled the development of new advanced and highly sensitive spectroscopic techniques like SERS. By the end of twentieth and start of twenty-first century, it was becoming clear that catalysis, surface science, solid-state physics, and nanoscience are strictly interconnected (Bell (2003); Zecchina, Groppo, and Bordiga, (2007)). This realization was reinforced by the simultaneous accumulation of new knowledge in the field of molecular metallic clusters and of metalloenzymes, whose structure began to be clarified in the same period. This strengthened the idea that the full comprehension of catalysis science is the result of a multidisciplinary approach.

With the advent, after 1970, of new physical methods, the complexity of the surface structure and of active center distribution in each catalytic system was fully emerging. For this reason efforts toward the formulation of unified and general theories of catalysis decreased and were substituted by the study of each system treated as a separate case. This evolution was especially favored by the progressive introduction of computing facilities of increasing power, such as the early studies, based on extended Hückel methods (1984) by Roald Hoffmann (1981 Nobel Prize) and coworkers (Sung and Hoffmann, 1985; Wong and Hoffmann, 1985) (Figure 2.6). These studies have revealed that the molecular orbitals centered on the carbon end of the carbon monoxide molecule overlap in a constructive way with the surface orbitals emerging on the surface. In other words, the formation of a bond between the CO and the surface atoms is the result of orbital recognition. After the early Hoffmann approach, the introduction of sophisticated DFT codes gave increasingly precise information on the catalytic site structure and on the species formed by interaction with reactants. For these reasons today most studies concerning specific catalysts and their surface structures are characterized by the synergic presence of both experimental and computational modeling data. Starting from the first years of the twenty-first century, the number of papers containing at least a computational part is increasing exponentially.

References

Bell, A. (2003) The impact of nanoscience on heterogeneous catalysis. *Science*, **299**, 1688–1691.

Benson, J.E. and Boudart, M. (1965) Hydrogen oxygen titration method for the measurement of supported platinum surface areas. *Journal of Catalysis*, **4**, 704–710.

Bodenstein, M. (1913) Eine theorie der photochemischen reaktionsgeschwindigkeiten. *Zeitschrift fur Physikalische Chemie*, **85**, 329–397.

Bodenstein, M. (1922) Chemische Kinetik. *Ergebnisse der exakten Naturwissenschaften*, **1**, 197–209.

Bodenstein, M. and Dux, W. (1913) Photochemische Kinetik des Chlor-knallgases. *Zeitschrift fur Physikalische Chemie*, **85**, 297–306.

Boreskov, G.K. (2003) *Heterogeneous Catalysis*, English translation by K.I. Zamaraev e A. V. Kzasin, Nova Science publisher, Hauppauge, New York.

Boudart, M. (1950) Pauling's theory of metals in catalysis. *Journal of the American Chemical Society*, **72**, 1040.

Boudart, M. (1985) Heterogeneous catalysis by metals. *Journal of Molecular Catalysis*, **30**, 27–38.

Boudart, M. and Djéga-Mariadassou, G. (1982) *La cinétique des réactions en catalyse hétérogène*. Elsevier-Masson, Paris.

Boudart, M., Aldag, A.W., Benson, J.E. *et al.* (1966) On the specific activity of platinum catalysts. *Journal of Catalysis*, **6**, 92–99.

Boudart, M., Vannice, M.A., and Benson, J.E. (1969) Adlineation, portholes and spillover. *Zeitschrift für Physikalische Chemie*, **64**, 171–177.

Boutonnet, M., Kizling, J., Stenius, P., and Maire, G. (1982) The preparation of monodisperse colloidal metal particles from microemulsions. *Colloids and Surfaces*, **5**, 209–225.

Brunauer, S., Emmett, P.H., and Teller, E. (1938) Adsorption of gases in multi molecular layers. *Journal of the American Chemical Society*, **60**, 309–319.

Cashion, K. and Polanyi, J.C. (1958) Infrared chemiluminescence from the gaseous reaction atomic H plus Cl_2. *Journal of Chemical Physics*, **29**, 455–456.

Che, M. and Bennett, C.O. (1989) The influence of particle-size on the catalytic properties of supported metals. *Advances in Catalysis*, **36**, 55–172.

Cremer, E. and Polanyi, M. (1933) The conversion of ortho into para hydrogen in the solid state. *Zeitschrift fur Physikalische Chemie*, **821**, 459–68.

Daniel, M. and Astruc, D. (2004) Gold nanoparticles: assembly, supra-molecular chemistry, quantum-size-related properties, and applications toward biology, catalysis and nanotechnology. *Chemical Reviews*, **104**, 293–346.

Dowden, D.A. (1950) Heterogeneous catalysis, part I. Theoretical basis. *Journal of the Chemical Society*, 242–265.

Eley, D.D. and Polanyi, M. (1935) Catalytic interchange of hydrogen with water and alcohol. *Transactions of the Faraday Society*, **32**, 1388–1397.

Eley, D.D. and Rideal, E.K. (1940) Parahydrogen conversion on tungsten. *Nature*, **146**, 401–402.

Eley, D.D. and Rideal, E.K. (1941) The catalysis of para-hydrogen conversion by tungsten. *Proceedings of the Royal Society of London A*, **178**, 429–451.

Emmett, P.H. and Brunauer, S. (1933) The adsorption of nitrogen by Iron synthetic ammonia catalysts. *Journal of the American Chemical Society*, **55**, 1738–1739.

Emmett, P.H. and Brunauer, S.J. (1937) The use of low temperature van der Waals adsorption isotherms in determining the surface area of iron synthetic ammonia catalysts. *Journal of the American Chemical Society*, **59**, 1553–1564.

Evans, M.G. (1938) Thermodynamical treatment of transition state. *Transactions of the Faraday Society*, **34**, 49–57.

Evans, M.G. and Polanyi, M. (1935) Some applications of the transition state method to the calculation of reaction velocities, especially in solution. *Transactions of the Faraday Society*, **31**, 875–894.

Evans, M.G. and Polanyi, M. (1937) On the introduction of thermodynamic variables into reaction kinetics. *Transactions of the Faraday Society*, **33**, 448–453.

Evans, M.G. and Polanyi, M. (1938) Inertia and driving force of chemical reactions. *Transactions of the Faraday Society*, **34**, 11–28.

Eyring, H. (1935) The activated complex in chemical reactions. *Journal of Chemical Physics*, **3**, 107–115.

Eyring, H. (1936) Viscosity, plasticity and diffusion as examples of absolute reaction rates. *Journal of Chemical Physics*, **4**, 283–291.

Eyring, H. and Polanyi, M. (1931) Über einfache Gasreaktionen. *Zeitschrift für Physikalische Chemie*, **B12**, 279–311.

Eyring, H., Walter, J., and Kimball, G. (1944) *Quantum Chemistry*. John Wiley and Sons Inc., Hoboken, NJ.

Farkas, A. (1931) Aktivierungsenergie der p-H_2-Umwandlung an nickel. *Zeitschrift für Physikalische Chemie B*, **14**, 371–379.

Farkas, L. and Sachsse, H. (1933) Über die homogene Katalyse der Para-Orthowasserstoff umwandlung. *Zeitschrift für Physikalische Chemie B*, **23**, 19–26.

Farkas, A., Farkas, L., and Harteck, P. (1934) Experiments on heavy hydrogen. II. The ortho-para conversion. *Proceedings of the Royal Society of London A.*, **144**, 481–493.

Gadsby, J., Hinshelwood, C.N., and Sykes, K.W. (1946) The kinetics of the reactions of the steam-carbon system. *Proceedings of the Royal Society of London*, **A187**, 151–187.

Gibson, C.H. and Hinshelwood, C.N. (1928) The influence of nitrogen peroxide on the union of hydrogen and oxygen-A problem of trace catalysis. *Transactions of the Faraday Society*, **24**, 559–566.

Halperin, W.P. (1986) Quantum size effects in metal particles. *Reviews of Modern Physics*, **58**, 3.

Hauffe, K. (1955) *Reaktionen in und an festen Stoffen*, vol. **2**. Springer, Berlin, p. 696.

Hendriksen, B.L.M. and Frenken, J.W.M. (2002) CO oxidation on Pt(110): scanning tunneling microscopy inside a high-pressure flow reactor. *Physical Review Letters*, **89**, 046101.

Hendriksen, B.L.M., Ackerman, M.D., Van Rijn, R. *et al.* (2010) The role of steps in surface catalysis and reaction oscillations. *Nature Chemistry*, **2**, 730–734.

Heyne, W. and Polanyi, M. (1928) Adsorption aus Losungen. *Zeitschrift für Physikalische Chemie*, **132**, 384–398.

Hinshelwood, C.N. (1940) *The Kinetics of Chemical Change*. Clarendon Press, Oxford.

Hinshelwood, C.N. (1952) *The Chemical Kinetics of the Bacterial Cell*. Clarendon Press, Oxford.

Hinshelwood, C.N. (1953) Autosynthesis. *Journal of the Chemical Society*, 1947–1956.

Horiuti, J. and Polanyi, M. (1933–1934) On the mechanism of ionization of hydrogen at a platinum electrode. *Memoirs and Proceedings of the Manchester Literary and Philosophical Society*, **47**.

Horiuti, J. and Polanyi, M. (1933a) A catalyzed reaction of hydrogen with water. *Nature*, **132**, 819.

Horiuti, J. and Polanyi, M. (1933b) Catalytic reaction of hydrogen with water and the nature of overvoltage. *Nature*, **132**, 931–932.

Horuti, J. and Polanyi, M. (1934) Direct introduction of deuterium into benzene. *Nature*, **134**, 847–849.

Langmuir, I. (1916) The constitution and fundamental properties of solids and liquids. I. Solids. *Journal of the American Chemical Society*, **38**, 2221–2295.

Langmuir, I. (1917) The constitution and fundamental properties of solids and liquids. II Liquids. *Journal of the American Chemical Society*, **39**, 1848–1902.

Langmuir, I. (1918) The adsorption of gases on plane surfaces of glass, mica and platinum. *Journal of the American Chemical Society*, **40**, 1361–1402.

Langmuir, I. (1919a) The arrangement of electrons in atoms and molecules. *Journal of the American Chemical Society*, **41**, 868–934.

Langmuir, I. (1919b) The structure of atoms and the octet theory of valence. *Proceedings of the National Academy of Sciences of the United States of America*, **5**, 252–264.

Langmuir, I. (1920) The octet theory of valence and its applications with special reference to organic nitrogen compounds. *Journal of the American Chemical Society*, **42**, 274–292.

Langmuir, I. (1922) Heterogeneous reactions. Chemical reactions on surfaces. *Transactions of the Faraday Society*, **17**, 607–620.

Langmuir, I. and Orange, J.A. (1913) Tungsten lamps of high efficiency. *General Electric Review*, **16**, 956–964.

Libuda, J. (2004) Molecular beams experiments on model catalysts: activity and selectivity of specific reaction centres on supported nanoparticles. *ChemPhysChem*, **5**, 625–631.

London, F. and Polanyi, M. (1930) Über die atom theoretische Deutung der Adsorptions-kräfte. *Naturwissenschaften*, **18**, 1099–1100.

Mars, P. and van Krevelen, D.W. (1954) Oxidations carried out by means of vanadium oxide catalysts. *Chemical Engineering Science*, **8**, 41–59.

Nernst, W. (1918a) Zur Anwendung des Einsteinschen photochemischen Äquivalentgesetzes. I. *Zeitschrift für Elektrochemie*, **24**, 335–336.

Nernst, W. (1918b) *des* neuen Wärmesatzes, Verlag von Wilhelm Knapp, Halle.

Ogawa, J. (2002) Eley-Rideal type mechanism for formate synthesis on a Cu(III) surface. *Nippon Kagakkai Koen Yokoshu*, **81**, 270–276.

Pauling, L. (1946) Molecular architecture and biological reactions. *Chemical & Engineering News*, **24**, 1375–1377.

Pauling, L. (1947) Atomic and interatomic distances in metals. *Journal of the American Chemical Society*, **69**, 542–553.

Pelzer, H. and Wigner, E. (1932) Über die Geschwindigkeitskonstante von Austausch-reaktionen. *Zeitschrift für Physikalische Chemie*, **B15**, 445–447.

Pisarzhevsky, L.V. (1955, in Russian) *Selected Works on Catalysis*. Izd. Acad. Nauk. USSR, Kiev.

Polanyi, M. (1913) Eine neue thermodynamische Folgerung aus der Quantenhypothese. *Verhandlungen der Deutschen Physikalischen Gesellschaft*, **15**, 156–161.

Polanyi, M. (1917) Gazokabsorptioja Szilard, nemillanoadszorbensen (Adsorption of Gases by a Solid Non-Volatile Adsorbent). PhD thesis. University of Budapest.

Polanyi, M. (1920) Zur Theorie der Reaktionsgeschwindigkeit. *Zeitschrift für Physik*, **1**, 90–101.

Polanyi, M. (1921a) Die chemische Konstitution der Zellulose. *Naturwissenschaften*, **9**, 288–293.

Polanyi, M. (1921b) Über Adsorptionskatalyse. *Zeitschrift fur Electrochemie*, **27**, 142–150.

Polanyi, M. (1928) Application of Langmuir's theory to the adsorption of gases on charcoal. *Zeitschrift für Physikalische Chemie*, **A138**, 459–464.

Polanyi, M. (1932a) Developments of the theory of chemical reactions. *Naturwissenschaften*, **20**, 289–296.

Polanyi, M. (1932b) *Atomic Reactions*. Williams & Norgate, Ltd., London.

Polanyi, M. (1934) Über eine Art Gitterstörung,die einem Kristall plastisch machen könnte. *Zeitschrift für Physik*, **89**, 660–664.

Polanyi, J.C. (1963) Infrared chemiluminescence. *Journal of Quantitative Spectroscopy and Radiative Transfer*, **3**, 471–476.

Polanyi, J.C. (1967) Dynamics of chemical reactions. *Discussions of the Faraday Society*, **44**, 293–297.

Polanyi, M. and Eyring, H. (1930) Zur Berechnung der Aktivierungswärme. *Naturwissenschaften*, **18**, 914–915.

Polanyi, M. and Schmid, E. (1929) Problems of plasticity. Deformation at low temperatures. *Naturwissenschaften*, **17**, 301–304.

Polanyi, M. and Wigner, E. (1925) Bildung und Zerfall von Molekülen. *Zeitschrift für Physik*, **33**, 429–434.

Polanyi, M. and Wigner, E. (1928) Über die Interferenz von Eigenschwingungen als Ursache von Energieschwankungen und chemischer Umsetzungen. *Zeitschrift für Physikalische Chemie*, **A139**, 439–444.

Polanyi, J.C. and Zewail, A.H. (1995) Direct observation of the transition state. *Accounts of Chemical Research* (Holy Grail Issue), **28**, 119–132.

Rideal, E.K. (1939) Parahydrogen conversion on tungsten. *Proceedings of the Cambridge Philosophical Society*, **35**, 130.

Rideal, E.K. and Taylor, H.S. (1919) *Catalysis in Theory and Practice*. Macmillan, London.

Schwab, G.M. (1931) *Katalyse vom Standpunkt der Kemischen Kinetik*. Verlag J. Springer, Berlin.

Schwab, G.M. (1950) On compact- dispersed silver. *Journal of Physical Chemistry*, **54**, 576–580.

Schwab, G.M. and Pietsch, E. (1926) Thermische Spaltung von Methan am Glühdraht. *Zeitschrift für Physikalische Chemie*, **32**, 430–434.

Schwab, G.M. and Pietsch, E. (1929) Zur Topoehemie der Kontakt-Katalyse. II, Über einen experimentellen Fall der Adlineation. *Zeitschrift für Physikalische Chemie B*, **2**, 262–264.

Semyonov, N. (1928a) Die Kinetik der Dissoziation von Zweitomigen Moleculen. *Zeitschrift für Physik*, **48**, 216–230.

Semyonov, N. (1928b) Zur Theorie des Verbrennungsprozesses. *Zeitschrift für Physik*, **48**, 571–582.

Semyonov, N. (1929) Kinetics of chain reactions. *Chemical Reviews*, **6**, 347–379.

Semyonov, N. (1934; translated in English in 1935, Oxford Press) *Chain Reactions* (in Russian), Goskhimizdat, Leningrad.

Semyonov, N. (1940) Study of properties of solutions of kinetic equations. *Uspekhi Fizicheskikh Nauk*, **23**, 251–257.

Stubbs, F.J. and Hinshelwood, C.N. (1951) The thermal decomposition of hydrocarbons. *Discussions of the Faraday Society*, **10**, 129–136.

Subramanian, V., Wolf, E.E., and Kamat, P.V. (2004) Catalysis with TiO_2/Au nanocomposites. Effect of metal particle size on the Fermi level equilibration. *Journal of the American Chemical Society*, **126**, 4943–4950.

Sung, S.-S. and Hoffmann, R. (1985) How carbon monoxide bonds to metal surfaces. *Journal of the American Chemical Society*, **107**, 578–584.

Taylor, H.S. (1925) A theory of the catalytic surface. *Proceedings of the Royal Society of London A*, **108**, 105–111.

Thompson, H.W. and Hinshelwood, C.N. (1929) The mechanism of the homogeneous combination of hydrogen and oxygen. *Proceedings of the Royal Society of London*, **A122**, 610–621.

Vol'kenshteĭn, F.F. (1963) *Élektronnayateoriyakatalizanapoluprovodnikakh, (M. Fizmatgiz)*, 1960; (*The Electronic Theory of Catalysis on Semi-conductors*), English translation by N. G. Anderson. Pergamon Press, Oxford.

Wigner, E.P. (1933) Über die paramagnetische Umwandlung von Para-Orthowasserstoff. *Zeitschrift für Physikalische Chemie B*, **23**, 28–32.

Wigner, E.P. (1938) The transition state method. *Transactions of the Faraday Society*, **34**, 29–41.

Wong, Y.-T. and Hoffmann, R. (1985) Chemisorption of carbon monoxide on three metal surfaces: Ni(11I), Pd(11I) and Pt(111). A comparative study. *Journal of Physical Chemistry*, **107**, 578–584.

Xu, Z., Xiao, F.S., Purnell, S.K. *et al.* (1994) Catalysis with TiO_2/gold nanocomposites. Effect of metal particle size on the Fermi level equilibration. *Nature*, **372**, 346–348.

Zecchina, A., Groppo, E., and Bordiga, S. (2007) Selective catalysis and nanoscience: an inseparable pair. *Chemistry – A European Journal*, **13**, 2440–2460.

3

Catalytic Processes Associated with Hydrocarbons and the Petroleum Industry

3.1 Petroleum and Polymer Eras

While the Fischer–Tropsch reaction between CO and hydrogen (both derived from carbon) to produce hydrocarbons was the fingerprint of carbon-based economy and catalysis at the end of nineteenth and initial part of twentieth century, the period starting around 1930 is named the era of petroleum. This definition is justified not only because in the twentieth century petroleum gradually became the major source of energy production but also because it is an invaluable reservoir for a multitude of carbon-containing compounds.

The refinery industry started to grow in the 1930s following the increased demand for gasoline for cars as well as for aviation. During World War II, oil-refining technology and capacity started to be boosted worldwide to support the war effort. After the war, the petrochemical industry developed very rapidly, one of the main driving forces being the necessity to lower the production costs of monomers necessary for the plastics and rubber industries, and other fine chemical products. For this reason the catalytic processes, associated with the petroleum industry (in particular, refinery and desulfurization) and with the chemistry of petroleum-derived compounds like olefins and aromatics, are of paramount importance and have stimulated much research and many discoveries. Eighty-five percent of industrial organic chemicals are currently produced by catalytic processes from petroleum and natural gas sources. As far as olefins are concerned, it is sufficient to mention their extensive use in the production of synthetic polymers and plastics, which are fingerprint products characterizing the second half of the twentieth century. The world of polymers and plastics is very large and represents a wide topic on its own, which cannot be thoroughly illustrated in this book. In this chapter only the processes leading to polymers deriving from monomers obtained from refinery and reforming (like polyethylene, polypropylene, acrylonitrile, PVC, and polystyrene) will be discussed. Other important polymers like nylon, NAFION, and polyurethane are mentioned in other chapters.

The Development of Catalysis: A History of Key Processes and Personas in Catalytic Science and Technology,
First Edition. Adriano Zecchina and Salvatore Califano.
© 2017 John Wiley & Sons, Inc. Published 2017 by John Wiley & Sons, Inc.

In connection with the need for selective processes to produce economically new chemicals from the cornucopia of carbon-containing molecules derived from refinery and reforming processes, fundamental and industrial studies of new catalysts have received unprecedented input.

3.2 Catalytic Cracking, Isomerization, and Alkylation of Petroleum Fractions

Hydrocarbon cracking is the process whereby long-chain molecules are broken down into simpler molecules such as light hydrocarbons via the breaking of carbon–carbon bonds. The cracking can be thermal or catalytic. In the second case heterogeneous and homogeneous catalysts with strongly acidic character are used.

A prototypical example of cracking reaction is illustrated below and concerns the formation of propene and heptane from decane:

Decane

$$CH_3-CH_2-CH_2-CH_2-CH_2-CH_2-CH_2-CH_2-CH_2-CH_3$$

$$\downarrow$$

$$CH_3-CH = CH_2 \; + \; CH_3-CH_2-CH_2-CH_2-CH_2-CH_2-CH_3$$

Propene   Heptane

In the example above linear heptane is formed. However, a branched isomer is formed as well. Of course, as petroleum is a very complex mixture of hydrocarbons, the this reaction must be considered as only one example.

Catalytic cracking of petroleum fractions with heterogeneous catalysts (in particular, over zeolite-based acidic catalysts) is today responsible for the manufacture of a high percentage of gasoline in the world and is the source of light alkanes, light olefins, diolefins, and aromatic molecules, like benzene and toluene (Kissin, 2001). However, acid zeolites are the final catalytic product of a long history of innovations over the twentieth century. Indeed in the first half of the twentieth century, several industries attempted to produce catalysts for petroleum cracking, and impure amorphous silica–aluminas were widely used as solid acid catalysts (Corma, 1995; Busca, 2007). Later, the process was stymied by the discovery that the available catalysts cease to work after a time because of poisoning by carbonaceous compounds.

Eugéne Jules Houdry understood the origin of the problem and developed a process to regenerate the catalyst.

The first solid acid catalysts, which could be regenerated and commercially used on a large scale (~1935), were various types of clays (acid-leached natural aluminosilicates), which can be considered as impure silica–aluminas

($SiO_2–Al_2O_3$). They were the first cracking catalysts in the so-called Houdry's process named after the inventor Eugène Jules Houdry.

Eugène Jules Houdry (1892–1962) was French and later became a naturalized American citizen. Houdry graduated from the École Nationale Supérieure des Arts et Métiers in 1911 and invented the catalytic cracking of petroleum feedstocks, which is the process of breaking long-chain hydrocarbons into shorter molecules (alkanes and alkenes). As a participant in the sport of automobile racing, Houdry was aware of the relevance of high-performance fuels for optimal machine performance. In 1922, while still in France, Houdry became informed about an exceptional gasoline produced catalytically from lignite by the French scientist Eugene Albert Prudhomme, who had patented in 1924 a procedure for the catalytic production of a liquid fuel similar to petrol. Houdry persuaded Prudhomme to set up a laboratory at Beauchamp near Paris. For the next few years, they worked together to develop a lignite-to-gasoline process according to the procedure originally invented by Prudhomme (1926).

The blooming automobile industry of the early 1930s sparked a strong demand for light gasoline with high octane content. Houdry had been alerted to this new need when he took a trip to the United States in 1922. At that time the obstacle to catalytic cracking was the regeneration of the catalyst. Houdry managed to demonstrate that the alumina catalyst he was using could be regenerated at a given temperature and in an air–hydrogen environment but was unable to convince European refiners that his discovery was worthwhile. Both the Anglo-Iranian Oil Company and the French Saint Gobain Company rejected his proposals. In 1930, however, Houdry found in the United States the sponsors for his invention. When World War II broke out in 1939, the Houdry process could produce the 100-octane gasoline required by the US Air Force.

In 1930 Houdry moved to America, and in 1941, during the Nazi occupation of France, he opposed the collaboration of the French government with the Germans. For this reason on May 1941, the French government declared that Houdry had lost the French citizenship.

He became the president of the US unit of "France Forever," an organization devoted to the support of Charles de Gaulle, the head of the French government in exile. In January 1942 was granted US citizenship. The first Houdry unit was built at Sun Oil's Pennsylvania oil refinery in 1937, and more units were built by the 1940s, all instrumental in US wartime aviation gasoline production (Figure 3.1).

In the following years the Houdry process was further developed by two engineers, Warren Kendall Lewis (1882–1975) and Edwin Richard Gilliland (1909–1973), both professors of engineering at MIT. They solved the problem of coke formation on the catalysts by using a continuously circulating fluidized catalyst (fluid catalytic cracking). Also the catalyst was changed as acidic zeolites became dominant; this process is still in widespread use.

Figure 3.1 Title page of *The Houdry Process* commemorative booklet produced by the National Historic Chemical Landmarks program of the American Chemical Society in 1996, with the portrait of Eugene Jules Houdry. The most dramatic benefit of the earliest industrial unit exploiting the Houdry process was the production of 100-octane aviation gasoline, which was of decisive importance in the Battle of Britain in 1940. *Source*: Courtesy of Sunoco, Inc.

Houdry's contributions to catalytic technology were recognized by numerous awards, including the Perkin Medal of the Society of Chemical Industry.

The discoveries of two great scientists, Vladimir Nikolayevich Ipatieff and Herman Pines, concerning the catalysis of converting paraffins into isoparaffins represented one of the cornerstones of the petroleum industry (Ipatieff, Pines, and Schmerling, 1940). They developed the new chemistry of acid-catalyzed reactions and investigated several acid catalysts, including sulfuric acid, phosphoric acid, hydrofluoric acid, $AlCl_3$, and other halides. A nice presentation of this novel and impressive chemistry can be found in *The Chemistry of Catalytic Hydrocarbon Conversions* (Pines, 1981) and in several other papers published with Ipatieff (Pines, Grosse, and Ipatieff, 1942; Ipatieff *et al.*, 1936; Pines and Ipatieff, 1945).

The biography of the Russian Vladimir Nikolayevich Ipatieff (1867–1952) is particularly interesting and merits detailed illustration (Figure 3.2). As a student at the famous Russian military academy *Mikhailovskaia artilleriiskaia akademiia* and coworker of Dmitry Konstantinovich Chernov (discoverer of the iron–carbon phase diagram), he became expert in the properties and analysis of steel and reached the highest ranks in the Czar's Nicholas II army up to the position of general. In 1896, Ipatieff studied in Munich under the guidance of Adolph von Baeyer. In 1907, he presented a dissertation on "Catalytic Reactions under High Pressure and Temperature" at St. Petersburg

University and received the degree of Doctor of Chemistry. In 1909, Ipatieff discovered an important phenomenon, the "promoter effect" of additives on catalysts. A promoter is a substance that when added to a solid catalyst is able to improve its activity by interacting with active components of the catalyst. Incorporation of promoters in catalysts is today widely practiced in synthetic chemistry and is the basis of many patents for the production of efficient industrial catalysts.

Despite his connection with high ranks of the Czar's army, Ipatieff succeeded, after the October Revolution, in remaining in the Soviet Union without political difficulties and devoted most of his time to studying chemical and catalytic problems. He rapidly reached high competence in this field up to the point that, in 1927, he was able to create in the Soviet Academy the High Pressure Institute, later called Ipatieff Institute for High Pressure Physics, and he was awarded the prestigious Lenin Prize.

After Lenin's death in 1924, Ipatieff felt threatened because of his past connections with the Czarist army, and fearing that he would be victimized by Stalin purges, he took advantage of a trip abroad on the occasion of an international meeting in Berlin to escape to the United States. Following his narration, he invited his wife to come with him, and as the train came to the Poland border, he announced to his shocked wife, "*Dear, look back at Mother Russia. You will never see her again.*" In the United States he became soon research director at Universal Oil Products (UOP) at Des Plaines as well as professor of chemistry at Northwestern University (Ipatieff, 1959).

Ipatieff received many awards, including the Berthelot Medal and the Gibbs Medal of the American Chemical Society.

A young Polish chemist, Herman Pines (1902–1996), joined the Ipatieff research group at UOP and became Ipatieff closest coworker until his death, when, at Northwestern University, he was nominated director of the Ipatieff High Pressure and Catalytic Laboratory.

Herman Pines (1902–1996) was born in Lodz, Poland. Since he was Jewish, he was prohibited from attending university in Poland. He consequently moved to France and attended the École Supérieure de Chimie Industrielle in Lyon, where, after completing his undergraduate studies in chemistry, he received his Chemical Engineering degree in 1927. After immigration to the United States in 1930, he accepted a position at UOP in Chicago, a position he held until 1952 (Figure 3.2).

From 1941 to 1952 Pines held several part-time positions at the Chemistry Department of Northwestern University. During this period he became associated with Vladimir Ipatieff with whom he conducted numerous experiments and obtained valuable results in the creation of a high octane aviation fuel. Pines authored numerous publications, two books on hydrocarbon conversion catalysis, held many patents, and was coeditor of *Advances in Catalysis* for more than 20 years. He received numerous awards based on his

(a)　　　　　　　　　　　(b)

(c)

Figure 3.2 (a) Vladimir Ipatieff (1867–1952), (b) Herman Pines (1902–1996), and (c) Vladimir Haensel (1914–2002). The highly innovative discoveries in the field of hydrocarbon transformation made by these researchers have been all achieved in industrial laboratories. Ipatieff (who had an adventurous life) and Pines jointly extended the knowledge of acid catalysis mechanisms in the chemistry of petrol. Haensel developed the reforming processes based on Pt particles supported on chlorinated alumina. *Source*: (a) The image of Ipatieff (of public domain) was taken before the October Revolution when he was general lieutenant of the Russian army and member of the Russian Academy of Sciences. (b) Courtesy of Northwestern University. (c) The Haensel image is from a drawing by A. Zecchina.

groundbreaking research, including the Fritsche Award of American Chemical Society and the Houdry Award in Applied Catalysis. Herman Pines also studied the mechanism of dehydration of alcohols on alumina, the mechanism of aromatization of alkanes over chromia, and hydrogen transfer reactions involving aromatic hydrocarbons.

While there is unanimous consensus that Pines was one of the towering scientists of this century, he always remained modest, and when his trendsetting discoveries of the 1930s were mentioned, he always referred them to Ipatieff. In his honor, the Herman Pines Award for outstanding research in the field of catalysis was established after his death.

The Pines–Ipatieff collaboration led to a series of discoveries that changed the world of industrial chemistry and petrol chemistry. Due to their collaborative work, knowledge of many fundamental aspects of acid catalysis based on the use of strong Brønsted acids like H_2SO_4 and Lewis acids like $AlCl_3$ in solution made a substantial leap. The role of carbocationic species was for the first time clearly elucidated. In the 1930s an unchallenged dogma of chemistry was that paraffins would not react with anything at low temperature; even the name of this class of compounds, "parum affinis," was based on this assumed lack of reactivity. It must have been quite a shock to scientists of those days when Pines and Ipatieff showed, in 1932, that in the presence of a strong acid (concentrated sulfuric acid), the paraffin isobutane would react, even at $-35\,°C$, with olefins. This was the basis of the alkylation process, patented in 1938 and industrially developed soon after.

The alkylation reactions can be described as follows. The first step is alkene (propene and butene) protonation by sulfuric acid, with formation of a carbenium ion intermediate:

$$CH_3 - CH = CH_2 + H^+ \leftrightarrow CH_3 - CH_2 - CH_2{}^+$$
$$CH_3 - CH_2 - CH = CH_2 + H^+ \leftrightarrow CH_3 - CH_2 - CH_2 - CH_2{}^+$$

The second step is alkylation of isobutane by the carbenium cations:

$$(CH_3)_3 - CH + CH_3 - CH_2 - CH_2{}^+ \rightarrow (CH_3)_3 - C - CH_2 - CH_2$$
$$- CH_3 + H^+$$
$$(CH_3)_3 - CH + CH_3 - CH_2 - CH_2 - CH_2{}^+ \rightarrow (CH_3)_3 - C - CH_2$$
$$- CH - (CH_3)_2 + H^+$$

with formation of isoheptane and isooctane. Notice that to form isooctane, the carbenium ion must be transformed into a branched isomeric form.

Its most spectacular application is in the synthesis of isooctane from *n*-butene and isobutane, which are hydrocarbons derived from petroleum cracking. Isooctane improves the quality of gasoline fuel and airplane fuel and has played a decisive role in the victory of the Royal Air Force during the Battle of Britain in 1941. The main results of this collaboration are summarized by Pines in 1981 (Pines, 1981). Also of great interest is the contribution of another UOP scientist, Louis Schmerling (1912–1991), who was the first to summarize the opinions of the UOP scientists on the role of carbocations in the mechanisms of acid-catalyzed reactions involving alkenes (Schmerling, 1953). His book, *Organic and Petroleum Chemistry for Non-Chemists*, has educated generations of industrial chemists in the comprehension of catalytic processes (Schmerling, 1983).

These major discoveries led to new processes for the isomerization of paraffins and the alkylation of aromatic compounds via carbenium intermediates. The alkylation of benzene with ethylene with formation of alkylbenzenes is a prototype reaction, which can be schematized as follows:

where the active intermediate is the $CH_3-CH_2^+$ carbenium ion formed by interaction of the proton with ethylene following the reaction:

$$CH_2 = CH_2 + H^+ \leftrightarrow CH_3-CH_2^+$$

Unfortunately when sulfuric acid is used as catalyst, this reaction leads to the formation of several isomers. In other words the catalysts are not selective.

The UOP group knowledge on paraffin isomerization was not limited to sulfuric acid. In fact it was found that $AlCl_3$ in the presence of small amounts of HCl was also a good catalyst. The role of HCl as promoter in forming $H^+(AlCl_4)^-$ (which is the true catalyst) from $AlCl_3$ can be found in Bloch *et al.* 1946). Herman Bloch (1912–1990), born in Chicago to a family who emigrated from Ukraine, graduated in 1933 and received his doctorate in 1936 at the University of Chicago under the guidance of Professor Julius Stieglitz. After his doctorate, he joined the UOP. All the discoveries described so far, which are based on the hypothesis of the elusive carbenium ion intermediates, had certainly a great influence on the work of George Olah (1994 Noble Prize) concerning the properties of superacids. The Olah contribution in the elucidation of the carbonium ion intermediate properties will be described in detail in the following chapter, which is more specifically devoted to homogeneous catalysis.

The Ipatieff and Pines advancements in acid catalysis certainly taken drew on the work of Brønsted (1928; Bell, 1936) on acid and base catalysis and after him of Meredith G. Evans and Michael Polanyi, when the latter joined the University of Manchester. We documented in the preceding chapter that these authors tackled the problem in terms of the potential energy surface (Evans and Polanyi 1937, 1938; Marcus, 1968).

A further development that arose from the studies performed at UOP laboratories was an appreciation of the fact that acid sites on the surfaces could catalyze reactions in a similar way to acids in solution. This led to the utilization of more convenient heterogeneous catalysts constituted by phosphoric acid supported on kieselguhr. This is the case where the permanent dialogue between homogeneous and heterogeneous catalysis is clearly emerging.

3.3 Reforming Catalysts

Another important discovery due to the UOP Russian scientist Vladimir Haensel (1914–2002) is the development of reforming catalysts made of platinum nanoparticles dispersed on chlorinated alumina (Figure 3.2).

Haensel, born in Freiburg, Germany, spent most of his youth in Moscow, where his father was an important professor of economics. After the Bolshevik Revolution, his father was nominated director of the Institute of Economic Research (financial section) from 1921 to 1928; there he authored Lenin's first 5-year economic plan.

In 1930 the Haensel family emigrated to the United States where Haensel's father accepted a position at Northwestern University. Haensel entered Northwestern University where he received a BS degree in general engineering.

In 1937, Haensel joined the "cracking" research division at UOP as a research chemist. In 1951, he was appointed director of refining research, and in 1969, he became vice president and director of research. In 1972, he was appointed vice president, a position he held until 1979.

The most important Haensel discovery was the reforming catalyst that is constituted by finely dispersed platinum on chlorinated alumina. Haensel was among the first to realize that platinum was in the form of nanoparticles and for this was awarded the Perkin Medal in 1967 (Figure 3.3).

The Pt component of the catalyst is responsible for the hydrogenating–dehydrogenating reaction. The important and fundamental aspect of Haensel's discovery is the bifunctional character of the reforming catalyst, which contains simultaneously both acidic and hydrogenating/dehydrogenating functions. Haensel's process was commercialized by UOP in 1949 for producing a high octane gasoline from low octane naphthas, and the UOP process became known as the "platforming process." In this process, which occurs in

Electron micrograph of a typical Pt–Al$_2$O$_3$
reforming catalyst

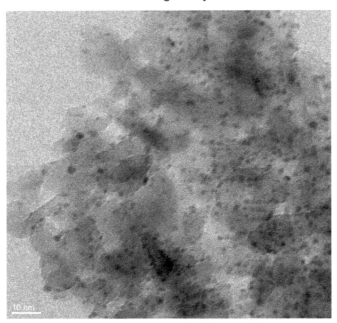

Figure 3.3 Electron micrograph of a catalyst constituted by Pt particles (black dots) supported on γ-Al$_2$O$_3$. This image is similar to that of a reforming catalyst reported in the literature. *Source*: Garcia-Dieguez *et al.*, 2010. Courtesy of E. Groppo and M. Manzoli.

the presence of H$_2$, the structures of hydrocarbon molecules rearrange to form highly branched and more aromatic molecules. The importance of this process also depends on the fact that H$_2$ is produced at the end of the reaction. The aromatics formed in the process became important intermediates for the plastics, textile, and synthetic fiber industries. The significance of the entire petrochemicals industry was greatly enhanced by the operation of the platforming process. After the Haensel discovery the design of more advanced reforming catalysts continued leading to the introduction of the new generation Pt–Re bimetallic catalysts, which are more resistant to poisoning (Sinfelt, 1983).

Haensel was a multifaceted person and a patron of the arts.

In his obituary he was quoted from a 1995 interview:

> *Work to produce something important. Do something new. Do something interesting, something that makes you want to shout out loud when you've got it. Life is too darn amazing – and too short – for anything less.*

3.4 Hydrodesulfurization (HDS) Processes

A process of paramount importance directly associated with petroleum industry is hydrodesulfurization (HDS). This process consists in the hydrogen-assisted removal of sulfur from sulfur-containing compounds present in oil and tar and is the first process that must be used to purify petroleum fractions. An example of sulfur removal from a thiol (a common contaminant of oil, which can be considered as a prototype of the plethora of sulfur-containing contaminants) with formation of aliphatic hydrocarbons and H_2S is given by:

$$C_2H_5 - SH + H_2 \rightarrow C_2H_6 + H_2S$$

Investigations into the HDS process go back to 1928. One of the problems concerning the utilization of carbon and oil derivatives in combustion engines is caused by the presence of sulfur contaminants whose concentration is typically between 0.05 and 5.0 wt%, which, during combustion, produce SO_2, a toxic substance of air. Another important reason for removing sulfur from the coal tar and naphtha streams within a petroleum refinery is that sulfur, even in extremely low concentrations, poisons the noble metal catalysts (platinum and rhenium) in the catalytic reforming units used to upgrade the octane rating. A third and more recently realized reason lies in the fact that the poisoning effect of sulfur on Pt-based catalyst prevents the full abatement of CO from engine exhaust gases.

The use of sulfur-containing fuels has been subjected to progressively stringent legislation. The first US federal air pollution law on defining acceptable levels of SO_2 was passed in 1955 (Stern, 1982). Since this date legislation in the United States and Europe and other countries has become progressively more stringent. All these facts taken together forced the search for catalysts that could be used for deep desulfurization of oil and gasoline. Research in this area forms a very rich chapter of industrial catalysis history. The notion that metal sulfides are, in presence of hydrogen, effective in the removal of sulfur compounds was laid down to 1928 in a patent of IG Farbenindustrie (1928). While in Germany the focus was on the development of optimum catalysts for the hydro-refining of coal tar, in the United States where petroleum prevailed, catalysts were developed for hydro-refining kerosenes and gas oils (Chianelli *et al.*, 2006). As early as 1943, an article by Byrns *et al.* (1943) appeared on the use of bentonite-supported Co-doped MoS_2 sulfide catalysts for industrial HDS. In the following decades, research focused on the continuous improvement of existing catalysts by optimizing the carrier, the properties (composition, texture, acidity, and porosity), and promoters.

Since the legislation was becoming more and more stringent, the necessity to design catalysts for deep hydro-treating processes became mandatory. It is for this reason that efficient catalysts containing the transition metal sulfides

MoS_2 and WS_2 operate today in almost every refinery in the world to upgrade fuels by removing contaminants, as well as by increasing the hydrogen content of the feedstock through hydrogenation of aromatic molecules. According to Toulhoat and Raybaud (2013), world production is currently more than 6×10^7 barrels per day, and the annual catalyst consumption is approximately 10^5 tons.

Present-day hydro-treating catalysts are mixed sulfides of mostly Co/MoS_2 or Ni/MoS_2 supported on high surface area carriers such as γ-alumina. Unsupported MoS_2 directly formed *in situ* in the oil batch has been also patented (Exxon Mobil's slurry catalyst hydro-conversion process: MRC process (Chianelli *et al.*, 2006); Eni-EST, hydro-conversion technology (Panariti *et al.*, 2000)).

For more than 50 years, research on HDS catalysts was based on experience and trial and error, which is an approach not rare in industrial research (Weisser and Landa, 1973). Only more recently surface science methods like HRTEM (Bahl, Evans, and Thomas, 1968), AFM (Helveg, *et al.*, 2000), UV–vis, FTIR, Raman, and XAS spectroscopies have been employed (Cesano *et al.*, 2011). The favorable optical properties of exfoliated MoS_2 single layers make this material suitable for photocatalytic applications using visible light.

3.5 Hydrocarbon Hydrogenation Reactions with Heterogeneous Catalysts

Catalytic hydrogenation of olefins, acetylenic compounds, and aromatic molecules derived from petrol cracking and reforming is fundamental to a group of important reactions for the synthesis of fine chemistry products.

As far as alkene reactivity is concerned, it is useful to recall that the C=C double is quite reactive and can be broken, allowing reagents to be added to the double bond.

After Sabatier, the hydrogenation of olefins (or even more complex substrates) in heterogeneous phase was generally performed with metal-supported particles (particularly Pd, Pt). Common supports are γ-alumina, silica, and amorphous carbon, all characterized by having a high surface area. In a widely accepted mechanism of olefin hydrogenation, confirmed after 1970 with the advent of modern physical methods for surface science studies, the olefin (or the substrate, in general) and the hydrogen molecules are simultaneously adsorbed on the surface (Figure 3.4).

Hydrogen is dissociated by the metal, and the adsorbed hydrogen atoms move freely on the surface and then hydrogenate the olefin in the adsorbed state. Of course, in more recent times the problem of the hydrogenation mechanism has attracted the efforts of several theoretical chemists. An early

The hydrogenation reaction

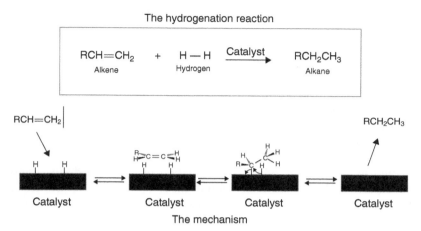

Figure 3.4 Schematic representation of the widely accepted hydrogenation mechanism of an olefin molecule on a flat idealized Pt surface. Adsorbed hydrogen atoms, which move freely on the surface, hydrogenate the olefin in the adsorbed state with the formation of an adsorbed alkyl intermediate. The final alkane product goes into the gas phase. *Source*: Adapted from Delbecq, Loffreda, and Sautet, 2010.

study based on a crystal orbital Hamilton population formalism in the framework of the extended Hückel molecular orbital theory was carried out in 2000 by Roald Hoffmann (1981 Chemistry Nobel Prize) and coworkers (Papoian, Nørskov, and Hoffmann, 2000), confirming that hydrogen moves freely on the Pt(111) surface. Since then, many studies based on more sophisticated codes have been performed (Delbecq, Loffreda, and Sautet, 2010).

On the basis of the results of these calculations, it can be readily understood why, being the adsorbed olefin surrounded by a multitude of mobile hydrogen atoms, the attack is not specific and hence enantioselectivity cannot be attained. For instance, when the $CRR'=CH_2$, olefin molecules are hydrogenated following the reaction

$$CRR' = CH_2 + H_2 \rightarrow HCRR' - CH_3$$

The resulting molecules, which contain optically active HCRR' groups, produce only racemic mixtures of D and L isomers. In other words, metal particles are not enantioselective catalysts. To induce enantioselectivity, which is an important property for pharmaceutical applications, Shiro Akabori and coworkers (Akabori and Izumi, 1962; Izumi *et al.*, 1963) discovered in 1963 that adsorption of a chiral organic component (D- or L-tartaric acid) on the surface of Raney nickel could induce asymmetric hydrogenation properties. To understand the origin of this theory, it is appropriate to briefly describe the Akabori biography.

Shirō Akabori (1900–1992) gained his PhD in 1931 under the supervision of R.K. Majima at the University of Tohoku. In 1932 he went abroad, and on his return to Japan in 1935, he became assistant professor at Osaka University, where he reached the position of full professor in 1939 and from 1953 on–off director of the newly built Institute of Applied Microbiology later named the Institute of Molecular and Cellular Biosciences. Finally, in 1967 he was called to the University of Tokyo where he became president of the Institute for Chemical and Applied Researches. Akabori studied the chemistry of amino acids and proteins as well as the biochemistry of oxidation processes that later became his principal research activity. In particular, he invented in 1931 a method for the reduction of amino acids to amino aldehydes (known as the Akabori reduction method) and in 1943 described the synthesis of the amino alcohols (Akabori synthesis). In 1952 he reported on a method of determination of C-terminal amino acids in proteins through reaction with hydrazine (Akabori, Ohno, and Narita, 1963). With his expertise in organic and amino acid chemistry, he was well informed about chiral organic compounds' preparation and properties. Consequently the use of chiral organic promoters as surface ancillary ligands capable of inducing enantioselectivity in a manner that will be later extensively adopted in homogeneous hydrogenation catalysis is not unexpected.

The use of organic chiral modifiers on metal surfaces has been the subject of several studies even in recent times. Of special mention is the progress made in the chiral modification of supported Pt and Pd catalysts by means of adsorbed cinchona alkaloids. However, as will be described in greater detail in the following chapter, the enantioselectivity obtained in this way is lower than that obtained with well-designed homogeneous catalysts and with enzymes (Barlow and Raval, 2003; Humblot, Barlow, and Raval, 2004).

In the framework of hydrogenation processes, particularly relevant has been the hydrogenation of benzene to obtain cyclohexane and cyclohexene, starting materials for the nylon synthesis:

In the process of benzene hydrogenation by Pd- and Pt-supported particles, the only product obtained is usually cyclohexane (C_6H_{12}) because the adsorbed benzene, lying flat on the surface, is massively attacked by the multitude of mobile of hydrogen atoms present on the surface. Although this is one of the desired intermediates for nylon production, in some cases it is preferable to obtain only partially hydrogenated products such as cyclohexene (C_6H_{10}), which is a useful molecule for the production of fine chemicals and also of nylon 6. Also, in order to induce selectivity, other components must be added to the pristine catalyst surface, following a strategy similar to that discussed previously. The strategy adopted in this case, after trial-and-error attempts, is the building of a three-dimensional structure partially covering the metal surface. The so-called Asahi catalyst based on Ru, patented in 1988 (Nagahara, Ono, and Konishi, 1997) using $ZnSO_4$ as the modifier, is an innovative example. The metal surface is covered by nanocrystals of $ZnSO_4$ (or other salts). The interstices between the nanocrystals constitute preferential paths, which can orient the benzene molecules during their interaction with the adsorbed hydrogen atoms so preventing total hydrogenation. Of course, as much of the surface is covered (poisoned) by salt, the increment of selectivity is obtained at the expense of hydrogenating activity, which is obviously depressed (Macleod, Isaac, and Lambert, 2001; Liotta, Martin, and Deganello, 1996). Other strategies are based on the coadsorption of organic molecules. Several authors have also suggested first modifying the proportion of surface hydrogen species by employing Pd/Au (or Pt/Au) alloys, instead of the pure metal. In fact, as Au is not active in hydrogen dissociation, the adsorbed hydrogen atoms located on Pt form a diluted adlayer, which prevents a multiple attack of the benzene molecule.

Another famous prototype hydrogenation reaction (the partial hydrogenation of the triple bond) is shown below:

The employed catalyst (the mythic Lindlar catalyst by the name of the discoverer) is palladium in finely divided form, supported on calcium carbonate. To avoid multiple hydrogenation and formation of the alkane, the surface of palladium is poisoned with Pb, which is not active in hydrogenation reactions. Further poisoning with quinoline enhances its selectivity. The catalyst was discovered in the 1970s (Lindlar and Dubuis, 1973) by Herbert Lindlar (1909–1984). Lindlar (of British origin) arrived in 1919 in Switzerland and studied chemistry at ETH Zurich and at the University of Bern where he obtained his PhD. He then joined the pharmaceutical company Hoffmann-La Roche where he worked, with the exception of a four-year hiatus, until his retirement in 1974. During these four years, Lindlar was in Zurich and Basel as an English vice-consul.

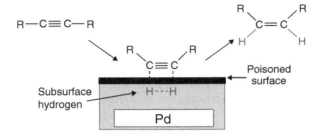

Figure 3.5 The mechanism of semihydrogenation of alkynes on the Lindlar. The surface is poisoned by Pb and by quinoline.

The Lindlar catalyst is highly interesting for three reasons:

1. Alkyne reduction is stereoselective, occurring via syn addition, that is, by addition of two substituents to the same side of a double or triple bond, to give the *cis*-alkene. It is supposed that the involved hydrogen diffuses from the interior of palladium (whose surface is nearly fully poisoned by Pb and quinoline).
2. It is able to reduce the triple bond leaving the double bonds unaltered.
3. It has several practical applications including the organic synthesis of vitamin A.

More recently the hydrogenation mechanism of these "mythic" catalysts has been investigated by means of density functional methods (Garcia-Mota *et al.*, 2010) (Figure 3.5).

3.6 Olefin Polymerization: Ziegler–Natta, Metallocenes, and Phillips Catalysts

An extremely important reaction involving ethylene and propylene is the that of polymerization, to which the German Karl Ziegler and the Italian Giulio Natta made a fundamental contribution (Figure 3.6).

They illuminated, with their stimulating activity and with their ability to disseminate new ideas, the cultural background of polymer chemistry, paving the route to new fundamental industrial processes that have completely changed the economy of the world.

Karl Waldemar Ziegler (1898–1973), born in Helsa near Kassel, graduated in Marburg in 1920 under the supervision of professor Karl Friedrich von Auwers (Ziegler, 1923). In 1926–1936 he was a professor at the University of Heidelberg.

In 1927 Ziegler made a discovery that influenced all his future scientific activity. Adding stilbene to an ethyl ether solution of phenylisopropyl potassium, he observed the first reaction of addition of an organo-alkali metal (Ziegler and Bähr, 1928). Further work showed that by adding butadiene

(a)   (b)

Figure 3.6 (a) Karl Ziegler (1892–1973) and (b) Giulio Natta (1903–1979). They were two of the most significant personalities of the time, who paved the route to new fundamental industrial processes (polymers production) that have completely changed the economy of the world. They were awarded the 1963 Nobel Prize *for their discoveries in the field of the chemistry and technology of high polymers.* Ziegler discovered that $TiCl_3$ in the presence of aluminum alkyl catalyzes the production of polyethylene at low pressure. Natta succeeded in selectively preparing macromolecules (polypropylene) having a spatially regular structure, an achievement not possible before. Images in the public domain.

to a solution of phenyl-isopropyl potassium, long-chain hydrocarbon with the reactive organo-potassium end still intact (living polymers) could be obtained. During World War II, he contributed to the German industrial effort to produce synthetic rubber (Ziegler, 1938). At the end of the war, he was nominated professor at the University of Halle (East Germany). To fully exploit and utilize Ziegler's expertise for the West, the American occupation forces convinced him to accept a position in West Germany at Mülheim where in 1953 his research team discovered that aluminum alkyl could catalyze the production of polyethylene under very mild conditions. The importance of the catalytic synthesis can be appreciated when it is taken into account that, before 1950, polyethylene was produced by a difficult high pressure process. In this process (discovered in the 1930s by ICI, Britain), ethylene gas was converted into a white solid polymer by heating it at very high pressures ($\sim$1000 bars) in the presence of minute quantities of oxygen. The polymerization reaction is a random one, producing a wide distribution of branched species and of molecule sizes. In addition Ziegler discovered that his polyethylene did not have the branched structure like the polyethylene made by the old process but consisted of very long, straight chains. Finally, in 1953 Ziegler developed a revolutionary polymerization process that in the following year was extended by Giulio Natta leading to the production of stronger plastic obtained by catalytic polymerization of propylene. Ziegler and Natta shared the 1963 Nobel Prize in Chemistry. The use of Ziegler–Natta catalytic systems soon made it possible to understand how simple hydrocarbons were polymerized into large molecular substances, so opening the way toward the extensive production of plastic.

The Ziegler adventure leading to the synthesis polyethylene was preceded by experiments concerning ethylene polymerization with alkyl aluminum catalysts. During these studies he found that in the presence of traces of nickel salts, 1-butene was almost the only product. This result convinced Ziegler to explore the effects of different metal salts. He found that in the presence of $TiCl_3$ and Et_2AlCl dissolved in alkanes, there occurred the immediate formation of high molecular weight polyethylene ($MW > 30,000$) at low ethylene pressures (Ziegler *et al.*, 1952). Further hydrocarbon polymerization studies continued to be extensively carried out by Ziegler (Ziegler *et al.*, 1955a, b, c, 1974; Ziegler, 1955).

We therefore come to the second hero of this fantastic transformation of synthetic industrial chemistry that transformed in a few years research in industrial and applied chemistry and contributed to a change in the economies of all developed countries worldwide: the Italian industrial chemist Giulio Natta (1903–1979). Giulio Natta graduated at the Polytechnic University of Milan in 1924 and by 1927 qualified to teach there. The interest of Giulio Natta in polymers started in 1932 when he met Hermann Staudinger (1953 Nobel Prize), the true father of macromolecular chemistry, who had been the first to put forward the hypothesis of a linear structure for both natural and synthetic polymers (Staudinger, 1920). In 1933 Natta became full professor at Pavia University where he stayed till 1935. At the University of Pavia, he performed his first research on the determination of crystal structure by X-ray spectroscopy (Natta, 1933). Later he used the same methods for studying catalysts and the structure of some high-organic polymers (Bruni and Natta, 1934).

In 1936 Natta took up the position of professor of Chemical Physics at the Polytechnic University of Turin where he was the director of the Institute of Industrial Chemistry. In 1938 he moved to Milan Polytechnic, where he worked for the remainder of his life. In the same year he began to investigate the polymerization of olefins and to study the production of synthetic rubber.

After World War II, he met engineer Piero Giustiniani, president of the Montecatini Chemical Industry, and began to collaborate with him to produce new kinds of polymeric materials. The link with Montecatini was one of the most fruitful collaborative efforts in Italy between university and industry in the postwar period, leading to the production of some 4000 patents. In 1953, with financial aid from Montecatini, he extended the research conducted by Ziegler to stereospecific polymerization, thus discovering polymers with a sterically ordered structure. Natta and coworkers employed a very useful terminology to indicate unambiguously the position of the substituents in a polymeric chain. They used the term "isotactic" macromolecules if all the substituents were located on the same side of the macromolecular backbone, the term "syndiotactic" macromolecules if the substituents had alternate positions along the chain, and "atactic" macromolecules if the substituents are placed randomly along the chain (Natta, 1956; Natta and Corradini, 1955; Natta and Danusso, 1959). By X-ray investigation Natta also succeeded in determining the exact arrangement of chains in the lattice of the new crystalline polymers he discovered (Natta and Corradini, 1955, 1956, 1960).

These studies led to the realization of isotactic polypropylene that Montecatini was the first to produce on an industrial scale, in 1957, in the Ferrara Chemistry Plant (Natta *et al.*, 1957). His later research led him to the synthesis of completely new elastomers by polymerization of butadiene and by copolymerization of ethylene with propylene (Natta *et al.*, 1955; Pino and Lorenzi, 1960).

In conclusion, Ziegler and Natta realized the importance of the selectivity of special catalysts able to specifically increase, often by many orders of magnitude, the rate of formation of one particular reaction product relative to that of other possible but unwanted by-products. Control of the absolute rate and the product distribution of chemical reactions has since been a challenge of paramount importance in synthetic organic chemistry, and many efforts were made to discover new stereoselective and chiral catalysts (i.e., catalysts favoring the preferential formation of one stereoisomer over another) and shape-selective catalyst (i.e., catalysts where the shape of active sites is tailored to induce the transformation of reactants into products with specific size). The importance of the role of selectivity was highlighted at the Natta presentation during the Nobel ceremony:

> *Professor Natta. You have succeeded in preparing by a new method macro molecules having a spatially regular structure. The scientific and technical consequences of your discovery are immense and cannot even now be fully estimated.*

The structure of the stereo-active centers of Ziegler–Natta catalysts and the polymerization mechanism are still the object of studies today. The active centers of heterogeneous catalysts are typically "supported" on the surface of solid particles, since the catalyst is dispersed on a second material that enhances the effectiveness or minimizes the costs. Sometimes the support is merely a surface on which the catalyst is spread to increase the surface area. More often the support and the catalytic center interact, affecting the catalytic reaction. This is the case of the original Ziegler–Natta catalysts where the active centers

Ti centers on MgCl$_2$ (001) plane cut along the (104) and (110) faces.

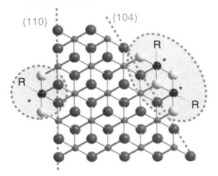

Electron microscopy of MgCl$_2$ support platelets showing (001)/(104) or (001)/(110) edges.

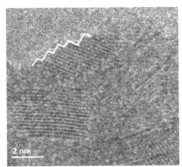

R = growing chain on titanium center
The asterisk is the coordination vacancy.

Figure 3.7 Idealized structure of an active center containing a fivefold-coordinated Ti with a growing polymer chain (CH$_2$)$_n$CH$_3$ and a coordination vacancy where an olefin molecule can be adsorbed and be inserted into the alkyl chain R. *Source*: Courtesy of Elena Groppo.

(coordinatively unsaturated Ti^{3+}) are located at the surface of TiCl$_3$. The role of support in enhancing the catalytic activity and selectivity is very important in olefin polymerization. In fact, when the Ziegler–Natta catalyst was supported on crystalline magnesium dichloride, two orders of magnitude increment of activity were observed. This important discovery was made independently half a decade after the award of the Nobel Prize by scientists at Mitsui Petrochemical and at Montecatini (Figure 3.7).

Due to this improvement these supported catalysts had a dramatic impact on the polyolefin industry and culminated in the commercial production of many high quality, reasonably priced polyolefins such as high performance HDPEs, linear low density polyethylenes (LLDPEs), highly isotactic PPs, and polyolefinic elastomers. However, as MgCl$_2$ support exposes various faces, the resulting supported catalysts are still heterogeneous and multisite, a fact that can negatively influence the selectivity. To improve the selectivity associated with the multisite structure of MgCl$_2$-supported catalyst (likely associated with the presence of different faces), the next step was consequently the modification of the surface by means of ancillary ligands (the so-called internal and external donor molecules) acting as additional ligands able to reduce the sites, heterogeneity and to tailor the structure of the active ones (Kashiwa, 1983). This achievement was obtained via a substantial amount of empirical optimization of these catalysts via a trial-and-error selection of the best ancillary ligands necessary to obtain polymers of relatively uniform molecular weight, composition, and stereochemistry. It is a widespread opinion that these modified $MgCl_2$ catalysts are real "single-site" catalysts even if experimental proof does not yet exist. A discussion on the structure of active sites in $MgCl_2$-supported Ziegler–Natta catalysts in the presence of electron donors can be found in the contribution of Liu *et al.* (2003).

Soon after Ziegler and Natta discovered heterogeneous olefin polymerization catalysts in the mid-1950s, efforts were directed toward devising homogeneous catalyst model systems that would prove more amenable to study. The use of metallocenes as polymerization catalysts was the start of an extraordinary adventure. This scientific achievement should be more appropriately treated in the chapter devoted to homogeneous catalysis. However, due to the close connection with Ziegler–Natta polymerization catalysis, discussion is anticipated. To fully understand the richness of this scientific area, a diversion onto so-called sandwich compounds (the metallocenes) in which a metal atom is situated between two aromatic ring systems is required. These compounds were discovered by Ernst O. Fischer and Geoffrey Wilkinson in 1952 (Fischer, 1952; Wilkinson and Birmingham, 1954):

They were both awarded the Nobel Prize in 1973 for this achievement, which is of great importance in the history of catalysis.

Ernst Otto Fischer (1918–2007), born near Munich in 1918, graduated in 1937 at Technical University of Munich, and, when the World War II broke out, served the army in Poland, France, and Russia.

At the end of the war in 1945, he studied chemistry in Munich and graduated in 1949. He then became an assistant to Professor Walter Hieber and worked on his PhD thesis on the carbon monoxide reactions catalyzed by Ni^{2+} salts (Hieber and Fischer, 1952). After his doctorate in 1952, he worked on transition metal and organometallic chemistry. In 1957 he was appointed professor at the University of Munich (Fischer, 1955; Fischer and Pfab, 1952).

In the 1960s he worked on metal π-complexes of cyclopentadienes and of six-ringed aromatics, on metal alkylidene and alkylidyne complexes, since then known as Fischer carbenes and Fischer carbynes.

The second protagonist in metallocene chemistry was Sir Geoffrey Wilkinson (1921–1996) (1973 Nobel prize winner) who graduated in 1941 at Imperial College of Science and Technology. As it was wartime, he was recruited for the nuclear energy project and sent to Canada in January 1943 where he remained until 1946, when he was accepted by Professor Glenn Seaborg at Berkeley for research in nuclear chemistry.

In 1949, back in England, Wilkinson became interested in transition metal complexes such as carbonyls and olefin complexes. His interest in organometallic chemistry developed in 1951, while he was at Harvard visiting the Nobel Prize awardee Roald Hoffmann and heard about the compound called dicyclopentadienyliron (ferrocene) (Kealy and Pauson, 1951). In June 1955, he was appointed to the chair of Inorganic Chemistry at Imperial College in the University of London where he worked on the complexes of transition metals (Ru, Rh, and Os), in particular complexes of unsaturated hydrocarbons and with catalytic activity in hydrogenation reactions. He remained at the Imperial College until his retirement. In 1962 he wrote, together with Frank Albert Cotton, the classic textbook *Advanced Inorganic Chemistry* (1962) that has been a reference book for generations of students all over the world.

Wilkinson correctly deduced that $Fe(Cp)_2$ compound's structure consists of a single iron atom sandwiched between two five-sided carbon rings:

Wilkinson synthesized a number of other "sandwich" compounds, or metallocenes, and his research earned him the Nobel Prize. In catalysis history, of particularly importance was his discovery in 1966 with John Osborn of the famous Wilkinson's catalyst, a homogeneous hydrogenation catalyst that had widespread significance for organometallic chemistry (Osborn *et al.*, 1966). We will return to this in the chapter devoted to homogeneous catalysis.

The catalytic properties of metallocenes had been, however, already studied in 1957 by the industrial chemist David S. Breslow, who worked at Hercules Inc., a Wilmington chemical company (Breslow and Newburg, 1957). An expert in polymer chemistry, he reported that the metallocene Cp_2TiCl_2 (Cp: cyclopentadienyl), which is soluble in aromatic hydrocarbons, could be active for olefin polymerization once activated with Et_3Al or Et_2AlCl. These soluble catalysts polymerized ethylene but were inactive for propylene and exhibited much lower activities than the heterogeneous systems.

This situation improved dramatically in the early 1980s, when Hansjorg Sinn, chief of the research group at the University of Hamburg and future minister of Science and Research of the City of Hamburg, together with his former student Walter Kaminsky discovered that titanium complexes $Cp_2Ti(CH_3)_2$ activated with Me_3Al containing small amounts of water showed a high polymerization rate, which reached a maximum when the molar ratio of water to trimethylaluminum was 1:1. They were able to isolate the product formed in the 1:1 mixture of water and trimethylaluminum as a white solid compound (named methylaluminoxane (MAO)) soluble in aromatic solvents. Even today the exact structure of MAO is not known. According to Sinn and Barron (Sinn, 1995; Koide, Bott, and Barron, 1996), it consists mainly of units of the basic structure $[Al_4O_3Me_6]$, which contains four aluminum, three oxygen atoms, and six methyl groups. It is generally assumed that the function of MAO is first to undergo a fast ligand-exchange reaction with the metallocene with formation of metallocene methyl and dimethyl compounds. In the next step, CH_3^- is abstracted from the metallocene by an Al center in MAO, thus forming a catalytically active metallocene cation and an MAO anion. The two moieties form an anion–cation pair.

A prominent contribution to this area is that of Walter Kaminsky (1941–) who discovered that when MAO was used to activate the organo-zirconium catalyst for ethylene and propylene polymerization, extremely high activities are obtained (Sinn, Kaminsky, and Vollmer, 1980; Sinn and Kaminsky, 1980). At such high activities the catalyst can remain in the product. MAO has close similarity to the Ziegler–Natta catalyst, and after its initiation, the active centers become supported on the insoluble polymer phase. Kaminsky has published more than 400 papers, mostly devoted to olefin chemistry, and received many honors. Today he is emeritus professor of the Technical University of Hamburg.

This result was very important because organo-zirconium complexes were thought to be totally inactive for olefin polymerization. A patent application was written covering these exciting results. The insertion time of one molecule of ethene into the growing chain is so short that a comparison with enzymes is fully appropriate.

Subsequently Kaminsky made the important discovery that Cp_2ZrCl_2, which can easily be synthesized and is more stable than $Cp_2Zr(CH_3)_2$, and is in combination with MAO a very active catalyst precursor (Figure 3.8).

Nearly contemporary to the Sinn and Kaminsky discoveries, a further milestone was reached when Hans-Herbert Brintzinger succeeded (Wild *et al.*, 1982) in synthesizing chiral-bridged metallocenes in 1982 at the University of Konstanz (Figure 3.8). Ewen (1984), at the Exxon Company (the United States), was able to demonstrate that appropriate chiral-bridged titanocenes catalyze the formation of partially isotactic polypropylene. A little later, highly isotactic material was obtained with analogous zirconocenes in the Kaminsky laboratory. After this discovery, a valuable development of industrial and scientific research in the metallocene sector was commenced by introducing the Zr coordination sphere, a variety of bridged metallocenes of different structure. A few examples of bridged metallocenes are shown below:

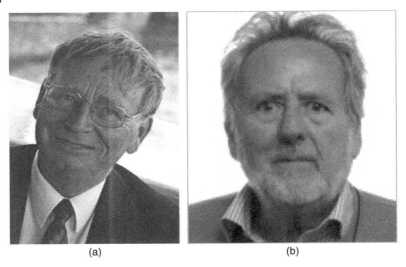

(a) (b)

Figure 3.8 (a) Walter Kaminsky (1941–) and (b) Hans-Herbert Brintzinger (1935–). They discovered metallocene polymerization catalysts. The main contribution of Kaminsky was the discovery that Cp_2ZrCl_2 activated by partially hydrolyzed Me_3Al (MAO) is an active olefin polymerization catalyst. Britzinger discovered that chiral-bridged titanocenes and zirconocenes catalyze the formation of partially isotactic polypropylene. The Kaminsky and Britzinger discovery opened the way to an unprecedented selectivity in polymer synthesis through a rational modification of the ligands sphere of the active metallic center. *Source*: (a) Reproduced with permission of Walter Kaminsky (University of Hamburg). (b) Reproduced with permission of Hans-Herbert Brintzinger (University of Kostanz).

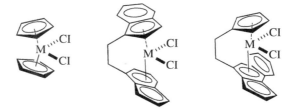

The rational modification of the ligands sphere of the active metallic center forms the basis of this development. Up until today, this fascinating research direction has not been concluded. It is, however, evident that the contributions of Kaminsky and Britzinger were not only contemporary but also highly synergic. For this reason they were jointly awarded the Heinz Beckurts Prize, the Alwin Mittasch Medal, and the Walter Ahlström Prize. Catalysts formed by MAO and properly modified metallocenes or other transition metal-containing complexes opened up the possibility of designing the microstructure of polyolefins in a way that was not possible previously.

These real "single-site" catalysts show only one kind of active site and are used today mainly for the rapidly increasing production of LLDPE and of some special copolymers and to elucidate the elementary steps of the polymerization reaction, which is simpler in homogeneous than in heterogeneous systems.

Despite the discovery of homogeneous metallocene catalysts, the classical $MgCl_2$-supported Ziegler–Natta catalysts did, however, not lose ground, and today they are still responsible for the production of the bulk of polyolefins.

The metallocene catalysts discovery has stimulated new research. Among this, the discovery of other weakly coordinating cocatalysts, such as tetra(perfluorophenyl)borate anions $[(C_6F_5)_4B]^-$, needs a specific mention because they have been successfully applied in place of MAO (Sishta, Hathorn, and Marks, 1992).

A convenient heterogeneous catalyst for olefin polymerization is the Phillips catalyst prepared by anchoring chromium trioxide on high surface area silica and reduced by hydrogen (Clark *et al.*, 1956). The Phillips catalysts were discovered in 1951 and patented in 1953 by Phillips research laboratories in Bartlesville, Oklahoma. In 1951, using as catalyst nickel oxide on silica–alumina, John Paul Hogan and Robert L. Banks invented "crystalline polypropylene" and high density polyethylene (HDPE), initially known by the name Marlex, following a previous patent obtained in 1945 by Grant C. Bailey and James A. Reid from the same company. In 1987 Hogan and Banks received the Perkin Medal for these discoveries. They were inducted into the National Inventors Hall of Fame. Both were given a *Heroes of Chemistry* Award by the American Chemical Society in 1989. In the following years the catalyst was gradually improved and found industrial application (Banks, 1980; Davis and Hettinger, 1983; Banks and Kukes, 1985) (Figure 3.9).

The biographies of the American chemists Robert Banks (1921–1989) and John Paul Hogan (1919–2012) show how new discoveries can originate from a propitious research environment in the industry. This fact has already been documented for the scientists involved in acid and reforming catalysts at UOP laboratories. Banks attended Southeast Missouri State University in 1940 and joined the Phillips Petroleum Company in 1946 where he worked until he retired in 1985. In 1956 Banks contributed to the discovery of the famous Cr/SiO_2 ethylene polymerization catalysts, for which he later received the Perkin Medal together with John Paul Hogan. Additional studies with different olefins brought convincing evidence to him that a new reaction was discovered (metathesis). For this discovery he received several awards: 1974 Oklahoma Chemist Award, 1979 American Chemical Society Petroleum Chemistry Award, and 1981 Pioneer Chemistry Award.

John Paul Hogan, born in Kentucky and educated at Oklahoma State University, was employed from 1944 to 1947 by the Phillips Petroleum Company as research chemist, promoted in 1947 to the position of project leader, from 1954

(a)   (b)

Figure 3.9 (a) Robert Banks (1921–1989) and (b) J. Paul Hogan (1919–2012). These industrial researchers developed new catalysts for polypropylene and high density polyethylene synthesis (Phillips catalysts for olefin polymerization). This Cr-based catalyst is still responsible for industrial production of a high fraction of polyethylene. The reuse of these images is free because they are of uncertain origin and can be classified as orphan works.

to 1960 to group leader, and finally in 1977–1985 to senior research associate. In 1985 he retired.

Thanks to their discovery, the Phillips Company obtained a patent on polypropylene and polyethylene and entered the plastic business. These plastics were initially known by the trademark name of Marlex (Banks, 1980; Davis and Hettinger, 1983; Banks and Kukes, 1985).

In 1998 the American Chemical Society established an award given to Banks (posthumously) and Hogan as *Heroes of Chemistry*. The development of a new high density polyethylene process was designated as a National Historic Chemical Landmark. The designation was conferred by the American Chemical Society. A plaque marking the designation was presented to Phillips Petroleum Company, Bartlesville, Oklahoma, on 12 November 1999. The inscription reads:

> *In 1951, while attempting to convert propylene into gasoline, J. Paul Hogan and Robert L. Banks of Phillips Petroleum Company discovered polypropylene, a high-melting crystalline aliphatic hydrocarbon. This discovery led to the development of a new catalytic process for the making of high-density polyethylene. Now, billions of pounds of polypropylene and high-density polyethylene are used each year in packaging of every*

shape and size, from milk jugs to large chemical drums; in toys, tools, furniture, and fibers; in water, sewer and gas pipes; and in auto parts. These polymers have become an integral part of our everyday lives.

The Cr/SiO_2 catalyst is very stable and efficient in gas-phase polymerization of ethylene. It is, however, unable to polymerize propene to isotactic polymers. Still today the polyolefins produced by means of the Phillips and Ziegler–Natta catalysts taken together comprise 95% of the polyethylene market. The importance of this catalyst is twofold. On one hand, it is responsible for the production of a large fraction of polyethylene, and on the other hand, it is very simple (and hence economic) because it does not require alkyl derivatives or MAO as activators. This old catalyst is an early example of heterogeneous "single-site" catalyst where the active sites (Cr^{x+}) are chemically anchored to the amorphous silica support. However, due to the amorphous character of the anchoring support, the isolated and grafted Cr^{x+} sites have slightly different structures, a fact that makes this catalyst heterogeneous (McDaniel, 2010; Groppo *et al.*, 2005). The real center is formed during the reaction with ethylene with surface chromates or by oxidative addition of ethylene to the previously reduced center. The study of the structure of these highly coordinated unsaturated sites represents an example of the combined and synergic application of FTIR, resonant Raman, EXAFS, XANES, and computational methods. These investigations have illustrated, for the first time, the coordination of up to three ligands (CO, NO, C_2H_4) and the mobility of metal centers under the action of adsorbates. The classical surface science methods (LEED, HREEL, STM, etc.) could not be used because of the fully insulating character of the material. This difficulty is commonly encountered for oxide-based catalysts.

3.7 Selective Oxidation Reactions

3.7.1 Alkane Oxidation

Partial oxidation of *n*-butane to maleic anhydride is the only successful example of the use of an alkane as reactant for the synthesis of a bulk chemical:

$$CH_3\text{-}CH_2\text{-}CH_2\text{-}CH_3 + 7/2\ O_2 \rightarrow C_4H_2O_3 + 4H_2O$$

The discovery of an active and selective catalyst, vanadyl pyrophosphate, made possible the development of the technology involved in this reaction. Today several firms are using this technology, including Mitsubishi, Sohio, Monsanto, and DuPont. The numerous contributions and problems associated with the catalyst structure and reaction mechanism have been thoroughly reviewed by Albonetti *et al.* (Albonetti, Cavani, and Trifiro, 1996).

An important problem related to partial oxidation of light paraffin is the methane utilization. As we know, today methane total combustion is mainly used for heat production and the generation of electricity. Methane is an ideal fuel for these purposes because it can be easily purified from sulfur compounds and because it has the largest heat of combustion relative to the amount of CO_2 formed. Yet methane is a greatly underutilized resource for chemical production (Lunsford, 2000). Among the methods that can be used to utilize methane for the production of useful chemicals, partial oxidation to methanol and formaldehyde and oxidative coupling are the most challenging and investigated. A detailed analysis of the abundant literature on this subject is outside the scope of this book. The main obstacle is the low selectivity at high conversions because the products are more reactive than methane. So far the direct and economically convenient partial oxidation of methane to methanol over a solid catalyst has not been successful. Similarly oxidative coupling of methane to C_2 hydrocarbons has not yet produced definite results. In conclusion, the direct catalytic conversion of methane remains a great challenge for the catalysis community as thoroughly reviewed by Labinger (2004) and Holmen (2009).

3.7.2 Olefin Oxidation

An important partial oxidation process involving olefins is ethylene epoxidation discovered in 1931 by the French chemist Theodore Lefort, using supported silver as catalyst (Lefort, 1931). The process was first commercialized by Union Carbide, and since 1940 almost all industrial production of ethylene oxide has relied on this process. Ethylene oxide is important or critical to the production of detergents, thickeners, solvents, plastics, and various organic chemicals such as ethylene glycol, ethanolamines, simple and complex glycols, polyglycol ethers, and other compounds. This places ethylene oxide as the 14th most produced organic chemical. The mechanism of this reaction and the state of oxygen on the surface of silver have been the subject of many investigations, particularly after the advent, several decades later, of new advanced physical and computational methods (Kestenbaum *et al.*, 2002; Campbell and Paffett, 1984; Bukhtiyarov and Knop-Gericke, 2011).

Another process of great industrial interest is epoxidation of propylene. In fact propylene oxide is the starting material for polyurethane, a much used plastic. Furthermore it used to produce solvents, chemical intermediates, flame retardants, synthetic lubricants, chemicals for oil field drilling, and textile surfactants.

Today the suitable oxidant is not oxygen but hydrogen peroxide, and the innovative and selective catalyst is titanium silicalite (TS-1), a crystalline microporous material discovered at the end of the twentieth century by EniChem, which will be described in the chapter devoted to zeolites and zeolitic materials.

The discovery of this catalyst has changed completely the processes for the production of propylene oxide. The commercialization of an integrated hydrogen peroxide–propylene oxide (HPPO) process, in which the hydrogen peroxide required for the epoxidation of propylene is produced *in situ*, started in 2008 (BASF, Dow Chemical).

Another important process was the epoxidation of olefins using organic hydroperoxides as oxidants. In 1969 Harald P. Wulff and Freddy Wattimena (Shell Company) showed that silica–titania was an effective heterogeneous catalyst for the reaction of olefins with organic hydroperoxides to form olefin oxides. However, because of the well-known high instability of hydroperoxides, the epoxidation of olefins with organic hydroperoxides remains an unresolved and difficult problem. The search of a selective epoxidation catalyst based on SiO_2–TiO_2 is also complicated by its amorphous nature, a fact that is not favoring selectivity.

As a final comment regarding catalytic hydrocarbon oxidation reactions, it must be underlined that this area not only spans over a century but is also very broad because many unsaturated molecules are involved. For more information, two important reviews can be consulted (Lucke *et al.*, 2004; Centi, Cavani, and Trifiro, 2001).

3.7.3 Aromatic Compounds Oxidation

Even today the direct catalytic oxidation of benzene to phenol:

is a challenging reaction in oxidation chemistry. More than 100 years ago Friedel and Crafts examined the conversion of benzene to phenol using oxygen. The experiments, carried out in 1888, were conducted in the gas phase in the presence of aluminum chloride (Friedel and Crafts, 1888). Later studies, made by the IG Farben chemists Burckhardt Helferich, Rolf Streeck, and Erich Günther, used molybdenum, tungsten, copper, and vanadium oxides as catalysts (Helferich, Streeck, and Günther, 1938).

After more than 100 years of research, only a small increase in yield was obtained with mixed Mo–V oxides supported on silica (Yamanaka *et al.*, 2000). The very low selectivity to hydroxylated products is due to the opening of the benzene ring with total oxidation associated with formation of CO_2 and water. These examples justify why the direct oxidation of benzene to phenol using oxygen as oxidant is considered one of the 10 "challenges for catalysis" (Cornils and Herrmann, 2003). As will be illustrated in the chapter devoted to zeolites, it has been recently demonstrated that the situation is improved when N_2O is used as oxidant and Fe–silicalite as catalyst (Panov *et al.*, 1992).

Vanadium compounds were applied as catalysts for the hydroxylation of benzene using hydrogen peroxide. In these catalysts the vanadium-active species are located in the channels of microporous systems; their discussion will therefore be postponed.

Partial oxidation of aromatic compounds has been studied by several authors. The partial oxidation of naphthalene to produce phthalic anhydride, which is an important industrial chemical, was obtained following the "Gibbs process" from the name of the inventor Harry D. Gibbs:

This historically important process, initiated in 1920, started to decline when petroleum products (in particular, *ortho*-xylene) became progressively available (Gibbs, 1922).

Today partial oxidation of *ortho*-xylene to produce phthalic anhydride, especially important for the large-scale production of plasticizers, is the alternative and dominating process.

In 2000, the worldwide production of phthalic anhydride was estimated to be about 3,000,000 tons per year. In all cases the active phase is supported V_2O_5, which is a good catalyst in other hydrocarbon partial oxidation reactions (Grasselli, Burrington, and Brazdil, 1981; Grasselli and Trifirò, 2011).

The most likely mechanism is of the Mars–van Krevelen type as discussed in the previous chapter. Other oxides like MoO_3 have been also used with success.

The shift from naphthalene to the xylene oxidation process was due to the availability of abundant and inexpensive *ortho*-xylene from the 1960s, a fact that has prompted research programs to discover catalysts that could selectively oxidize *ortho*-xylene to phthalic anhydride. This research effort culminated in the discovery of supported V_2O_5/TiO_2 catalysts, consisting of a two-dimensional surface vanadium oxide phase strongly interacting with the titania support (sometimes referred to as "monolayer catalyst") (Weissermel and Arpe, 1978; Bond, 1997). A few years later fundamental information about the structure of active sites was obtained by means of advanced spectroscopic methods like FTIR and Raman. For more information, the papers by Bert Weckhuysen *et al.* and by Israel E. Wachs can be consulted (Weckhuysen and Keller, 2003; Wachs, 2013).

Figure 3.10 Robert Grasselli (1930). Grasselli has greatly contributed to the catalytic oxychlorination and to the pioneering mechanistic analyses and deductions for selective oxidation in terms of the Mars–van Krevelen mechanism. *Source*: Courtesy of R. Grasselli.

3.8 Ammoximation and Oxychlorination of Olefins

Together with olefin epoxidation, catalytic ammoxidation of propene and oxychlorination of ethene comprise approximately one-quarter of the value produced by all catalytic processes worldwide (see, for instance, propene ammoximation to produce acrylonitrile) (Burrington *et al.*, 1982; Grasselli, 1999) (Figure 3.10).

The ammoxidation reaction of propylene with ammonia and air to acrylonitrile (Sohio process) is represented in the following:

$$CH_2 = CH - CH_3 + NH_3 + \frac{3}{2}O_2 \rightarrow CH_2 = CHCN + 3H_2O$$

This process accounts for most of the acrylonitrile currently annually produced worldwide. Acrylonitrile, first synthesized in 1893 (Moureu, 1894) by the French chemist Charles Moureu (1863–1929), did not become of practical utility until the 1930s, when the industry began using it in new applications such as polyacrylonitrile (PAN) polymer, rubbers, and adiponitrile. Recently it has been discovered that PAN carbon fibers can be produced that find industrial application as light and robust materials, which maintains the importance of PAN. This process is the result of the contribution of many industrial researchers. Everything started in 1953 when Sohio (a Petroleum Company founded by John D. Rockefeller) decided under the advice of Franklin Veatch (an innovative research supervisor) that converting light refinery gases such as propane, propene, and so on, to compounds containing oxygen could be profitable. At the time, partial oxidation of aliphatic and olefinic hydrocarbons was primitive and expensive. The breakthrough for the synthesis of acrylonitrile from propene, ammonia, and oxygen came when bismuth phosphomolybdate was used as the catalyst in an experiment designed by Idol in March 1957 (Veatch *et al.*, 1960).

The chronological details of PAN synthesis are particularly interesting. In 1955 the first major discovery that bismuth phosphomolybdate catalyzed propylene to acrolein transformation in a single catalytic reaction step was made. The second improvement came when it was found that acrylic acid could be made in a subsequent step.

The third step came when it was realized that acrylonitrile (a derivative of acrylic acid) could be successfully performed by feeding acrolein, ammonia, and air over the catalyst (phosphomolybdate) that produced acrylic acid from acrolein.

The discovery of the synthesis of acrylonitrile is important not only economically but also because it is the result of the joint efforts of several researchers in an industrial and risky program and in an industrial milieu. Today, nearly all acrylonitrile is produced by the Sohio process (1996), and catalysts developed at the Sohio Laboratory are used around the world. The whole process was based on the working hypothesis that the oxidation step could be better performed by lattice oxygen coming from a multivalent solid metal oxide rather than by O_2 gas, whose function was only to fill the oxygen vacancies created by the oxidation process. The influence of Mars–van Krevelen model here is evident. Many oxide mixtures have been tested following a mixture of intuition, of knowledge on lattice oxygen mobility of various oxidic systems, and of trial and error. Subsequent studies have established that the most active and selective heterogeneous catalysts for ammoxidation of olefins to their corresponding nitriles contain at least three essential components. In fact the catalytically active sites are composed of elements that chemisorb the olefins, abstract α-hydrogens from olefins to generate allylic species, and selectively insert nitrogen into the allylic surface intermediates. Furthermore the active sites in the immediate surroundings must provide a means for the facile transport of electrons oxygen ions and vacancies (Callahan *et al.*, 1970; Grasselli, Burrington, and Brazdil, 1981). We are clearly in the presence of an *ante litteram* example of complex nanomachinery for molecule manipulation.

Important contributions to the understanding of the ammoxidation mechanism were made by the industrial chemist Robert K. Grasselli who, after having studied chemistry at Harvard, graduated at Case Western Reserve University. He was appointed as research fellow at the Standard Oil Company (Sohio) and in 1959 became director of the Catalysis and Solid State Science section, a position that he held until 1986, when he moved to the Office of Naval Research as director of the Chemistry division until 1989. From 1989 to 1995 he was research scientist at the Mobil Research and Development Corporation in Princeton, and from 1983 to 1990 full professor of Chemical Engineering and Material Science at the University of Delaware. He has produced 175 patents, had more than 150 scientific publications, and received many honors. He was also the founder of the European Gordon Research Conferences.

Another important reaction involving an olefin (ethene) deriving from the oil cracking of petrol is oxychlorination, which leads to the intermediate (vinyl chloride) for the synthesis of polyvinyl chloride (PVC), which is the third commodity plastic in the world after polyethylene and polypropylene. In the early history of vinyl chloride and PVC synthesis, Henri Victor Regnault (1810–1878), best known for his careful measurements of the thermal properties of gases, played an important role because the first synthesis of both compounds must be attributed to him (Fox, 1911). This French chemist and physicist deserves specific attention, since he is an important example of an interdisciplinary scientist, not commonly found today. In 1830 he entered the École Polytechnique and passed in 1832 to the École des Mines, developing an important aptitude for experimental chemistry. Only a few years later he was appointed as professor of Chemistry University of Lyon.

At that time the laboratory of Liebig at Giessen was one of the most important centers for the promotion of experimental chemistry. Its international renown attracted the attention of many students from foreign countries, in particular, from France. Due to Liebig's association with Joseph Louis Gay-Lussac, Regnault in 1835 reached the Giessen laboratory, where he synthesized several chlorinated hydrocarbons (e.g., vinyl chloride, dichloromethane) (Regnault, 1835, 1839).

The results were published in the *Annales de Chimie et de Physique* and earned him membership of the Academy of Sciences. In 1838 he observed, under illumination, the formation from vinyl chloride of a white solid, which only one century later was recognized as a polymer. We know now that the white powder was PVC and that the reaction catalyzed by sunlight is one of the oldest observations of photocatalysis.

In the following years he developed a great interest in physical chemistry (Regnault, 1854) and designed sensitive thermometers (Regnault, 1842) and hygrometers (Regnault, 1845) and measured the specific heats of many substances (Regnault, 1841) as well as the thermal expansion coefficient of several gases. Two laws governing the specific heat of gases are named after him. In the course of this work, he discovered that at temperatures near a substance's boiling point, Boyle's law is only an approximation. The second occasion was in 1872 when the German Eugen Baumann (1846–1896) observed the same effect (formation of a white solid inside flasks) on the newly discovered vinyl chloride gas exposed to sunlight. However, as the material was difficult to work with, no commercial applications were found.

The vinyl chloride story continues when, in 1912, Frans, a German chemist working for Griesheim-Elektron, patented a means to produce vinyl chloride from acetylene and hydrogen chloride using mercuric chloride as a catalyst.

Today the vast majority of vinyl chloride monomer (VCM) is produced in two steps. The first step is oxychlorination of ethene using $CuCl_2$ as catalyst as follows:

$$CH_2 = CH_2 + 2HCl + \frac{1}{2}O_2 \rightarrow ClCH_2CH_2Cl + H_2O$$

The second step is thermal cracking (dehydrohalogenation) of dichloro-ethane:

$$Cl - CH_2 - CH_2 - Cl \rightarrow CH_2 = CH - Cl + HCl$$

The initial choice of $CuCl_2$ as catalyst (usually supported on alumina) is more the result of empirical observations made by industrial researchers than of a fully rational design, even if the testing of a metal chloride to perform reactions where chlorine is involved is certainly intuitive. It is difficult to single out the person responsible for the first observation concerning the catalytic properties of $CuCl_2$ in oxychlorination reactions. This situation is not rare where industrial catalysis is concerned. It is known today that, as observed for active oxygen in partial oxidation reactions catalyzed by V_2O_5, in ammoximation reactions the Cl does not come directly from HCl but from the $CuCl_2$ catalyst itself and that the role of HCl is only to restore the chlorine concentration. On this basis it can be concluded that we are dealing with a Mars–van Krevelen-type mechanism, whose general validity in many processes is, once more, clearly demonstrated. The detailed mechanism and the structure of catalytic sites are today still under investigation (Leofanti *et al.*, 2002).

As briefly mentioned, VCM is an intermediate in the production of PVC, which is one of the oldest synthetic materials.

The first patent on PVC was submitted in 1913 by Friedrich Heinrich August Klatte (1880–1934). His method again used polymerization of vinyl chloride with sunlight (Klatte and Rollett, 1917). The most significant breakthrough occurred, however, in the United States in 1926 when the industrial scientist Waldo Lonsbury Semon (1898–1999), employed at the BF Goodrich Company, discovered that when the powder of this hard and brittle solid was dissolved in certain solvents, it swelled to form a gel that was moldable. The result was an elastic, but not adhesive substance that, in combinations of up to 50% with plasticizer, is now familiar in floor tiles, garden hoses, imitation leather, shower curtains, fabric for umbrellas, raincoats, shoe heels, automobile seats, and dozens of other commonly used objects. He also made pioneering contributions in polymer science, including new rubber antioxidants, and his technical leadership led to the discovery of three major new polymer families: thermoplastic polyurethane, synthetic "natural" rubber, and oil-resistant synthetic rubbers. He patented his discovery in 1933 (Semon, 1933). Semon made more than 5000 other synthetic rubber compounds. He was awarded the Charles Goodyear Medal in 1944 and the Elliott Cresson Medal in 1964.

Today PVC is produced from VCM via a radical mechanism by using a radical initiator. The low cost and excellent durability of PVC make it the material of choice for health care, transport, textiles, and construction industries.

3.9 Ethylbenzene and Styrene Catalytic Synthesis

Ethylbenzene is important industrially because it is the intermediate for the synthesis of styrene by catalytic dehydrogenation from which polystyrene (a widely employed resin) is obtained.

In the early days ethylbenzene was produced via Friedel–Crafts synthesis from benzene and ethylene. In classical organic chemistry the Friedel–Crafts reaction (discovered in 1877) is the usual way to attach alkyl substituents to an aromatic ring (Friedel and Crafts, 1877). The traditional catalysts are $AlCl_3$ and $FeCl_3$, the former being the most used and powerful Lewis acid. The reaction mechanism is based on the reaction between $AlCl_3$ ($FeCl_3$) and RCl (R = alkyl group):

$$RCl + AlCl_3 \rightarrow R^+ - (AlCl_4)^-$$

with formation of R^+ intermediate (the true active species). Then R^+ is attacking the benzene ring with elimination of HCl (Olah, 1963).

Today ethylbenzene (one of the top 50 chemicals produced in the world) is produced by the alkylation of benzene with ethane, using an acidic zeolite as catalyst. This process, which is very efficient and with extremely high selectivity (>98–99%), will be described in the chapter devoted to zeolites and microporous catalysts.

From ethylbenzene, styrene is obtained by catalytic dehydrogenation. Most ethylbenzene dehydrogenation catalysts are based on iron(III) oxide, promoted by potassium oxide or potassium carbonate. Understanding the promoting effect of potassium has been the goal of many researchers. There is now consensus that potassium and iron oxide forms the phase $KFeO_2$, which is the active catalyst (Hirano, 1986).

In 2010 about 27 million tons of styrene monomer was produced. Snamprogetti and Dow are developing a process for the synthesis of styrene from ethane and benzene. The complex catalyst is composed of gallium oxide, platinum, and potassium on alumina. From previous considerations it is emerging that Fe_2O_3 (already documented to play a role in Fischer–Tropsch synthesis) and Ga_2O_3 oxide are fundamental in this dehydrogenation process. As usual, this fact has for decades fueled the interest of surface scientists working on Fe_2O_3 and Ga_2O_3.

3.10 Heterogeneous Metathesis

Observation of olefin metathesis based on heterogeneous catalysts goes back to the 1950–1960 period. In 1956 Herbert S. Eleuterio (1972-), working at the DuPont's Petrochemicals department in Wilmington, Delaware, obtained a propylene–ethylene copolymer from a propylene feed, passed over a molybdenum oxide on aluminum catalyst that produced a mixture of propylene, ethylene, and 1-butene, a fact that indicated that a metathesis reaction was taking place. However, DuPont did not commercialize these results. When Eleuterio tried the experiment with cyclopentene (Eleuterio, 1960a, b), he realized that he was observing a reaction in which the cyclopentene ring had been opened and joined up again (Koch and Eleuterio, 1969). The seminal contribution of Eleuterio was recognized in 2006 when the Nobel Laureates Robert H. Grubbs and Richard R. Schrock, recipients of the 2005 Nobel Prize in Chemistry for the development of olefin metathesis, visited the DuPont Experimental Station to celebrate the contributions of Eleuterio in the early stage development of this novel technology.

Around the same period several chemists at other petrochemical companies had similar baffling results. In 1960, for instance, researchers working at the Standard Oil Company (Edwin F. Peters and Bernard L. Evering) observed that propylene in contact with molybdenum oxide on alumina treated with tri-isobutyl aluminum yields ethylene and butenes (Peters and Evering, 1957). In 1964, Robert L. Banks and Grant C. Bailey, of the Phillips Petroleum Company, reported the disproportionation of propylene to ethylene and butenes using molybdenum hexacarbonyl or molybdenum oxide supported on alumina (Banks and Bailey, 1964).

In the following years, the catalyst was gradually improved and found industrial application (Banks, 1980; Davis and Hettinger, 1983; Banks and Kukes, 1985). This strange chemistry of the olefins had some almost magical behavior that could not be explained by the reactions of olefins known at the time.

Today the heterogeneous olefins metathesis catalysts are usually transition metal oxides such as Re_2O_7 and MoO_3 supported on several inorganic matrices (SiO_2, Al_2O_3, ZrO_2, and TiO_2) (Ivin, 1997; Schuster and Blechert, 1997; Thayler, 2007). As for Re_2O_7 supported on Al_2O_3, it is generally agreed that monomeric tetrahedral ReO_4 structures are grafted and stabilized over the alumina surface. This species is the precursor of the active one with single-site character. It is supposed that the metal center of the precursor species has three equivalent Re=O moieties pointing outward from the surface and a Re–O–Al bond formed via *an anchoring process* involving an acidic OH group of the support (Coverdale, Dearing, and Ellison, 1983; Oikawa *et al.*, 2004). Less is known about the coordination and valence state of the real working center after interaction with the reactants.

As we will discuss in the chapter specifically devoted to homogeneous catalysts, alkene metathesis will be very efficiently performed a decade later by means of homogeneous complexes with well-defined molecular structure, whose introduction certainly received input from the discoveries of Eleuterio and Banks. We can state without uncertainty that throughout the history of catalysis, homogeneous and heterogeneous catalysis influenced each other in a very fruitful way. Both types still need further improvements because while homogeneous catalysts typically suffer from several drawbacks such as bimolecular deactivation processes and metal recovery (separation of products from the catalyst as well as metal contamination of the product), the heterogeneous counterparts suffer from the presence of multiple sites and ill-defined structures. A line of research has been suggested (Coperet *et al.*, 2003; Coperet, 2007; Wischert *et al.*, 2012) that by combining the advantages of both disciplines, optimal systems could be obtained. This approach is called surface organometallic chemistry (SOMC), which not only is useful for designing new metathesis catalysts by anchoring alkylidyne complexes to surface silanols of SiO_2 but also can allow a better comprehension of the Re_2O_7/Al_2O_3 traditional ones.

3.11 Catalytic Synthesis of Carbon Nanotubes and Graphene from Hydrocarbon Feedstocks

In recent times the attention of chemists, material scientists, and physicists has been attracted by novel carbon structures like fullerenes (Kroto *et al.*, 1985), carbon nanotubes (Iijima, 1991), and graphene (Rao *et al.*, 2009), as well as by the fabrication of carbon nanotubes and graphene from hydrocarbon feedstock.

While fullerene- and fullerene-type molecules are prepared by incomplete combustion of carbon feedstocks (Mojica, Alonso, and Méndez, 2013), carbon nanotubes and graphene are fruitfully prepared by means of a catalytic "bottom-up" approach.

In particular, while the formation of carbon filaments on nickel (a steam-reforming catalyst) as well as on group VIII metal and metal alloys is known since the early 1970s (Baker *et al.*, 1972), only with the work of Jijima in 1990, the synthesis of filaments in form of nanotubes was elucidated (Iijima, 1991). The catalytic growth of filaments, already observed in the 1970s, has certainly inspired the catalytic way toward the synthesis of these hollow structures. In the catalytic processes now used, they grow at about 700°C on a substrate of metal catalyst particles, usually nickel, cobalt, and iron.

To grow the nanotubes, two gases are bled on the metal particles: a process gas (such as nitrogen or hydrogen) and a carbon-containing gas (such as acetylene, ethylene, or methane). Nanotubes grow at the "active" sites of the

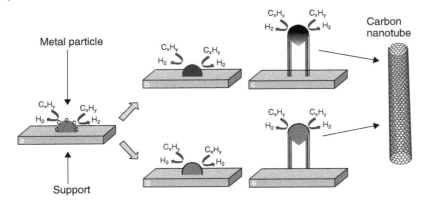

Figure 3.11 Mechanism of catalytic formation of carbon nanotubes. *Source*: Tessonnier and Su, 2011. Reproduced with permission of Wiley.

metal catalyst. The diameters of the nanotubes are related to the size of the metal particles. Note that nickel and cobalt are among the metals that are active in Fischer–Tropsch synthesis.

Two mechanisms have been proposed. Following the first one after the initial deposition of carbon on the metal particle, the formation of solid solution occurs, which then creates the carbon nanotubes via C—C bond formation (upper part of the figure). Following the second mechanism (which is most likely when Ni, Co, and Fe particles are involved), the initially formed carbon atoms migrate to the surface and then form C—C bonds (Figure 3.11).

During the catalytic process a mixture of single-walled or multiwalled nanotubes is usually formed, whose relative proportion depends on many factors, like the composition of the gas phase, temperature, metal particle size, and support. The optimization of all parameters is still stimulating many investigations. The carbon nanotubes formed in the catalytic way are impure because they often contain metal particles. For this reason also many purification methods have been explored. A survey of the brief history of the carbon nanotube synthesis can be found in the Tessonnier and Su (2011) review.

Graphene has been prepared in many different ways and was first exfoliated mechanically from graphite in 2004 (Geim, 2009).

This simple preparation technique has led the explosive growth of interest in graphene. Unfortunately, the graphene particles thus obtained are very small (only several microns or tens of microns at best) and are of irregular shape. Their application to electronic devices is therefore inhibited.

Different preparation techniques were required, and it is at this point that catalysis went into the game. "Bottom-up" methods of graphene synthesis on metals surfaces are now becoming increasingly widely used (Edwards and Coleman, 2013).

The surface growth mechanism of graphene is a catalytic process where the metal catalyzes the formation of an ordered bidimensional structure of C—C bonds (graphene). The process is based on the exposure of the metal surface to acetylene and other carbon feedstocks at high temperature. Following the Fischer–Tropsch lesson, the useful metals should not form stable carbides and promote the C–C formation. So it is not unexpected that Cu is a useful metal. In order to achieve favorable surface-catalyzed reactions, the number of graphene nucleation sites must be controlled not only by the reduction of surface defects of the metal but also by the rate of carbon exposure, slower rates resulting in fewer nucleation sites and hence larger graphene domains. The choice of the best processes is still a matter of intensive study.

References

Akabori, S. and Izumi, Y. (1962) Method of preparing a raney nickel optically active hydroxy acid hydrogenation catalyst, Hyogo-ken, US 3,203,905.

Akabori, S., Ohno, K., and Narita, K. (1963) Asymmetric Hydrogenation with Modified Raney Nickel. I. Studies on Modified Hydrogenation Catalyst. II, *Bulletin of the Chemical Society of Japan*, **36**, 21–25.

Albonetti, S., Cavani, F., and Trifiro, F. (1996) Key aspects of catalyst design for the selective oxidation of paraffins. *Catalysis Review Science and Engineering*, **38**, 413–438.

Bahl, O.P., Evans, E.L., and Thomas, J.M. (1968) The identification and some properties of point defects and non-basal dislocations in molybdenite surfaces. *Proceedings of the Royal Society of London*, **A306**, 53–65.

Baker, T.K., Barber, M.A., Harris, P.S. *et al.* (1972) Nucleation and growth of carbon deposits from the nickel catalyzed decomposition of acetylene. *Journal of Catalysis*, **26**, 51–62.

Banks, R.L. (1980) Industrial aspects of the disproportionation reaction. *Journal of Molecular Catalysis*, **8**, 269–276.

Banks, R.L. and Bailey, G.C. (1964) Olefin disproportionation. A new catalytic process. *Industrial and Engineering Chemistry Product Research and Development*, **3**, 37–70.

Banks, R.L. and Kukes, S.G. (1985) New developments and concepts in enhancing activities of heterogeneous metathesis catalysts. *Journal of Molecular Catalysis*, **28**, 117–131.

Barlow, S.M. and Raval, R. (2003) Complex organic molecules at metal surfaces, bonding, organisation and chirality. *Surface Science Reports*, **50**, 201–341.

Bell, R.P. (1936) The theory of reactions involving proton transfers. *Proceedings of the Royal Society of London. Series A, Mathematical and Physical Sciences*, **154A**, 414.

Bloch, H.S., Pines, H., and Schmerling, L. (1946) The mechanism of paraffin isomerization. *Journal of the American Chemical Society*, **68**, 153.

Bond, G.C. (1997) Preparation and properties of vanadia/titania monolayer catalysts. *Applied Catalysis A: General*, **157**, 91–103.

Breslow, D.S. and Newburg, N.R. (1957) Bis-(cyclopentadienyl)-titanium dichloride–alkylaluminum complexes as catalysts for the polymerization of ethylene. *Journal of the American Chemical Society*, **79**, 5072–5073.

Brønsted, J.N. (1928) Acid and basic catalysis. *Chemical Reviews*, **5**, 231–338.

Bruni, G. and Natta, G. (1934) Struttura del caucciù non stirato studiata con i raggi di elettroni. *Rendiconti Accademia Nazionale dei Lincei*, **XIX**, 536.

Bukhtiyarov, V.I. and Knop-Gericke, A. (2011) Ethylene epoxidation over silver catalysts, in *Nanostructured Catalysts, Selective Oxidations* (eds C. Hess and R. Schlögl). Royal Society of Chemistry, Cambridge, pp. 214–247.

Burrington, J.D., Craig, J.D., Kartisek, T., and Grasselli, R.K. (1982) Aspects of selective oxidation and ammoxidation mechanisms over bismuth molybdate catalysts: 4. Allyl amine as a probe for nitrogen insertion. *Journal of Catalysis*, **06**, 225–232.

Busca, G. (2007) Acid catalysts in industrial hydrocarbon chemistry. *Chemical Reviews*, **107**, 5366–5410.

Byrns, A.C., Bradley, W.E., and Lee, M.W. (1943) Catalytic desulfurization of gasolines by cobalt molybdate process. *Industrial and Engineering Chemistry*, **35**, 1160–1167.

Callahan, J.L., Grasselli, R.K., Milberger, E.C., and Strecker, H.A. (1970) Oxidation and ammoxidation of propylene over bismuth molybdate catalyst. *Industrial & Engineering Chemistry Process Design and Development*, **9**, 134–142.

Campbell, C.T. and Paffett, M.T. (1984) Model studies of ethylene epoxidation catalyzed by the Ag(110) surface. *Surface Science*, **139**, 396–416.

Centi, G., Cavani, F., and Trifiro, F. (2001) in *Selective Oxidation by Heterogeneous Catalysis* (eds M.V. Twigg and M.S. Spencer). Kluwer Academic/Plenum Publ., New York.

Cesano, F., Bertarione, S., Piovano, A. *et al.* (2011) Model oxide supported MoS_2 HDS catalysts: structure and surface properties. *Catalysis Science and Technology*, **1**, 123–136.

Chianelli, R.R., Siadati, M.H., Perez De la Rosa, M. *et al.* (2006) Catalytic properties of single layers of transition metal sulfide catalytic materials. *Catalysis Reviews*, **48**, 1–41.

Clark, A., Hogan, J.P., Banks, R.L., and Lanning, W.C. (1956) Marlex catalyst systems. *Industrial and Engineering Chemistry*, **48**, 1152–1155.

Coperet, C. (2007) Design and understanding of heterogeneous alkene metathesis catalysts. *Dalton Transactions*, 5498–5504.

Coperet, C., Chabanas, M., Petroff, S.-A.R., and Basset, J.-M. (2003) Bridging the gap between homogeneous and heterogeneous catalysis through surface organometallic chemistry. *Angewandte Chemie International Edition*, **42**, 156–181.

Corma, A. (1995) Inorganic solid acids and their use in acid-catalyzed hydrocarbon reactions. *Chemical Reviews*, **95**, 559–614.

Cornils, B. and Herrmann, W.A. (2003) Concepts in homogeneous catalysis: the industrial view. *Journal of Catalysis*, **216**, 23–31.

Coverdale, A.K., Dearing, P.F., and Ellison, A. (1983) The structure of rhenium oxide–alumina metathesis catalysts. *Journal of the Chemical Society, Chemical Communications*, 567–568.

Davis, B.H. and Hettinger, W.P. Jr., (eds) (1983) Discovery and development of olefin disproportionation (metathesis), in *Heterogeneous Catalysis: Selected American Histories*, vol. **222**. ACS Symposium Series, p. 403.

Delbecq, F., Loffreda, D., and Sautet, P. (2010) Heterogeneous catalytic hydrogenation: is double bond/surface coordination necessary? *Journal of Physical Chemistry Letters*, **1**, 323–326.

Edwards, R.S. and Coleman, K.S. (2013) Graphene film growth on polycrystalline metals. *Accounts of Chemical Research*, **43**, 23–30.

Eleuterio, H.S. (1960a) German Patent 1072811.

Eleuterio, H.S. (1960b) Perfluoropropylene POLYMERS. US Patent 2,958,685, 1 November.

Eleuterio, H.S. (1991) Olefin metathesis: chance favors those minds that are best prepared. *Journal of Molecular Catalysis*, **65**, 55–61.

Evans, M.G. and Polanyi, M. (1937) On the introduction of thermodynamic variables into reaction kinetics. *Transactions of the Faraday Society*, **33**, 448–452.

Evans, M.G. and Polanyi, M. (1938) Inertia and driving force of chemical reactions. *Transactions of the Faraday Society*, **34**, 11–28.

Ewen, J.A. (1984) Mechanisms of stereochemical control in propylene polymerizations with soluble group 4B metallocene/methylalumoxane catalysts. *Journal of the American Chemical Society*, **106**, 6355–6364.

I.G. Farbenindustrie A.G. (1928) British Patent 315.439.

Fischer, E.O. (1952) Über Cycopentadien-Komplexe des Eisen und des Kobalts. *Angewandte Chemie*, **64**, 620–621.

Fischer, E.O. (1955) Metallverbindungen des Cyclopentadiens und des Indens. *Angewandte Chemie*, **67**, 475–482.

Fischer, E.O. and Pfab, W. (1952) Zur Kristallstruktur der Di-Cyclopentadienyl-Verbindungen des zweiwertigen Eisens, Kobalts und Nickels. *Zeitschrift für Naturforschung*, **B7**, 377–379.

Fox, W. (1911) Henri Victor Regnault, The Catholic Encyclopedia, vol. **12**. Robert Appleton Company, New York.

Friedel, C. and Crafts, J.M. (1877) Sur une Méthode Générale Nouvelle de Synthèse d'Hydrocarbures, d'Acétones, etc. *Comptes Rendus Chimie*, **84**, 1392–1395;1450–1454, 85, 74–77.

Friedel, C. and Crafts, J.M. (1888) Sur une nouvelle methode generale de synthese, des combinaisons aromatiques. *Annales de Chimie et de Physique Paris*, **14**, 433–472.

García-Diéguez, M., Finocchio, E., Larrubia, M. *et al.* (2010) Characterization of alumina-supported Pt, Ni and Pt–Ni alloy catalysts for the dry reforming of methane. *Journal of Catalysis*, **274**, 11–20.

Garcia-Mota, M., Gomez-Diaz, J., Novell-Leruth, G. *et al.* (2010) A density functional theory study of the 'mythic' Lindlar hydrogenation catalyst. *Theoretical Chemistry Accounts*, **128**, 663–673.

Geim, A.K. (2009) Graphene: status and prospects. *Science*, **324**, 1530–1534.

Gibbs, H.D. (1922) Phthalic anhydride. V-The preparation of phthalic anhydride by the catalysis of the vapor phase reaction between naphthalene and atmospheric air. *Industrial and Engineering Chemistry*, **14**, 120–125.

Grasselli, R.K. (1999) Advances and future trends in selective oxidation and ammoxidation catalysis. *Catalysis Today*, **49**, 141–153.

Grasselli, R. and Trifiro, F. (2011) New insights into heterogeneous oxidation catalysis. *Topics in Catalysis*, **54**, 587.

Grasselli, R.K., Burrington, J.D., and Brazdil, J.F. (1981) Mechanistic features of selective oxidation and ammoxidation catalysis. *Faraday Discussions of the Chemical Society*, **72**, 203–223.

Groppo, E., Lamberti, C., Bordiga, S. *et al.* (2005) The structure of active centers and the ethylene polymerization mechanism on the Cr/SiO_2 catalyst: a frontier for the characterization methods. *Chemical Reviews*, **105**, 115–184.

Helferich, B., Streeck, R., and Günther, E. (1938, German Patent IG Farben, DE 501467, 1927) Ein neues Verfahren zur Darstellung aromatischer Aldehyde. *Journal für Praktische Chemie*, **151**, 251–256.

Helveg, S., Lauritsen, J.V., Laegsgaard, E. *et al.* (2000) Atomic-scale structure of single-layer MoS_2 nanoclusters. *Physical Review Letters*, **84**, 851.

Hieber, W. and Fischer, E.O. (1952) Über den Mechanismus der Kohlenoxydreaktion von Nickel(II)- und Kobalt(II)-Salzen bei Gegenwart von Dithionit. *Zeitschrifft fur anorganische und allgemeine Chemie*, **269**, 292–307.

Hirano, T. (1986) Active phase in potassium–promoted iron-oxide catalyst for dehydrogenation of ethylbenzene. *Applied Catalysis*, **26**, 65–79.

Holmen, A. (2009) Direct conversion of methane to fuels and chemicals. *Catalysis Today*, **142**, 2–8.

Humblot, V., Barlow, S.M., and Raval, R. (2004) Two-dimensional organisational chirality through supramolecular assembly of molecules at metal surfaces. *Progress in Surface Science*, **76**, 1–19.

Iijima, S. (1991) Helical nanotubes of graphitic carbon. *Nature*, **354**, 56.

Ipatieff, V.N. (1959) *My Life in the United States; The Memoirs of a Chemist.* Northwestern University Press, Evanston.

Ipatieff, V.N., Grosse, A.V., Pines, H., and Komarewsky, V.I. (1936) Alkylation of paraffins with olefins in the presence of aluminum chloride. *Journal of the American Chemical Society*, **58**, 913–920.

Ipatieff, V.N., Pines, H., and Schmerling, L. (1940) Isomerization accompanying alkylation II. The alkylation of benzene with olefins, naphthenes, alcohols and alkyl halides. *Journal of Organic Chemistry*, **5**, 252–260.

Ivin, J.K. (1997) *Olefin Metathesis and Metathesis Polymerization*, 2nd edn. Academic Press, Cambridge, MA.

Izumi, Y., Imaida, M., Fukawa, H., and Akabori, S. (1963) Asymmetric hydrogenation with modified Raney nickel. I. Studies on modified hydrogenation catalyst. II. *Bulletin of the Chemical Society of Japan*, **36**, 155–160.

Kashiwa, N. (1983) Patent: Catalyst component for alpha-olefin polymerization, Chisso Corporation Nov 9, EP0093494A1

Kealy, T.J. and Pauson, P.L. (1951) A new type of organo-iron compound. *Nature*, **168**, 1039–1040.

Kestenbaum, H., Lange de Oliveira, A., Schmidt, W. *et al.* (2002) Silver catalyzed oxidation of ethylene to ethylene oxide in a microreaction system. *Industrial & Engineering Chemistry Research*, **41**, 710–719.

Kissin, Y.V. (2001) Chemical mechanisms of catalytic cracking over solid acidic catalysts: alkanes and alkenes. *Catalysis Review*, **43**, 85–146.

Klatte, F. and Rollett, A. (1917; US Patent 1,241,738, filed: July 3, 1914; issued: Oct. 2, 1917) Plastic composition and process of producing it. *Journal of the Society of Chemical Industry, London*, **36**, 1185.

Koch, T.A and Eleuterio, H.S (1969) Cyclododecatriene-(1,5,9) process. US Patent **3**, 381 045.

Koide, Y., Bott, S.G., and Barron, A.R. (1996) Alumoxanes as co-catalysts in palladiu catalyzed co-polymerization of carbon monoxide and ethylene: genesis of a structure activity relationship. *Organometallics*, **15**, 2213–2226.

Kroto, H.W., Health, J.R., O'Brien, S.C. *et al.* (1985) C60: Buckminsterfullerene. *Nature*, **318**, 7779–7780.

Labinger, J.A. (2004) Selective alkane oxidation: hot and cold approaches to a hot problem. *Journal of Molecular Catalysis A: Chemical*, **220**, 27–35.

Lefort, T. (1931) Silver-catalyzed epoxidation of ethylene. French Patent 1931 and US Patent 1,998,878, 1935.

Leofanti, G., Marsella, A., Cremaschi, B. *et al.* (2002) Alumina-supported copper chloride – 4. Effect of exposure to O_2 and HCl. *Journal of Catalysis*, **205**, 375–381.

Lindlar, H. and Dubuis, R. (1973) Palladium catalyst for partial reduction of acetylenes. *Organic Synthesis*, **5**, 880–886.

Liotta, L.F., Martin, G.A., and Deganello, G. (1996) The influence of alkali metal ions in the chemisorption of CO and CO_2 on supported palladium catalysts: a Fourier transform infrared spectroscopic study. *Journal of Catalysis*, **164**, 322–333.

Liu, B., Nitta, T., Nakatani, H., and Terano, M. (2003) Stereospecific nature of active sites on $TiCl_4/MgCl_2$ Ziegler–Natta catalyst in the presence of an internal electron donor. *Macromolecular Chemistry and Physics*, **204**, 395–402.

Lucke, B., Narayana, K.V., Martin, A., and Jahnisch, K. (2004) Oxidation and ammoxidation of aromatics. *Advanced Synthesis & Catalysis*, **346**, 1407–1422.

Lunsford, J.H. (2000) Catalytic conversion of methane to more useful chemicals and fuels: a challenge for the 21st century. *Catalysis Today*, **63**, 165–174.

Macleod, N., Isaac, J., and Lambert, R.M. (2001) Sodium promotion of Pd/gamma-Al_2O_3 catalysts operated under simulated "three-way" conditions. *Journal of Catalysis*, **198**, 128–135.

Marcus, R.A. (1968) Theoretical relations among rate constants, barriers, and Brønsted slopes of chemical reactions. *Journal of Physical Chemistry*, **72**, 891–899.

McDaniel, M.P. (2010) A review of the Phillips supported chromium catalyst and its commercial use for ethylene polymerization. *Advances in Catalysis*, **53**, 123–606.

Mojica, M., Alonso, J.A., and Méndez, F. (2013) Synthesis of fullerenes. *Journal of Physical Organic Chemistry*, **26**, 526–539.

Moureu, C. (1894) Acrylonitrile. *Annales de Chimie et de Physique*, **2**, 187–192.

Nagahara, H., Ono, M., Konishi, M., and Fukuoka, Y. (1997) Partial hydrogenation of benzene to cyclohexene. *Applied Surface Science*, **121/122**, 448–451.

Natta, G. (1933) Struttura e polimorfismo degli acidi alogenidrici. *Gazzetta Chimica Italiana*, **63**, 425–439.

Natta, G. (1956) Stereospezifische Katalysen und isotaktische Polymere. *Angewandte Chemie*, **68**, 393–403.

Natta, G. and Corradini, P. (1955) Studi Roentgenografici sulla cristallinità e sulla struttura di idrocarburi ad alto peso molecolare. *Atti Accademia Nazionale dei Lincei, Memorie*, **4**, 73–76.

Natta, G. and Corradini, P. (1956) Über die Kristallstrukturen des 1,4-cis-Polybutadiens und des 1,4-cis-Polyisoprens. *Angewandte Chemie*, **68**, 615–616.

Natta, G. and Corradini, P. (1960) General considerations on the structure of crystalline polyhydrocarbons. *Nuovo Cimento*, **XV**, 9–39.

Natta, G. and Danusso, F. (1959) Nomenclature relating to polymers having sterically ordered structure. *Journal of Polymer Science*, **34**, 3–11, Nottingham Symposium.

Natta, G., Pino, P., Corradini, P. *et al.* (1955) Crystalline high polymers of α-olefins. *Journal of the American Chemical Society*, **77**, 1708–1710.

Natta, G., Pino, P., Mazzanti, G. *et al.* (1957) Diciclopentadienilici del titanio per lo studio dei catalizzatori di polimerizzazione dell'etilene a bassa pressione. *La Chimica e L'Industria*, **39**, 1032–1033.

Oikawa, T., Ookoshi, T., Tanaka, T. *et al.* (2004) A new heterogeneous olefin metathesis catalyst composed of rhenium oxide and mesoporous alumina. *Microporous & Mesoporous Materials*, **74**, 93–103.

Olah, G.A. (ed.) (1963) *Friedel–Crafts and Related Reactions*, vol. **1**. Interscience, New York.

Osborn, J.A., Jardine, F.H., Young, J.F., and Wilkinson, G. (1966) The preparation and properties of tris(triphenylphosphine)halogenorhodium(I) and some reactions thereof including catalytic homogeneous hydrogenation of olefins and acetylenes and their derivatives. *Journal of the Chemical Society A*, 1711–1732.

Panariti, N., Del Bianco, A., Del Piero, G., and Marchionna, M. (2000) Petroleum residue upgrading with dispersed catalysts: part 1. Catalysts activity and selectivity. *Applied Catalysis A: General*, **204**, 203–213.

Panov, G.I., Sheveleva, G.A., Kharitonov, A.S. *et al.* (1992) Oxidation of benzene to phenol by nitrous oxide over FeZSM-5 zeolites. *Applied Catalysis A: General*, **82**, 31–36.

Papoian, G., Nørskov, J.K., and Hoffmann, R. (2000) A comparative theoretical study of the hydrogen, methyl, and ethyl chemisorption on the Pt(111) surface. *Journal of the American Chemical Society*, **122**, 4129–4144.

Peters, E.F. and Evering, B.L. (1957) Assignors to standard oil company Chicago III, a corporation of Indiana, for the invention relates to novel catalysts which are useful in hydrocarbon conversions and to processes for the manufacture of said catalysts, US Patent 2,963,447 filled October 31, Ser. No. 693,453.

Pines, H. (1981) *The Chemistry of Catalytic Hydrocarbons Conversion*. Academic Press, New York.

Pines, H. and Ipatieff, V.N. (1945) Reaction of methylcyclopentane with olefins in the presence of sulphuric acid and hydrogen fluoride catalysts. *Journal of the American Chemical Society*, **67**, 1631–1638.

Pines, H., Grosse, A.V., and Ipatieff, V.N. (1942) Alkylation of paraffins at low temperatures in the presence of aluminum chloride. *Journal of the American Chemical Society*, **64**, 33–36.

Pino, P. and Lorenzi, G.P. (1960) Optically active vinyl Polymers. II. Optical activity of isotactic and block polymers of optically active a-olefins in dilute hydrocarbon solution. *Journal of the American Chemical Society*, **82**, 4745–7.

Prudhomme, E.A. (1926) Improvement in the manufacture of liquid fuel similar to petrol, British Patent 12 689/24; *Process for the manufacture of synthetic liquid fuels*, US Patent US1711856 A.

Rao, C.N.R., Sood, A.K., Subrahmanyam, K.S., and Govindaraj, A. (2009) Graphene: the new two-dimensional nanomaterial. *Angewandte Chemie International Edition*, **48**, 7752–7777.

Regnault, H.V. (1835) Sur la Composition de la Liqueur des Hollandais et sur une nouvelle Substance éthérée. *Annales de Chimie et de Physique*, Gay-Lussac & Arago, Tome 58, Paris, Crochard Libraire, 301–320.

Regnault, H.V. (1839) Sur les chlorures de carbone CCl et CCl2. *Annales de Chimie et de Physique*, **70**, 104–107.

Regnault, H.V. (1841) Recherches sur la chaleur spécifique des corps simples et des corps composés. *Annales de Chimie et de Physique*, **5**, 129–207.

Regnault, H.V. (1842) Note sur la comparaison du thermomètre à air et du thermomètre à mercure. *Annales de Chimie et de Physique*, **6**, 353–369.

Regnault, H.V. (1845) Etude sur l'hygrométrie. *Annales de Chimie et de Physique*, **15**, 129–236.

Regnault, H.V. (1854) *Course élémentaire de chimie*, vol. **4**. Mason, Paris.

Schmerling, L. (1953) Reactions of hydrocarbons. *Industrial and Engineering Chemistry*, **45**, 1447–1455.

Schmerling, L. (1983) *Organic and Petroleum Chemistry for Non-Chemists*, Pennwell Books, Tulsa, OK.

Schuster, M. and Blechert, S. (1997) Olefin metathesis in organic chemistry. *Angewandte Chemie International Edition in English*, **36**, 2036–2056.

Semon, W.L. (1933) For the synthetic rubber-like composition and method of making the same; method of preparing polyvinyl halide products. U.S. Patents 1, 929,453 and 2,188,396

Sinfelt, J.H. (1983) *Bimetallic Catalysts: Discoveries, Concepts and Applications*, Wiley, New York.

Sinn, H. (1995) Proposals for Structure and effect of methylaluminoxane, based on mass blances and phase separation experiments. *Macromolecular Symposium*, **97**, 27–52.

Sinn, H. and Kaminsky, W. (1980) Ziegler–Natta catalysis. *Advances in Organometallic Chemistry*, **18**, 99–149.

Sinn, H., Kaminsky, W., and Vollmer, H.J. (1980) Living polymers on polymerization with extremely productive Ziegler catalysts. *Angewandte Chemie International Edition in English*, **19**, 390–392.

Sishta, C., Hathorn, R.M., and Marks, T.J. (1992) Group 4 metallocene-alumoxane olefin polymerization catalysts CPMAS-NMR spectroscopic observation of cation-like zirconocene alkyls. *Journal of the American Chemical Society*, **114**, 1112–1114.

Sohio (1996) Acrylonitrile Process commemorative booklet, Acrylonitrile.

Staudinger, H. (1920) Über polymerisation. *Berichte der Deutschen Chemischen Gesellschaft*, **53**, 1073–1085.

Stern, A.C. (1982) History of air pollution legislation in the United States. *Journal of the Air Pollution Control Association*, **32**, 1–44.

Tessonnier, J. and Su, D.S. (2011) Recent progress in the growth mechanism of carbon nanotubes. *ChemSusChem*, **4**, 824–847.

Thayler, A.M. (2007) Making metathesis work. *Chemical and Engineering News*, **85**, 37–47.

Toulhoat, H. and Raybaud, P. (eds) (2013) *Catalysis by Transition Metal Sulphides from Molecular Theory to Industrial Application*, vol. **XXXI**. Ed. TECHNIP, Paris.

Veatch, F., Callahan, J.L., Idol, J.D. Jr., and Milberger, E.C. (1960) New route to acrylonitrile. *Chemical Engineering Progress*, **56**, 65–67.

Wachs, I.E. (2013) Catalysis science of supported vanadium oxide catalysts. *Dalton Transactions*, **42**, 11762–11769.

Weckhuysen, B.M. and Keller, D.E. (2003) Chemistry, spectroscopy and the role of supported vanadium oxides in heterogeneous catalysis. *Catalysis Today*, **78**, 25–46.

Weisser, O. and Landa, S. (1973) *Sulfide Catalysts, Their Properties and Applications*. Pergamon Press, New York.

Weissermel, K. and Arpe, H.-J. (1978) *Important Raw Materials and Intermediates*. Industrial Organic Chemistry, Verlag Chemie, Weinheim, pp. 335–339.

Wild, F.R.W.P., Zsolnai, L., Huttner, G., and Brintzinger, H.H. (1982) ansa Metallocene derivatives IV. Synthesis and molecular structure of chiral ansa-titanocene derivatives with bridged tetrahydroindenyl ligands. *Journal of Organometallic Chemistry*, **232**, 233–247.

Wilkinson, G. and Birmingham, I.M. (1954) Bis-cyclopentadienyl compounds of Ti, Zr, V, Nb and Ta. *Journal of the American Chemical Society*, **76**, 4281–4284.

Wischert, R., Laurent, P., Copéret, C. *et al.* (2012) γ-Alumina: the essential and unexpected role of water for the structure, stability, and reactivity of "defect" sites. *Journal of the American Chemical Society*, **134**, 14430–14449.

Yamanaka, I., Katagiri, M., Takenaka, S., and Otsuka, K. (2000) Studies in surface science and catalysis. *Studies in Surface Science and Catalysis*, **130**, 809–814.

Ziegler, K. (1923) Zur Kenntnis des dreiwertigen Kohlenstoffs, I: Über Tetra-aryl-allyl-Radikale und ihre Derivate (submitted as Thesis of Habilitation to the faculty of philosophy of the University of Marburg). *Liebigs Annalen der Chemie*, **434**, 34–78.

Ziegler, K. (1938) Über Butadienpolymerisation und die Herstellung des künstlichen Kautschuks. *Chemiker-Zeitung*, **62**, 125–127.

Ziegler, K. (1955) Das Mülheimer Normaldruck-Polyathylen-Verfahren. *Kunststoffe*, **46**, 506.

Ziegler, K. and Bähr, K. (1928) Über den vermutlichen Mechanismus der Polymerisationen durch Alkalimetalle. *Berichte der Deutschen Chemischen Gesellschaft*, **61**, 253–263.

Ziegler, K., Gellert, H.G., Kühlhorn, H. *et al.* (1952) Koordinative Ketten Transfer Polymerisation. *Angewandte Chemie*, **64**, 323–350.

Ziegler, K.K., Holzkamp, E., Breil, H., and Martin, H. (1955a) Das Mülheimer Normaldruck-Polyäthylen-Verfahren. *Angewandte Chemie*, **67**, 541–547.

Ziegler, K., Holzkamp, E., Breil, H., and Martin, H. (1955b) Polymerisation von Äthylen und anderen Olefinen. *Angewandte Chemie*, **67**, 426.

Ziegler, K., Holzkamp, E., Breil, H., and Martin, H. (1955c) Das Mülheimer Normaldruck-Polyäthylen-Verfahren. *Angewandte Chemie*, **67**, 541–547.

Ziegler, K., Breil, H., Martin, H., and Holzkamp, E. (1974) (Studien- und Verwertungsgesellschaft), Verfahren zur Homopolymerisation von Propylen und α-Butylen, DBP 1257430, patented 18.07.1974.

4

Surface Science Methods in the Second Half of the Twentieth Century

4.1 Real Dispersed Catalysts versus Single Crystals: A Decreasing Gap

Heterogeneous catalysts are typically constituted by an active species (the catalyst) dispersed on a second material, which enhances the effectiveness or minimizes the costs. As a first approximation, the support can be merely considered a surface on which the catalyst is spread to increase the surface area. Real catalysts often consist of mixtures of phases, some of which have catalytically active surfaces while others are catalytically silent, others having instead the function of supporting the small particles of active phases or of keeping them from sintering. More often the support and the catalyst interact, affecting the structure and morphology of the supported particles and hence the catalytic reaction. It is so concluded that real heterogeneous catalysts are actually very complex systems, exposing different phases, faces, edges, corners, and other defects. The complexity of the catalytic materials was already recognized in early times by several chemists including Langmuir, Taylor, Rideal, Schwab, and Boreskov. The situation is made even more complex and intriguing by the fact that the particles are often nanometric in size, a situation that is at the origin of further differences with respect to the bulk materials (quantum effect). Before the 1950s, the study of the surface species and structures at the molecular level was not possible, and all information on the surface properties was indirect (for instance, adsorption isotherms). The situation changed dramatically when the revolutionary development of new physical surface science experimental methods, initiated in the 1950s and 1960s, afforded new techniques for characterizating at the molecular level the surface structures of dispersed solids and single crystals.

These new methods also have recently made it possible to characterize the structure and other properties of the working catalysts *in situ* and under *operando* conditions.

The Development of Catalysis: A History of Key Processes and Personas in Catalytic Science and Technology, First Edition. Adriano Zecchina and Salvatore Califano.

Among the new applied techniques, particularly important were different types of vibrational spectroscopies (infrared and Raman spectroscopies) enriched with the availability of high power laser sources, nuclear magnetic resonance (NMR), electron spin resonance (ESR), microcalorimetry, electron microscopies, different forms of molecular beam techniques, and X-ray absorption spectroscopies (XAS), X-ray absorption near edge structure (XANES), and extended X-ray absorption fine structure (EXAFS) that, at the end of the twentieth century, drew on the availability of synchrotron radiation sources. As we will see next, this list is far from complete. The different types of methods and experiments (Hansen *et al.*, 2001) have provided us with a wealth of information on surface structures, adsorption geometries, bonds strengths, and elementary reaction steps.

Although well-defined single-crystal surfaces are only a crude ideal model of the catalyst's surfaces used in practice both in the laboratory experiments and in industrial processes, they represent an ideal playground for the understanding of the fundamental phenomena occurring on surfaces. As we will document more fully below, in the last quarter of the twentieth century, physical methods for investigating surface properties were developed and reached sufficiently high sensitivity as to make possible the study of single-crystal faces. Consequently study models of supported catalysts could be produced and investigated under realistic high pressure and high temperature conditions.

In summary, the application of the new physical methods followed two apparently different but later converging lines: the one centered on finely divided solids and the other centered on single crystals. The first line involved materials nearer to real catalysts, while the second tried to understand the fundamental phenomena occurring on well-defined surfaces. These two lines converged when the surface scientists of the second line started the study of stepped and defective surfaces and of particles and clusters deposited on metal single-crystal faces and when those of the first line became more and more involved in the study of surface properties of particles with a well-defined morphology or of crystalline microporous materials characterized by a well-defined structure at the molecular level. In the near future, these converging approaches are expected to potentially close the gap between the surface science and real catalysts studies.

4.2 Physical Methods for the Study of Dispersed Systems and Real Catalysts

IR spectroscopy is the technique that has produced the largest information on surface species and surface structures on finely dispersed systems. The application of IR spectroscopy to study the structure of the surface species goes back to 1940 when the future Russian academician Alexander Nikolaevich

Terenin (founder of the Soviet school of photochemistry) and his coworker, K. Kasparov, first succeeded to obtain the IR spectrum of surface species on porous glass (Terenin and Kasparov, 1940). His contribution was published about 15 years before the start of analogous studies by R. Eischens in the United States (Eischens, Pliskin, and Francis, 1954).The application of IR spectroscopy to surface characterization, initiated before the World War II, was continued by Terenin after the end of the war (Terenin *et al.*, 1955; Terenin and Roev, 1959). Unfortunately, Kasparov could not participate because he died on the Leningrad battlefield. A description of this exceptional scientist can be found in the *Biographical Memoir of Academician Alexander Terenin* (Tsyganenko, 2008; Terenin, 1975).

Terenin's career started when in 1919 he was chosen by Professor D. Rozhdestvensky, the first director of State Optical Institute, among the 20 best students of physics to complete the staff of the institute. Besides Terenin the group included Vladimir Fock, Evgeny Gros, Sergey Frish, and several other future brilliant scientists, whose names are now well known among the specialists (it suffices to mention the Hartree–Fock method). One of the first results of the work, carried out at the State Optical Institute, was a paper published on Nature (Terenin, 1921). In 1932, Terenin organized the Laboratory of Optics of surface phenomena, which in 1957 became the Laboratory of Photosynthesis and since 1960 is known as the Laboratory of Photocatalysis. The scientific merits of A. Terenin were recognized in the USSR by Stalin prize, and in 1954 he was awarded the S. Vavilov Gold Medal of the USSR Academy of Sciences, in 1959 the Ciamician Gold Medal of the University of Bologna, and in 1964 the Finzen Gold Medal (Figure 4.1).

Fifteen years later, R. Eischens and colleagues at Texaco were able to obtain the IR spectrum of carbon monoxide (CO) adsorbed on finely divided noble metal particles supported on silica (Figure 4.1). Although the samples were completely black, they were still sufficiently transparent to IR radiation to allow a good spectroscopic analysis in transmission. The observed frequencies were entirely similar to those of homogeneous metal carbonyls containing linear and bridged CO, indicating that the adsorption of CO on the finely divided metal particles, which were very close to those of an industrial hydrogenation catalyst, involves the formation of chemical bonds with surface atoms characterized by the $\sigma{-}\pi$ character (Eischens, Pliskin, and Francis, 1954). This result can be considered as the second milestone in the study of the structure of adsorbed species on solids of catalytic interest. Robert Philip Eischens (1920–2010), born in North Dakota, earned a scholarship at the University of Wisconsin–Madison. After his BS in chemistry, he spent two years at the University of Chicago and completed his doctorate in chemistry at Northwestern University. In the following years Eischens became a senior scientist at the Texaco Company in charge of catalysis research where he studied by IR many real catalysts (Mapes and Eischens, 1954; Pliskin and

(a)

(b)

(c)

Figure 4.1 (a) Alexander Nikolaevich Terenin (1896–1967), (b) Robert Philip Eischens (1921–2010), and (c) Norman Sheppard (1921–). They pioneered the application of vibrational spectroscopies to adsorbed species. Terenin was the first to use IR spectroscopy for surface studies, using a self-made spectrometer. Eischens was an industrial researcher who applied IR spectroscopy to the study of supported metal catalysts. Norman Sheppard, a recognized expert of vibrational molecular spectroscopy, following Terenin's research line, studied oxidic systems by IR and Raman spectroscopies. *Source*: (a) Courtesy of *Journal of Photochemistry and Photobiology*. Reproduced with permission of Elsevier. (b) The Eischens image is from a drawing of A. Zecchina. (c) The Sheppard photograph was taken by A. Zecchina.

Eischens, 1956; (Eischens and Pliskin, 1957). His interest in the study of surface species on finely dispersed solids can be understood if we consider his position in catalysis research at Texaco.

Later Eischens worked for the US Department of Energy (1976–1980) as program manager of catalysis research and at Lehigh University as professor of chemistry. Eischens received many professional awards and traveled the world as a distinguished researcher. In 1997, when proposed for the Nobel Prize in chemistry, his research (extending back four decades) was described as the basis for much of our current understanding of the science of heterogeneous catalysis. After the seminal Eischens contributions, the IR technique became quickly the single most widely used physical tool in studies of species adsorbed on catalysts.

Nearly at the same time, Norman Sheppard and colleagues at the Department of Colloid Science in Cambridge, utilizing the same porous glasses already investigated by Terenin, obtained the IR spectrum of adsorbed ammonia (Yates, Sheppard, and Angell, 1955) (Figure 4.1). The inspiration of Terenin's work is quite evident! N. Sheppard was also the first, 20 years later, to apply Raman spectroscopy to surface studies in his laboratory at East Anglia University (Nguyen and Sheppard, 1978), and he was the first in 1980 to observe by Raman spectroscopy a light-stimulated surface polymerization reaction (Rives-Arnau and Sheppard, 1980).

Before becoming involved in surface studies, N. Sheppard was a pioneer in the development of IR, Raman, and NMR spectroscopy techniques in the United Kingdom, and he was the internationally recognized expert in the vibrational spectroscopy of molecules. Born in Hull, he was educated at the University of Cambridge. In this period, he worked under the supervision of Sir Gordon Sutherland on the study of IR spectra of natural and synthetic rubber (Sheppard and Sutherland, 1945). Later, he went to the United States to study the vibrational spectra of hydrocarbons in the liquid and solid states at Pennsylvania State University with Professor D.H. Rank (Sheppard, 1988). Back to the United Kingdom in 1948, he spent, thanks to a Ramsay Memorial fellowship, 7 years at the department of Colloid Science of the University of Cambridge. In Cambridge, he worked mainly on the spectra of hydrocarbons (Cohen and Sheppard, 1959; Sheppard, 1948; Sheppard and Simpson, 1952).

In 1964, he moved to Norwich where at the University of East Anglia as professor of chemical physics, he worked on the vibrational spectroscopy of small molecules, especially of those adsorbed on surfaces of microcrystalline solids (Hussain and Sheppard, 1992). As for the surface studies, particularly important was his expertise in the assignment of the IR bands of hydrocarbons and small molecules gained in the Cambridge period, a fact that often allowed him to identify the structure of surface intermediates on catalysis surfaces. He retired in 1986 from his teaching duties but continued to serve as president of the Faraday Division of the Royal Society and to work on the spectroscopy of adsorbed molecules.

A few years after the first successful application of Raman spectroscopy to adsorbed species by Sheppard, also Israel E. Wachs, at the Exxon Research and Engineering Company (United States), started to systematically apply Raman techniques to catalytic materials (Chan *et al.*, 1984). The Raman technique in its resonance (Damin *et al.*, 2006; Guo *et al.*, 2012) and the surface-enhanced Raman spectroscopy (SERS) version are today increasingly applied in catalysis and surface science laboratories (Damin *et al.*, 2008).

Only a few years after the Terenin and Sheppard contributions, D.J.C. Yates (School of Mines, Columbia University, New York), continuing the porous glass line, obtained the IR spectrum of adsorbed methanol and water (Folman and Yates, 1958), while John T. Yates Jr, continuing the Eischens line at Department of Chemistry, Spectroscopy Laboratory of MIT, obtained the spectrum of CO adsorbed on dispersed Ni on alumina (Yates and Garland, 1961). After these pioneering applications of IR spectroscopy, the increasing worldwide diffusion of commercial spectrometers opened the way toward the extensive investigation of the vibrational properties of physically and chemically adsorbed species on a large variety of systems including dispersed metals, nontransition and transition oxides, halides, and porous materials. Among the research groups active in this area, we recall the group in Moscow directed by Alexander Kiselev (Kiselev and Lygin, 1959; Galkin, Kiselev, and Lygin, 1964), the University of Turin group headed by Enzo Borello and Adriano Zecchina (Borello, Zecchina, and Castelli, 1963; Borello, Zecchina, and Morterra, 1967), the Knözinger group in Munich (Stolz and Knözinger, 1971), the A.K. Kazansky group of the Russian Academy of Sciences (Michel, Kazansky, and Andreev, 1978), the J.C. Lavalley group at the Laboratoire Catalyse et Spectrochimie de Caen, France (Nguyen *et al.*, 1980), and the Guido Busca group of the University of Genova (Busca and Lorenzelli, 1982).

In the last 15 years of the twentieth century, with the diffusion of FTIR spectrometers, there became available diffuse reflectance infrared Fourier transform spectroscopy (DRIFT) in many universities and laboratories. Starting from the 1970s, the application of IR spectroscopy and of diffuse reflectance modes to surface studies of catalysts became very popular, and the number of published paper is now approaching to 10,000. The reason of this diffusion was the cheap instrumental investment, the high sensitivity, the simplicity of the apparatus, and the high penetration of IR radiation, which allows species to be detected also in micropores of microporous catalysts (a capacity exceeding other surface science techniques). All these features made this technique fully dominant in the studies of adsorbates and the intermediate supported catalysts and microporous materials like zeolites.

A partial limitation of IR spectroscopy in the transmission and diffuse reflectance mode was traditionally that it could be used only for *in situ* experiments on finely divided systems with surface areas $\geq 10\,\mathrm{m^2/g}$ because the particles are still transparent and still have acceptable scattering properties.

Figure 4.2 The high resolution transmission electron microscopy (HRTEM) image of MgO nanocubes exhibiting nearly perfect low index faces and adsorption properties comparable to those of (001) faces obtained by cleaving MgO single crystals under high vacuum. *Source*: Adapted from Spoto *et al.*, 2003, Elsevier.

MgO particles with perfect cubic shape obtained by burning Mg ribbon

Defect free (100) faces

50 nm

Of course, these materials, like the similar heterogeneous catalytic systems, expose a plethora of different surfaces, edges, corners, and defects. Consequently the surface species are also heterogeneous, a fact that often complicates the interpretation of the IR spectra. Only in recent time the use of FTIR spectrometer has allowed to obtain significant IR spectra of adsorbed species in transmission mode on well-defined microcrystals, characterized by surface areas as low as $2\,m^2/g$. These microcrystals, as fully documented by high resolution transmission electron microscopy (HRTEM), can expose defect-free faces that compare very well with those of single crystals or of films of microcrystals deposited on single-crystal metal substrates. An extraordinary example of MgO microcrystals exhibiting a nearly perfect morphology is reported in Figure 4.2. It is interesting that the IR spectrum obtained in transmission mode of CO adsorbed on the microcrystals of this dispersed system is nearly identical to that obtained by Heidberg *et al.* (1995) for CO adsorbed on ultrahigh vacuum cleaved MgO (001) single-crystal face.

This result clearly shows that the gap between dispersed and single-crystal spectroscopies can be closed (Spoto *et al.*, 2004). A limitation of IR spectroscopy concerns its difficult application to the study of high surface area carbonaceous solids and to the catalysts consisting of metal particles supported on them. This limitation is less severe for Raman spectroscopy.

In the 1970s diffuse reflectance spectroscopy in the UV-visible and near- IR regions started to be employed for the study of oxides surface states and surface species. The pioneering contribution in this area came from Kamil Klier at the J. Heyrovsky Institute of Physical Chemistry, Czechoslovak Academy of Sciences (Polàk and Klier, 1969). A few years later (1974) the technique was applied by Frank Stone in England (Zecchina and Stone, 1974) to the study of surface states and surface species on alkaline earth oxides and their solid solution with transition metal oxides. The information derived by this technique were numerous and concerned surface excitons, internal electronic transitions in surface species formed by interaction of the surface with gases, coordination and valence states of surface ions, and their modification upon adsorption of molecules, the plasmonic absorption of small metal particles. Today the technique is widely applied in surface studies of catalytic systems. The complementary luminescence spectroscopy applied to the study of oxide surfaces goes back to 1978 (Coluccia, Deane, and Tench, 1978).

Already in 1969 Kirill Ilyich Zamaraev, at the Institute of Catalysis in Novosibirsk, Siberian Branch of the USSR Academy of Sciences, started to utilize ESR spectroscopy for catalysis and surfaces, among these, in the study of catalytic activity of homogeneous Cu(II) complexes (Zhitomirski and Zamaraev, 1969). In 1984 Zamaraev, elected director of the Institute of Catalysis later named Boreskov Institute, became also the pioneer of the application of NMR spectroscopy to the study of heterogeneous catalysts (Stepanov *et al.*, 1984). Nearly at the same time, the ESR technique was applied by Jack. H. Lunsford at Texas A&M University (Lunsford, 1967). In the 1970s ESR spectroscopy of surface radicals gradually gained importance. It was in that period employed by Kazansky at the Zelinsky Institute in Moscow (Shvets, Vorotinzev, and Kazansky, 1969) and later by Michel Che at the Institut de Recherches sur la Catalyse of Villeurbanne (Shelimov and Che, 1978).

The energetic interaction of molecules with surfaces of dispersed systems had been traditionally investigated by calorimetry and the heat evolved during the interaction was measured. The first calorimeter for the study of gas–solid adsorption phenomena was constructed in 1854 by Pierre-Antoine Favre (1813–1880), professor of chemistry of the Faculty of Sciences of Marseilles (1854). The calorimetric methods were successively improved by Albert Tian, also professor of chemistry at the same university, who invented the isothermal heat-flux microcalorimeter. Tian described his compensation isothermal microcalorimeter for the first time in 1922 and then brought further improvements to this thermocouple instrument in 1924 and 1926. This type of calorimeter was further improved in 1947 by Tian's student Edouard Calvet, who introduced the differential setup and a rational construction of the two twinned calorimetric elements in the 1940s, transforming the Tian's appliance into a true laboratory instrument. This sensitive differential microcalorimeter is now widely used with the name of Calvet calorimeter in studies of catalytic and high surface area materials (Gravelle, 1985; Fubini, 1988). The information on adsorption phenomena given by calorimetry is very wide. From calorimetric data it is possible to study adsorption kinetics in order to distinguish between physical and chemical adsorption and, in the second instance, to obtain information on the chemical bonds formed on interaction of molecules with the catalytic surfaces (Cardona-Martinez and Dumesic, 1992). Today a Calvet calorimeter is present in practically all laboratories interested in surface science and catalysis.

To complete the survey of the most widely used physical methods finding application in surface studies, XAS must be mentioned. Although the first observation concerning the presence of a fine structure in the X-ray absorption spectra goes back to 1920 (Stumm von Bordwehr, 1989), only after the appearance of high intensity synchrotron radiation sources (about 1970), XAS in its XANES and EXAFS manifestations became widely utilized in surface studies. The importance of XANES and EXAFS lies first in the elemental specificity. In addition XANES and EXAFS spectra allow important information to be extracted on the valence and coordination state (e.g., octahedral, tetrahedral) of the absorbing metal center both before and after its interaction with the molecules. As valence state and coordination of surface elements can change upon adsorption, the application of XANES and EXAFS to surface studies and catalysis is straightforward. These techniques are fully complementary to IR and UV–vis–NIR spectroscopies, and they are often usefully utilized in this conjunction. The number of papers in this area is about 3500, and they are almost fully concentrated in the last 20 years, that is, in the period that has seen a progressive and worldwide increment and spreading of strong radiation sources. The most investigated materials are supported metals particles and metal-exchanged zeolites, that is, those systems characterized by a high concentration of surface centers whose coordination and valence states can be perturbed by an adsorption of molecules (Bordiga *et al.*, 2013).

4.3 Surface Science of Single-Crystal Faces and of Well-defined Systems

In the previous section we documented the vibrational, ESR, and micro-calorimetry that have been mainly employed for the study of highly dispersed systems and of real catalysts. However, it must be recalled that they have been used also for the study of single-crystal surfaces. In particular, in the last part of the twentieth century, reflectance absorption IR spectroscopy (RAIRS) on single crystals and films has added information on the vibrational properties of adsorbed species on well-defined surfaces, which could be usefully compared with the knowledge obtained concerning dispersed systems. Primarily noteworthy are the studies of the group of Hajo Freund and collaborators at the Fritz Haber Institute (FHI). Their studies were focused on RAIRS of well-defined thin oxide films grown on metal single-crystal supports. As these films are formed by small crystals of well-defined shape (similar to those presumably present on the parent catalytic materials), the resulting data provided new information on the reactivity of defects (edges, corners, etc.), so contributing to close the gap between surface science and catalysis. A description of the results obtained with this approach can be found in a recent review article (Kuhlenbeck, Shaikhutdinov, and Freund, 2013).

Recently ESR has found application in single-crystal surface science. In fact this technique, although more easily employed on finely divided solids, had been for the first time applied to well-defined single-crystal surfaces in the Hajo Freund laboratory at the FHI (Schlienz *et al.*, 1995).

Then again, although microcalorimetry was considered for a long time as only useful for the study of finely divided materials, it must be underlined that more recently David King at the University of Cambridge has constructed a single-crystal calorimeter operating under high vacuum (Borroni-Bird and King, 1991) and that this innovative calorimeter has added new strength to the single-crystal surface science research line.

For the study of single-crystal faces, the most used technique is the low-energy electron diffraction (LEED) technique, a sophisticated application of the discovery of electron diffraction made in 1927 by Clinton Joseph Davisson (1881–1958) and his assistant Lester Halbert Germer (1896–1971) (Davisson and Germer, 1927) and implemented by George Paget Thomson (Thomson and Reid, 1927). These two experiments opened a new era for the application of electron diffraction to the study of matter.

Electron diffraction did not, however, become a popular tool for the study of surfaces until the early 1960s. In the early 1960s ultrahigh vacuum became widely available and better detection methods were devised, and thus LEED experienced intensive use, thanks also to the realization of new techniques for producing clean single-crystal metal surfaces. Consequently its operability as a true research instrument in chemical physics for the determination of bond angles and bond lengths of single-crystal metal surfaces and of molecular systems adsorbed on them became a true reality.

In particular, starting from the 1970s, Philip Woodruff (University of Warwick), Philip Hofmann Alexander Bradshaw and Gerhard Ertl (FHI), Gabor Somorjai (Berkeley), D.W. Goodman (Texas University), Y. Iwasawa (University of Tokyo), and David King (University of Cambridge) gave invaluable contributions to the development of the technique. A detailed documentation of all the contributions is, however, beyond the scope of this book. The expectations created by the LEED technique, which could be applied only to single-crystal faces of metals and semiconductors, were immense in the surface science community because it was expected to allow the achievement of conclusive information about the surface interactions and reactions occurring at catalytic sites located in well-defined structures.

Since Pt is one of the most active catalysts in many reactions, it can be understood why one of the first contributions of LEED analysis was concerned just with the platinum single-crystal surface with (100) orientation (Hagstrom, Lyon, and Somorjai, 1965). Yet, surprisingly, it was found that flat low index faces of Pt (which was known from a century to be the most active hydrogenation catalyst) did not dissociate hydrogen at all and that only stepped surfaces containing a high concentration of edges, where Pt atoms are in low coordination state, are active (Somorjai, 2000). After the first studies on Pt, the LEED technique was gradually extended to many other metals and confirmed that metal centers characterized by low coordination state, like those located on edges, corner, and other defects, which are very abundant on highly dispersed materials, play a fundamental role in surface science. This fact gradually convinced the surface scientists that the distance between models and real catalytic systems was still remaining very large. It is from this period that the awareness of a "persisting" material gap between single crystals (even properly stepped) and dispersed particles became evident. The material gap is even larger for other reasons. The fact is that particles ranging in size from 1 to 10 nm (i.e., a common range for supported particles in catalysts) exhibit physical and chemical properties that are intermediate between atomic and molecular size regimes, on the one hand, and the macroscopic bulk, on the other (quantum size effect). This fact implies that the activity and selectivity of supported metal nanoparticles are strictly connected with the particle size, not only because the number of defect sites of nanocrystals increases markedly with the decrease in size but also because their electronic properties are altered with respect to the bulk.

The results obtained by LEED on surface structures and reconstructions of single crystals were so important that it can be stated with confidence that LEED fueled the introduction of many other surface techniques, which became gradually available in the 1970–2000 period.

One of these techniques is the scanning tunneling microscopy (STM), developed in 1981 by the German Gerd Binnig (1947–) and the Swiss Heinrich Rohrer (1933–) at the IBM research laboratories of Zurich (Binnig and Rohrer, 1986). The tunneling microscope technique is based on the tunnel effect realized by letting a small conducting tip approach very closely a surface so that electrons can tunnel through the vacuum when a voltage difference is applied among them. With this kind of microscope, a resolution of Angstroms can be easily obtained, well suited to image individual atoms and to explore the presence of impurities attached to a surface. An important advantage of this technique with respect to LEED is that it can be applied not only in ultrahigh vacuum but also in air or in a liquid and even at high temperatures. Another advantage of STM is the possibility to do conductance and photoemission spectroscopy.

Another technique introduced in the same period is the high resolution electron energy loss (HREEL) spectroscopy pioneered by Harald Ibach in 1985 (Franchy and Ibach, 1985; Ibach, 1991) at the Forschungszentrum Jülich of the Institut für Bio- und Nanosysteme, GmbH. This technique gives information on the vibrational properties of molecules adsorbed on clean (low index) metal surfaces. As the only information about the vibrational properties of adsorbed species was previously given by IR studies on dispersed and hence highly defective metals particles, the new technique provided fresh understanding of the basic phenomena occurring on well-defined surfaces.

To end the enumeration of the most important spectroscopic methods employed today in surface studies, sum–frequency generation (SFG) must be mentioned. This technique is widely utilized today in many laboratories (Shen, 1989; Rupprechter, 2002). We will shortly return to the histories of these techniques in our discussion of the Somorjai contributions to surface science and catalysis.

With the aid of all these new methods, the basic chemistry occurring on surfaces upon exposure to reactive gases is now better understood. However, it must be constantly kept in mind that characterization of the catalyst surface itself constitutes only a small step toward our understanding of catalytic processes. In fact the surface science methods often are unable to distinguish spectator from active species. Furthermore, under technologically relevant working conditions, very significant changes of the catalyst structure induced by reactants can occur, and this is why spectroscopic methods under *operando* conditions are needed (Arean, Weckhuysen, and Zecchina, 2012).

The judicious application of some of the previously mentioned techniques (in particular, LEED) characterized the research of German physicist Gerhard Ertl and Hungarian chemistry professor Gabor Somorjai, the two people who have more than anyone else contributed to the development of the application of modern surface science to catalysis science.

Gerhard Ertl (1936–), born in Stuttgart, graduated in 1961 at the Münich Technische Universität where he completed his PhD thesis in 1965 (Figure 4.3). In his research he undertook the study of the surface structure of metallic catalysts and semiconductors and developed innovative techniques for observing how atoms and molecules absorb on very clean surfaces. The main question that engrossed him (his own words) was "How do chemical reactions proceed?" The novelty of his approach lies in the fact that he "always attempted to tackle chemical questions with physical methods." But unlike many of his colleagues, there was no one method to which Ertl remained devoted.

In 1974, by coupling LEED measurements of electron diffraction to desorption techniques and to theoretical models, he succeeded in explaining how hydrogen bonds to palladium (Conrad, Ertl, and Latta, 1974) and nickel (Christmann *et al.*, 1974) surfaces. In 1977 he worked out a theory of the molecular mechanisms of the ammonia synthesis (Carrà and Ugo, 1969), discovered by Fitz Haber 70 years before, concentrating his attention on the process that he considered the determining step for the reaction rate constants, namely the dissociation of the N_2 molecules into single atoms of nitrogen (Ertl, 2012). Ertl first showed that nitrogen atoms, derived from the breaking of the N≡N triple bond, are present on the iron catalyst surface (Bozso *et al.*, 1977) and proposed a model of the iron–nitrogen bond. He proved also that the activation energy (Ertl, 2005) is different for different crystallographic planes of the iron, pointing out the greater efficiency of the (111), (110), and (100) planes, and clarified the details of the reaction mechanism (Ertl and Huber, 1980).

Figure 4.3 (a) Gerhard Ertl (1936–), (b) Gabor Somorjai (1935–), and (c) John Meurig Thomas (1932–), ambassadors of the relations between surface science and catalysis. Ertl (2007 Nobel Prize) clarified the mechanism of the ammonia synthesis. Somorjai introduced new technologies such as LEED for the study of surfaces, while Thomas made extensive contributions to relations between structure of surface sites and catalysis. *Source*: (a) The Ertl image is in the public domain. (b) Reproduced with permission of Somorjai. (c) Reproduced with permission of Thomas.

In the following years Ertl studied the nonlinear mechanisms involved in the classical oxidation reaction of CO on platinum and palladium catalysts (Skottke *et al.*, 1987). He identified the conditions that give rise to oscillating reactions (Cox, Ertl, and Imbihl, 1985). The relevance of CO-oxidation reaction on noble metals can be appreciated when its relation with catalytic exhaust pipes for cars and with combustion cells is considered. Ertl was appointed director of Physical Chemistry Department at the FHI on 1985. The presence of Gerhard Ertl has attracted at FHI top-notch scientists like Robert Schlögl (1994) and Hans-Joachim Freund (1996) who, since then, have played an important role in the fundamental study of catalytic systems, contributing to keep the FHI tradition at the highest level.

The influence of Gerhard Ertl was not limited to FHI. He has contributed to the foundation of a worldwide experimental school of thought by showing how reliable results can be attained in this difficult area of research in applying the new physical methods. For his research Gerhard Ertl received in 2007 the Nobel Prize for chemistry. His insights have provided the scientific basis of modern surface chemistry of single-crystal faces (Pettinger *et al.*, 2007).

A coworker of Gerhard Ertl at the University of Munich was Helmut Knözinger (1935–2014) who studied physics at the Ludwig-Maximilians-Universität of Munich, where he received his PhD in physical chemistry in 1967. The activity of Helmut Knözinger was primarily in the characterization of finely dispersed systems, spanning from surface spectroscopy (Knözinger and Rumpf, 1978; Knözinger, 1979) and surface chemistry to heterogeneous catalysis (Knözinger, Thornton, and Wolf, 1979). Knözinger also is a coeditor of the handbook of heterogeneous catalysis in collaboration with Gerhard Ertl, Jens Weitkamp (University of Stuttgart) and Ferdi Schüth (Frankfurt University).

Of the same importance, and with many equivalent aspects, is the research of the Hungarian Jewish Gabor Arpad Somorjai (1935–), who as a child escaped the holocaust, thanks to the help of the Swedish diplomat Raoul Wallenberg who in 1944 procured for him and his mother false Swedish passports (Figure 4.3). Somorjai studied chemical engineering at Budapest but in 1956 escaped to the United States where he obtained a PhD in 1960 (Somorjai, 1960) at the University of California, Berkeley, and where, after a short period at the IBM research laboratories in New York, he has been a professor and researcher since 1964. At Berkeley Somorjai became aware that many catalytic reactions occur preferentially at surfaces defects.

Since 1965, Somorjai has directed a research project concerning the study of the molecular bases of heterogeneous catalysis by characterizing the structure of the crystal surfaces (Lyon and Somorjai, 1967) and determining the nature of the bonds with adsorbed molecules (Somorjai, 1994). In this research Somorjai has utilized single crystals of transition metals in a large number of catalytic reactions both in high pressure cells and under high vacuum, applying advanced techniques to control *in situ* the catalyst surface during the reactions. Since 1994 he has been investigating nanostructures (Somorjai and Park, 2008; Li and Somorjai, 2010; Somorjai, Frei, and Park, 2009) produced either by lithography through bombardment with electrons or by epitaxial growth (Habas *et al.*, 2007), as well as in photo lithography using either atomic force microscopy (AFM). In the 1990s Somorjai together with Yuen-Ron Shen (1935–), an expert of nonlinear optics, developed the technique of overtone generation by frequency sum or difference (Cremer *et al.*, 1996) that Somorjai has applied to the study of catalytic reactions on surfaces (Somorjai and McCrea, 2000).

As for catalysis science, it is worth to recall that Somorjai also investigated the activation of C—H bonds for several alkenes and alkanes on the (111) crystal face of platinum as a function of temperature and pressure in the absence and presence of hydrogen. He applied vibrational SFG to characterize the adsorbate structures and high pressure scanning tunneling microscopy (HP-STM) to monitor their surface mobility under reaction conditions (Somorjai and Marsh, 2005).

Somorjai, interested in the Haber–Bosch process, reached different conclusions from those of Ertl as far as the slow stage of the reaction is concerned. According to Somorjai the slow stage is the one in which the ammonia molecules are detached from the catalyst. He derived this idea from the fact that under industrial conditions, excess NH_3 act as poison (Spencer, Schoonmaker, and Somorjai, 1982). As it is emerging from the quoted references, the Somorjai research was mainly focused on model metal catalysts, which, being conductive, are more suitable for the application of LEED and other surface science methods. This fact makes the Ertl and Somorjai approaches very similar. Nevertheless, he did not ignore that many industrial processes are catalyzed by insulating oxides and chlorides. Hence he decided to deposit the Ziegler–Natta catalysts on a gold support following an ingenious multistep method involving the initial deposition of $MgCl_2$, followed by on top deposition of $TiCl_4$, then by electron irradiation, and finally by activation with $AlEt_3$. With this procedure he obtained an active polymerization catalyst in an ultrahigh vacuum chamber (Magni and Somorjai, 1996). Unfortunately, the employed methodology was not sensitive enough to allow a complete characterization of the structure of the catalytic centers, which, even today, is not fully known. The *in situ* methodology to create on the metal surface a film of insulating catalyst has been extensively used by the Hans–Joachim Freund group at the FHI, which also exploited RAIRS as a characterizing technique.

A third scientist who has greatly contributed to heterogeneous catalysis studies, keeping at a very high level the tradition of catalytic studies established at Cambridge by Rideal and Eley, is John Meurig Thomas (1932–) whose main interest was focused on the structure of catalytic sites on well-defined systems (Figure 4.3).

Born in Llanelli (South Wales) from a family of miners, Thomas graduated in 1958 from the University of Wales, completing his PhD work through collaboration with Queen Mary College, University of London. After a year he joined the University of Wales, where he started his academic career. In 1969, he became professor of chemistry at the University College of Wales, Aberystwyth.

In 1978, he was called at the University of Cambridge as head of the Physical Chemistry Department, and in 1986 he was nominated director of the Royal Institution of Great Britain. In 1987, the BBC televised his Royal Institution Christmas Lectures on crystals, continuing the tradition started by Faraday in 1826. The Faraday personality and contributions were celebrated by Thomas in 1991 in a beautiful book entitled *Michael Faraday and the Royal Institution: The Genius of Man and Place* (Taylor & Francis, 1991). He resigned as director in 1991 but remained associated with the Davy–Faraday Research Laboratory of the Royal Institution until 2006.

In 1991, he was knighted "for services to chemistry and the popularization of science." In his research Thomas has used different chemical–physical techniques, in particular, high resolution electron microscopy, synchrotron radiation, high resolution X-ray diffraction, and spectroscopy to investigate the nature and the location of active sites of solid catalysts with well-defined structure (oxides, zeolites, microporous materials) under operating conditions (Wright *et al.*, 1986) and to devise new mesoporous (Raja *et al.*, 2003), microporous, and molecular sieve catalysts. He has also contributed to the development of industrial applications of heterogeneous catalysis (Thomas and Thomas, 1997), inventing the solvent-free (Thomas *et al.*, 2001) catalytic synthesis of ethyl acetate, which has reached today a production of more than 200,000 ton/year in the United Kingdom (Ballantine, Purnell, and Thomas, 1984). He later devised a single-step, solvent-free process for the production of caprolactam, an intermediate for nylon-6 (Thomas and Raja, 2005).

In 2012 Sir John published an important illustration of the concept of single-site heterogeneous catalysis (Thomas, 2012), emphasizing the importance of nanoporous materials to assemble a wide range of new, well-defined catalytically active centers and their interesting regioselective and enantioselective catalysis.

Very recently he has published an important review on heterogeneous catalysis entitled "The concept, reality and utility of single-site heterogeneous catalysts (SSHCs)" (Thomas, 2014). The concept of single site in catalysis can be applied to many heterogeneous, homogeneous, and enzymatic catalysts and can be usefully in closing the gap between these different branches of catalysis science. For instance, the presence on the surface of active single sites with well-defined structure is a common feature of heterogeneous and homogeneous polymerization catalysts (Groppo *et al.*, 2005).

The previously mentioned scientists have primarily discussed the interaction of gas molecules with well-defined surface sites, located either on single-crystal surfaces or on catalytic materials with well-defined crystallographic structures. However, as many "single-site" heterogeneous catalysts are synthesized under wet conditions, attention must be paid to the preparation methods. A scientist who has focused on understanding the state of the surface under wet conditions and the mechanism of anchoring of metal centers to the hydrated surfaces is Michel Che (Tougerti *et al.*, 2012; Dumond, Marceau, and Che, 2007). Michel Che, already cited as a pioneer of ESR application to catalytic materials (Che and Naccache, 1971) and president-founder of European Federation of Catalysis Societies (EFCATS), has pioneered a molecular approach to heterogeneous catalysis that is based on transition elements taken as probes of surface structures of the anchoring support. His work, complementary to that of John Meurig Thomas, has led to the emergence of interfacial coordination chemistry at the junction of colloidal and supramolecular chemistry.

References

Arean, C., Weckhuysen, B.M., and Zecchina, A.O. (2012) Operando surface spectroscopy – placing catalytic solids at work under the spotlight. *Physical Chemistry Chemical Physics*, **14**, 2125–2127.

Ballantine, J.A., Purnell, J.H., and Thomas, J.M. (1984) Sheet silicates: broad spectrum catalysts for organic synthesis. *Molecular Catalysis*, **27**, 157–167.

Binnig, G. and Rohrer, H. (1986) Scanning tunneling microscopy. *IBM Journal of Research and Development*, **30**, 355.

Bordiga, S., Groppo, E., Agostini, G. *et al.* (2013) Reactivity of surface species in heterogeneous catalysts probed by in situ X-ray absorption techniques. *Chemical Reviews*, **113**, 1736–1850.

Borello, E., Zecchina, A., and Castelli, M. (1963) IR spectra of butenes adsorbed on silica aerosil. *Annali di Chimica*, **53**, 69–74.

Borello, E., Zecchina, A., and Morterra, C. (1967) Infrared study of methanol adsorption on aerosil. I. Chemisorption at room temperature. *Journal of Physical Chemistry*, **71**, 2938–2915.

Borroni-Bird, C.E. and King, D.A. (1991) An ultrahigh vacuum single crystal adsorption microcalorimeter. *Review of Scientific Instruments*, **62**, 2177–2185.

Bozso, F., Ertl, G., Grunze, M., and Weiss, M. (1977) Interaction of nitrogen with iron surfaces.1. Fe(100) and Fe(111). *Journal of Catalysis*, **49**, 18–41.

Busca, G. and Lorenzelli, V. (1982) Infrared spectroscopic identification of species arising from reactive adsorption of carbon oxides on metal oxide surfaces. *Materials Chemistry*, **7**, 89–126.

Cardona-Martinez, N. and Dumesic, J.A. (1992) Applications of adsorption microcalorimetry to the study of heterogeneous catalysis. *Advances in Catalysis*, **38**, 149–244.

Carrà, S. and Ugo, R. (1969) A reinterpretation of the mechanism of the heterogeneous catalytic synthesis of ammonia. *Journal of Catalysis*, **15**, 435–438.

Chan, S.S., Wachs, I.E., Murrell, L.L. *et al.* (1984) *In situ* laser Raman spectroscopy of supported metal oxides. *Journal of Physical Chemistry*, **88**, 5831–5835.

Che, M. and Naccache, C. (1971) Nature of paramagnetic species produced by oxygen treatment of titanium dioxide. *Chemical Physics Letters*, **8** (1), 45–48.

Christmann, K., Schober, O., Ertl, G., and Neumann, M. (1974) Adsorption of hydrogen on nickel single crystal surfaces. *Journal of Chemical Physics*, **60**, 4128–4140.

Cohen, A.D. and Sheppard, N. (1959) High-resolution nuclear magnetic resonance spectra of hydrocarbon groupings: some dichloro-propenes. *Proceedings of the Royal Society of London A*, **252**, 488–505.

Coluccia, S., Deane, A.M., and Tench, A.J. (1978) Photoluminescent spectra of surface states in alkaline earth oxides. *Journal of the Chemical Society, Faraday Transactions 1: Physical Chemistry in Condensed Phases*, **74**, 2913–2922.

Conrad, H., Ertl, G., and Latta, E. (1974) Adsorption of hydrogen on palladium single-crystal surfaces. *Surface Science*, **41**, 435–446.

Cox, M.P., Ertl, G., and Imbihl, R. (1985) Spatial self-organization of surface-structure during an oscillating catalytic reaction. *Physical Review Letters*, **54**, 1725.

Cremer, P.S., Xingcai, S., Shen, Y.R., and Somorjai, G.A. (1996) The first measurement of an absolute surface concentration of reaction intermediates in ethylene hydrogenation. *Catalysis Letters*, **40**, 143–145.

Damin, A., Bonino, F., Bordiga, S. *et al.* (2006) Vibrational properties of Cr^{II} centers on reduced Phillips catalysts highlighted by resonant Raman spectroscopy. *ChemPhysChem*, **7**, 342–344.

Damin, A., Usseglio, S., Agostini, G. *et al.* (2008) Au nanoparticles as SERS probes of the silica surface layer structure in the absence and presence of adsorbates. *Journal of Physical Chemistry C*, **112**, 4932–4936.

Davisson, C. and Germer, L.H. (1927) Reflection of electrons by a crystal of nickel. *Nature*, **119**, 558–560.

Dumond, F., Marceau, E., and Che, M. (2007) A study of cobalt speciation in Co/Al2O3 catalysts prepared from solutions of cobalt-ethylenediamine complexes. *Journal of Physical Chemistry C*, **111**, 4780–4789.

Eischens, R.P. and Pliskin, W.A. (1957) Infrared study of the catalyzed oxidation of CO. *Advances in Catalysis*, **9**, 662–668.

Eischens, R.P., Pliskin, W., and Francis, S.A. (1954) Infrared spectra of chemisorbed carbon monoxide. *Journal of Chemical Physics*, **22** (10), 1786–1787.

Ertl, G. (2005) Activation of diatomic molecules at solid surfaces. *Philosophical Transactions of the Royal Society of London A*, **363**, 955–958.

Ertl, G. (2012) The arduous way to the Haber-Bosch process. *Zeitschrift für Anorganische und Allgemeine Chemie*, **638**, 487–489.

Ertl, G. and Huber, M. (1980) Mechanism and kinetics of ammonia decomposition on iron. *Journal of Catalysis*, **61**, 537–539.

Folman, M. and Yates, D.J.C. (1958) Infra-red and length-change studies in adsorption of H_2O and CH_3OH on porous silica glass. *Transactions of the Faraday Society*, **54**, 1684–1691.

Franchy, R. and Ibach, H. (1985) CO on W(100) at 100 K studied by a high resolution EELS spectrometer with multi-channel detector. *Surface Science*, **155**, 15–23.

Fubini, B. (1988) Adsorption calorimetry in surface chemistry. *Thermochimica Acta*, **135**, 19–29.

Galkin, G.A., Kiselev, A.V., and Lygin, V.I. (1964) Infra-red spectra and energy of adsorption of aromatic compounds on silica. *Transactions of the Faraday Society*, **60**, 431–439.

Gravelle, P.C. (1985) Application of adsorption calorimetry to the study of heterogeneous catalysts reaction. *Thermochimica Acta*, **96**, 365–376.

Groppo, E., Lamberti, C., Bordiga, S. *et al.* (2005) The structure of active centers and the ethylene polymerization mechanism on the Cr/SiO_2 catalyst: a frontier for the characterization methods. *Chemical Reviews*, **105**, 115–184.

Guo, Q., Sun, K., Feng, Z. *et al.* (2012) A thorough investigation of the active titanium species in TS-1 zeolite by *in situ* UV resonance Raman spectroscopy. *Chemistry – A European Journal*, **18**, 13854–138.

Habas, S.E., Lee, H., Radmilovic, V. *et al.* (2007) Shaping metal nanocrystals through epitaxial seeded growth. *Nature Materials*, **6**, 692–697.

Hagstrom, S., Lyon, H.B., and Somorjai, G.A. (1965) Surface structures on the clean platinum (100) surface. *Physical Review Letters*, **15**, 491.

Hansen, T.W., Wagner, J.B., Hansen, P.L. *et al.* (2001) Atomic-resolution *in situ* transmission electron microscopy of a promoter of a heterogeneous catalyst. *Science*, **294**, 1508–1510.

Heidberg, J., Kandel, M., Meine, D., and Wildt, U. (1995) The monolayer CO adsorbed on MgO(100) detected by polarization infrared spectroscopy. *Surface Science*, **331-333**, 1467–1472.

Hussain, G. and Sheppard, N. (1992) Infrared investigation of the surface species obtained by chemisorption of acetylene (ethyne), C2H2 and C2D2, on high-area zinc oxide. *Journal of the Chemical Society, Faraday Transactions*, **88**, 2927–2930.

Ibach, H. (1991) *Electron Energy Loss Spectrometers*, Springer, Berlin.

Kiselev, A.V. and Lygin, V.I. (1959) Infrared absorption spectra and structure of the hydroxyl layers on silicas of different degrees of hydration. *Colloid Journal (USSR)*, **21**, 561–568.

Knözinger, H. (1979) XPS photoelectron study of rhodium complexes attached to chemically modified silicas. *Inorganica Chimica Acta*, **37**, 537–538.

Knözinger, H. and Rumpf, E. (1978) Stabilization of hexarhodiumhexadecacarbonyl by attachment to chemically modified silica surfaces. *Inorganica Chimica Acta*, **30**, 51–58.

Knözinger, H., Thornton, E.W., and Wolf, M. (1979) Rhodium catalysts prepared by attachment of hexarhodiumhexadecacarbonyl onto chemically modified silicas. Characterization of the infrared spectra in the carbonyl stretching region. *Journal of the Chemical Society, Faraday Transactions I*, **75**, 1888–1899.

Kuhlenbeck, H., Shaikhutdinov, H.S., and Freund, H.J. (2013) Well-ordered transition metal oxide layers in model catalysis – a series of case studies. *Chemical Reviews*, **113**, 3986–4034.

Li, Y. and Somorjai, G.A. (2010) Nanoscale advances in catalysis and energy applications. *Nano Letters*, **10**, 2289–2295.

Lunsford, J.H. (1967) EPR study of NO adsorbed on magnesium oxide. *Journal of Chemical Physics*, **46**, 4347–4351.

Lyon, H.B. and Somorjai, G.A. (1967) Low energy electron diffraction study of the clean (100), (111), and (110) faces of platinum. *Journal of Chemical Physics*, **46**, 2539–2550.

Magni, E. and Somorjai, G.A. (1996) Preparation of a model Ziegler-Natta catalyst: electron irradiation induced titanium chloride deposition on magnesium chloride thin films grown on gold. *Surface Science*, **345**, 1–16.

Mapes, J.E. and Eischens, R.P. (1954) The infrared spectra of ammonia chemisorbed on cracking catalysts. *Journal of Physical Chemistry*, **58**, 1059–1062.

Michel, D., Kazansky, V.B., and Andreev, V.M. (1978) Study of the interaction between surface hydroxyls and adsorbed water molecules on porous glasses by means of infrared spectroscopy. *Surface Science*, **72**, 342–356.

Nguyen, T.T. and Sheppard, N. (1978) Raman spectra of molecules adsorbed on heavy metal oxides : Propylene on ZnO: the anionic nature of the adsorbed allyl species. *Journal of the Chemical Society, Chemical Communications*, (**20**), 868–869.

Nguyen, T.T., Lavalley, J.C., Saussey, J., and Sheppard, N. (1980) Infrared and Raman spectroscopic study of some aromatic acetylenic compounds adsorbed on zinc oxide. *Journal of Catalysis*, **61**, 503–514.

Pettinger, B., Domke, K.F., Zhang, D. *et al.* (2007) Direct monitoring of plasmon resonances in a tip-surface gap of varying width. *Physical Review B*, **76**, 113409.

Pliskin, W.A. and Eischens, R.P. (1956) Infrared spectra of chemisorbed olefins and acetylene. *Journal of Chemical Physics (Letter)*, **24**, 482–483.

Polák, R. and Klier, K. (1969) Spectra of synthetic zeolites containing transition metal ions-III. A simple model calculation of the system adsorbed molecule-NiA zeolite. *Journal of Physics and Chemistry of Solids*, **30**, 2231–2233.

Raja, R., Thomas, J.M., Jones, M.D. *et al.* (2003) Cascade reactions: asymmetric synthesis using combined chemo-enzymatic catalysts. *Journal of the American Chemical Society*, **125**, 14982–14983.

Rives-Arnau, V. and Sheppard, N. (1980) Raman spectroscopic study of the polymerization of acetylene on titanium dioxide (rutile). *Journal of the Chemical Society, Faraday Transactions 1: Physical Chemistry in Condensed Phases*, **76**, 394–402.

Rupprechter, G. (2002) Polarization–modulation infrared reflection absorption spectroscopy of functioning model catalysts from ultrahigh vacuum to ambient pressure. *Advances in Catalysis*, **51**, 133–263.

Schlienz, H., Beckendorf, M., Katter, U.J. *et al.* (1995) Electron spin resonance investigation of the molecular dynamics of NO_2 on Al_2O_3(111) under ultrahigh vacuum conditions. *Physical Review Letters*, **74**, 761.

Shelimov, B.N. and Che, M. (1978) An EPR study of the kinetics of oxygen isotopic exchange involving O^- adsorbed species on V_2O_5 SiO_2 catalysts. *Journal of Catalysis*, **51**, 143–149.

Shen, Y.R. (1989) Surface properties probed by second-harmonic and sum-frequency generation. *Nature*, **337**, 519–525.

Sheppard, N. (1948) Some characteristic frequencies in the Raman spectra of saturated aliphatic hydrocarbons. *Journal of Chemical Physics*, **16**, 690–697.

Sheppard, N. (1988) Vibrational spectroscopic studies of the structure of species derived from the chemisorption of hydrocarbons on metal single-crystal surfaces. *Annual Review of Physical Chemistry*, **39**, 589–644.

Sheppard, N. (2008) The U.K.'s contributions to IR spectroscopic instrumentation. From wartime fuel research to a major technique for chemical analysis. *Analytical Chemistry*, **64**, 877–882.

Sheppard, N. and Simpson, D.M. (1952) The infrared and Raman spectra of hydrocarbons Part I. Acetylenes and olefins. *Quarterly Reviews*, **6**, 1–33.

Sheppard, N. and Sutherland, G.B.B.M. (1945) Some infrared studies on the vulcanisation of rubber. *Transactions of the Faraday Society*, **41**, 261–271.

Shvets, V.A., Vorotinzev, V.M., and Kazansky, V.B. (1969) The oxygen anion-radicals on V_2O_5/SiO_2 (Letter). *Journal of Catalysis*, **15**, 214–218.

Skottke, M., Behm, R.J., Ertl, G. *et al.* (1987) LEED structure analysis of the clean and (2×1)H covered Pd(110) surface. *Journal of Chemical Physics*, **87**, 6191.

Somorjai G.A. (1960) Small angle x-ray study of metallized catalysts. PhD thesis. University of California, Berkeley.

Somorjai, G. (1994) Surface reconstruction and catalysis. *Annual Review of Physical chemistry*, **45**, 721.

Somorjai, G. (2000) The development of molecular surface science and the surface science of catalysis: the Berkeley contribution. *Journal of Physical Chemistry B*, **104**, 2969–2979.

Somorjai, G.A. and Marsh, A.L. (2005) Active sites and states in the heterogeneous catalysis of carbon–hydrogen bonds. *Philosophical Transactions of the Royal Society*, **A363**, 879–900.

Somorjai, G.A. and McCrea, K.R. (2000) Sum frequency generation: surface vibrational spectroscopy studies of catalytic reactions on metal single-crystal surfaces. *Advances in Catalysis*, **45**, 385–438.

Somorjai, J. and Park, Y. (2008) Evolution of the surface science of catalysis from single crystals to metal nanoparticles under pressure. *Journal of Chemical Physics*, **128**, 182504.

Somorjai, G.A., Frei, H., and Park, J.Y. (2009) Advancing the frontiers in nanocatalysis, biointerfaces and renewable energy conversion by innovations of surface techniques. *Journal of the American Chemical Society*, **131**, 16589–16605.

Spencer, N.D., Schoonmaker, R.C., and Somorjai, G.A. (1982) Iron single crystals as ammonia synthesis catalysts: effect of surface structure on catalyst activity. *Journal of Catalysis*, **74**, 129–135.

Spoto, G., Gribov, E., Damin, A. *et al.* (2003) The IR spectra of $Mg^{2+}(CO)$ complexes on the (0 0 1)surfaces of polycrystalline and single crystal MgO. *Surface Science*, **540**, L605–L610.

Spoto, G., Gribov, E.N., Ricchiardi, G. *et al.* (2004) Carbon monoxide MgO from dispersed solids to single crystals: a review and new advances. *Progress in Surface Science*, **76**, 71–146.

Stepanov, V.G., Shubin, A.A., Ione, K.G. *et al.* (1984) NMR study of the state of aluminium and silicon atoms in Y and ZSM type zeolites and mordenite as a function of the extent of decationization and the heat-treatment conditions. *Kinetics and Catalysis*, **25**, 1043–1050.

Stolz, H. and Knozinger, H. (1971) Adsorption properties of alumina-VI. An IR-investigation of the adsorption of pyridine. *Kolloid-Zeitschrift & Zeitschrift für Polymere*, **243**, 71–76.

Stumm von Bordwehr, R. (1989) A history of the X-ray absorption fine structure. *Annals of Physics*, **14**, 377–466.

Terenin, A. (1921) The normal orbit of the electron in the atom of mercury. *Nature*, **107**, 203.

Terenin, A.N. (1975) *Selected Works: Spectroscopy of Adsorbed Molecules and Surface Compounds*, vol. **3**. Nauka, Leningrad.

Terenin, K.A.N. and Kasparov, K. (1940) IR spectroscopy study of ammonia adsorption on pure or Fe-containing silica. *Russian Journal of Physical Chemistry*, **14**, 1362–1636.

Terenin, A.N., Jaroslavsky, N.G., Karjakin, A.W., and Sidorova, A.I. (1955) Spectroscopie infrarouge des molecules adsorbees sur verre poreux. *Mikrochimica Acta*, **2/3**, 467–470.

Terenin, A. and Roev, L. (1959) Infra-red spectra of chemisorbed nitric oxide molecules. *Spectrochimica Acta*, **15**, 275–276.

Thomas, J.M. (2012) The societal significance of catalysis and the growing practical importance of single-site heterogeneous catalysts. *Proceedings of the Royal Society of London A.*, **468**, 1884.

Thomas, J.M. (2014) The concept, reality and utility of single-site heterogeneous catalysts. *Physical Chemistry Chemical Physics*, **16**, 7647–7661.

Thomas, J.M. and Raja, R. (2005) Design of a "green" one-step catalytic production of ε-caprolactam (precursor of nylon-6). *Proceedings of the National Academy of Sciences (PNAS)*, **102**, 13732–13736.

Thomas, J.M. and Thomas, W.J. (1997) *Principles and Practice of Heterogeneous Catalysis*. VCH, Weinheim.

Thomas, J.M., Raja, R., Sankar, G. *et al.* (2001) Solvent-free routes to clean technology. *Chemistry – A European Journal*, **7**, 2972.

Thomson, G.P. and Reid, A.A. (1927) Diffraction of cathode rays by a thin film. *Nature*, **119**, 890.

Tougerti, A., Llorens, I., D'Acapito, F. *et al.* (2012) Surface science approach to the solid–liquid interface: surface-dependent precipitation of $Ni(OH)_2$ on a-Al_2O_3 surfaces. *Angewandte Chemie, International Edition*, **51**, 7697.

Tsyganenko, A. (2008) Science—this is my religion: biographical memoir of Academician Alexander Terenin (1896–1967). *Journal of Photochemistry and Photobiology A: Chemistry*, **196**, 123.

Wright, P.A., Thomas, J.M., Cheetham, A.K., and Nowak, A.K. (1986) Localizing active sites in zeolitic catalysts: Neutron-powder-profile analysis and computer simulation of deutero-pyridine bound to gallozeolite-L. *Nature*, **318**, 611–614.

Yates, J.T. Jr., and Garland, C.W. (1961) Infrared studies of carbon monoxide chemisorbed on nickel and on mercury-poisoned nickel surfaces. *Journal of Physical Chemistry*, **65**, 617.

Yates, D.J.C., Sheppard, N., and Angell, C.L. (1955) Infrared spectrum of ammonia adsorbed on porous silica glass. *Journal of Chemical Physics*, **23**, 1980.

Zecchina, A. and Stone, F.S. (1974) Reflectance spectra of CO chemisorbed on MgO, and evidence for the formation of cyclic adsorbed species. *Journal of Chemical Society, Chemical Communications*, **15**, 582–584.

Zhitomirski, A.N. and Zamaraev, K.I. (1969) Electron-paramagnetic resonance investigation of electronic structure in copper (II) complexes in relation to their catalytic activity in the hydrogen peroxide decomposition oxidation-reduction reaction. *Journal of Structural Chemistry*, **10**, 357–363.

5

Development of Homogeneous Catalysis and Organocatalysis

5.1 Introductory Remarks

A homogeneous catalytic process is a process in which catalysts and reactants are both in the same phase (solution), and the catalysts precursors are molecular species of a well-defined and known structure.

The history of homogeneous catalysis spans the entire twentieth century and is interconnected with the parallel history of heterogeneous catalysis. Homogeneous and heterogeneous catalyses have complementary positions in the chemical industry. For example, as documented in the previous chapters, many major industrial products such as ammonia, nitric acid, sulfuric acid, gasoline, and polymers are produced via heterogeneous catalysis. In this chapter we will show that many basic chemicals, fine chemicals, pharmaceuticals, and agrochemicals are produced via homogeneous catalysis. The synergic interactions between these two branches of catalysis were constant and productive throughout the twentieth century. The aim of this chapter is not to give a detailed and fully comprehensive report about all processes and reactions associated with homogeneous catalysts (a goal definitely outside the limited ambition of this book) but, more simply, to give information about the main catalysts classes and their evolution in the twentieth century and thus to illustrate the innovative character of this rich branch of catalysis science.

Depending on the nature of the involved catalysts, we can divide the homogeneous processes into several groups: processes based on acid–base catalysts, on organometallic catalysts, on organic catalysts, and on mixed organic–metallorganic catalysts.

In principle, homogeneous catalysis does not suffer from the drawbacks of heterogeneous catalysis, as is the case with heterogeneity catalysis, nor do we lack deep understanding of the structure of surface sites. Rather, a serious practical problem with homogeneous catalysis is the separation of reactants and products (which are all in the same phase) from the catalyst

The Development of Catalysis: A History of Key Processes and Personas in Catalytic Science and Technology,
First Edition. Adriano Zecchina and Salvatore Califano.
© 2017 John Wiley & Sons, Inc. Published 2017 by John Wiley & Sons, Inc.

(often a precious metal) and from the solvent. For the processes catalyzed by strong acids, corrosion and environmental problems are additional problems. These drawbacks are partly counterbalanced by the fact that the catalyst (or better the catalyst precursor) is unequivocally, chemically, and structurally characterized and based on a rational design. Efforts to recover the catalyst involve immobilizing or anchoring homogeneous catalysts to solid surfaces, but this often causes alterations in the structure of the originally anchored precursor. However, the solvents used in this process are subject to stringent environmental legislation that has generated increasing pressure for cleaner methods of chemical production, through reduction, or, preferably, elimination of poisonous solvents. It is indeed a fact that the best solvent is no solvent, or if a solvent (diluent) is needed, it should preferably be water, which is nontoxic, nonflammable, inexpensive, and abundantly available (Sheldon, 2000).

5.2 Homogeneous Acid and Bases as Catalysts: G. Olah Contribution

Acids and bases can act as powerful homogeneous catalysts and they operate in solution for a great variety of chemical reactions involving organic molecules, some even known from centuries (see, for instance, the Valerius Cordus discovery mentioned in Chapter 1). After 1920 homogeneous catalysts received a big stimulus from the acid-catalyzed refinery reactions developed by Houdry first and by Ipatieff and Pines later. The mechanism of acid- and base-catalyzed reactions is explained in terms of the Brønsted–Lowry concept of acids and bases as a process based on the transfer of protons from the acid catalyst to the substrate or from the substrate to the basic catalyst. In terms of the more general Lewis theory of acids and bases, the reaction proceeds via a donation of an electron pair by a base catalyst or acceptance by an acid catalyst. There are many possible chemical compounds that can act as sources for the protons to be transferred in an acid catalysis system. In the Ipatieff and Pines contributions described in Chapter 4, the acids used for acid catalysis included sulfuric acid, hydrofluoric acid, and phosphoric acid. Later on, toluenesulfonic acid, alkanesulfonic acid, heteropoly acids, the superacid trifluoromethanesulfonic acid also known as triflic acid, perchloric acid, and the "magic acid" FSO_3H-SbF_5 (which is even stronger) discovered by George Andrew Olah (1927–) were assayed for the generation of carbocations mainly *via* interaction with olefinic hydrocarbons. In agreement with Lewis's theory, also $AlCl_3$, $FeCl_3$, and $SbCl_5$ (which are Lewis acids) were shown to be effective as acid catalysts. In this application, however,

$AlCl_3$ and $FeCl_3$ behave as catalysts only in presence of HCl as an activator, a fact that suggests that the active species are actually $H[AlCl_4]$ and $H[FeCl_4]$.

Of course, the contribution of Ipatieff and Pines to the understanding of the mechanism of acid catalysis was devoted to the processes associated with the petroleum industry. They discovered that strong acid catalysts protonate the alkenes (for instance, propene or butene) to produce reactive and short-lived carbocations (carbenium ions) that can undergo double bond isomerization or, in turn, alkylate isobutene (Ipatieff, Pines, and Schmerling, 1940) and aromatic hydrocarbons. As for the much less abundant class of reactions catalyzed by bases, we only mentioned that in the 1930s Herman Pines had already found that solid base catalysts are more effective for double bond isomerization than acid catalysts (Pines and Stalik, 1977; Pines, 1981).

After the pioneering studies of Ipatieff and Pines, the study of carbocations was resumed at the beginning of the 1960s by the Hungarian chemist George Andrew Olah, who proved that the very short-lived carbocations can be stabilized by superacids (Figure 5.1). This important discovery allowed lengthening the carbocation's lifetime, making it possible to investigate their structure by chemical–physical techniques and to profit in their applications to complex reactions.

George Andrew Olah (1927–) was born in Budapest where he entered the Technical University of Budapest and graduated in 1949. Fascinated by the works of Fritz Seel (1915–1987), professor at the University of Saarland, a great expert of fluorine compounds, he became very early interested in fluorine

Figure 5.1 George A. Olah. Olah studied the generation of carbocations *via* interaction of strong Brønsted acids with olefinic hydrocarbons. He was awarded the 1994 Nobel Prize in Chemistry with the motivation *for his contribution to carbocation chemistry*. *Source*: Reproduced with permission of G. Olah.

chemistry, a field that strongly influenced his future life. In particular, he started to be interested first in acylation reactions with acyl fluorides R–COF and then in alkylation with alkyl fluorides, using BF_3 as a catalyst:

$$ArH + RCOF \xrightarrow{BF_3} ArCOR + HF$$

$$ArH + RF \xrightarrow{BF_3} ArR + HF$$

From these studies, his multiannual custom with the carbocation chemistry took origin. After the tragic events in Hungary in October 1956, Olah, with the family, emigrated first to London and then in 1957 to Canada. In 1957, he joined Dow Chemical, an important American chemical company where after a short while he was promoted to *scientist*. In 1964, he moved to the United States upon being appointed at the *Eastern Research Laboratories* of the Dow Chemical in Massachusetts. The following year, he joined the University of Cleveland, Ohio, as teacher and department director. In 1977, he moved to the University of California, Los Angeles (UCLA), where he still works today. In 1991, he assumed the office of director of the Loker Hydrocarbon Research Institute, an important organization in the field of hydrocarbons.

Already while he was in Hungary, Olah had started to synthesize the superacid FSO_3H and boron trifluoride BF_3. When he arrived in Canada at the Dow Chemical, he expanded this line of research by starting a systematic study of the carbocations, which, in the 1950s, were often indicated as intermediate products in several chemical reactions. It is a fact that their existence had been hypothesized several times, for instance, by Pines. However, owing to their reactivity, they had extremely short lifetimes (of the order of nanosecond). For this reason nobody had succeeded in proving their presence even with the most sophisticated spectroscopic techniques available at the time. Without experimental evidence of their existence, it was impossible to establish whether they were the product of the researchers' imagination or a true physical reality. In the 1960–1970 period, George Olah, with a series of brilliant experiments, definitively solved the problem, developing the methods to prepare stable carbocations with a relatively long lifetime, sufficient to study their structure and their physical properties with spectroscopic techniques. Olah realized that by using as solvents the superacids, with acidity greater of that of sulfuric acid up to 18 orders of magnitude, he was not only able to protonate any organic molecule but the formed carbocations remained stable in solution. Under these conditions they could not react with the solvent molecules supplying protons in rearrangement reactions following reactions of the type

$$(CH_3)C^+ \rightleftarrows (CH_3)_2C=CH_2 + H^+$$

and then could be studied by NMR spectroscopy.

Olah found that two types of carbocations exist, those trivalent named "carbenium ions" formed by interaction of the proton with olefins in which the positive carbon atom is connected to three atoms and those penta-coordinated called "carbonium ions" with five substituents formed by interaction of the proton with saturated hydrocarbons. On the basis of research that definitely proved the existence of the carbonium ions, Olah disproved the belief of the time that in organic compounds the carbon atoms could be at maximum fourfold coordinated. Looking for Lewis acids, Olah found that (Olah *et al.*, 1962) antimony pentafluoride SbF_5 was an extremely powerful Lewis acid, able to ionize the alkyl fluorides in solution, giving rise to stable alkyl cation–anion pairs with sufficiently long lifetimes, according to the equations

$$(CH_3)_3C-F + SbF_5 \rightleftarrows [(CH_3)_3C]^+[SbF_6]^-$$

He studied several important superacids such as FSO_3H (fluorosulfuric acid), CF_3SO_3H (triflic acid), as well as the very powerful fluoroantimonic acid $HF-SbF_5$ and the so-called magic acid FSO_3H-SbF_5, a mixture of antimony pentafluoride (Lewis acid) and fluorosulfonic acid. Also superacids based on fluorides such as AsF_5, TaF_5, NbF_5, and other powerful Lewis acids such as $B(O_3SCF_3)$ were successfully utilized. The name *magic acid* for the FSO_3H-SbF_5 mixture was conceived by his coworker Joachim Lukas, a German researcher that worked with him in the 1960s. An episode about him is often reported that he was able to demonstrate, during a party in the laboratory, that a piece of candle wax could be slipped into the FSO_3H-SbF_5 acid. The candle wax dissolved completely quite rapidly, giving rise to a solution that, when examined with ^1H-NMR, showed to contain the tert-butyl cation (Olah *et al.*, 1967). One of the significant results of the carbocation stabilization with superacids was the preparation of protonated methane (Olah and Schlossberg, 1968) using the magic acid following the reaction

$$CH_4 + H^+ \rightarrow CH_5^+$$

so showing that methane can behave as a weak Lewis base. Olah and coworkers also proved that penta-coordinated methane can be used to prepare long hydrocarbon chains. However, this interesting observation has not found any practical application. The research on penta-coordinated carbonium ions showed that these species can produce new reactions, a result that has greatly contributed to the development of petrol-chemistry. It allows, in particular, the linear chains of saturated hydrocarbons (low octane number) to be transformed into branched chains (high octane number). The concept of penta-coordinated carbonium ions proved later its utility also in reactions occurring in the cavities of acidic zeolites. Olah's group disclosed also an original method for producing branched aliphatic ketones in hydrocarbon mixtures of isoalkanes by means of superacid catalyzed formylation–rearrangement reactions. The method can be used to simultaneously isomerizes and formylate hydrocarbons in complex hydrocarbon mixtures such as refinery streams, alkylate mixtures, and natural gas liquids. Liquefied natural gases are upgraded and oxygenated by addition or by direct production of branched aliphatic ketones. Due to Olah's work, the chemistry of carbenium and carbonium ions became familiar to the chemist community. For his research on superacids and carbocations, Olah was awarded with the 1994 Nobel Prize in Chemistry.

More recently Olah has dedicated his attention to the study of methanol as the ideal fuel for the future, since it can be produced via catalytic methods (Olah, 2005). He is convinced that this could lead to a new economy, more sustainable and less polluting. We will return to this point in the final chapter.

Earlier in this chapter we mentioned that to favor separation from reactants, efforts have been made to anchor strong acids to heterogeneous supports like silica or silica alumina. Along this line a special position is occupied by Nafion, a fluorinated microporous polymer that branches ending with ($-SO_3H$) acidic groups, which can act as strongly acidic heterogeneous catalysts. Nafion was discovered in 1960 by Walther Grot, a researcher working at DuPont. The high concentration of F in the polymer chain confers a highly acidic character to the sulfonic groups, which are so often considered as superacidic (Figure 5.2). In presence of water the SO_3H groups form droplets of highly concentrated acid. For this reason this group occupies a position intermediate between homogeneous and heterogeneous systems. Nafion has various applications. It has been

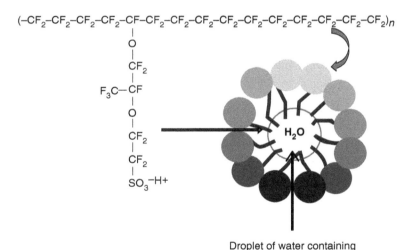

Droplet of water containing sulfonic groups

Figure 5.2 Structure of the $-CF_2-$ polymer with pendant branches carrying sulfonic groups at the end, and idealized structure of a membrane tubular channel containing pendant chains and a droplet of water. *Source*: Adapted from Schmidt-Rohr and Chen (2008).

demonstrated that, like superacids in solution, it can perform alkylation and acylation of substrates.

5.3 Organometallic Catalysts

Apart from homogeneous acid–base catalysts, the "classical" homogeneous organometallic catalyst consists of a central metal atom or cluster of metal atoms, surrounded by a ligand sphere and by an outer solvent sphere. Their interplay governs activity, productivity, and selectivity of the catalytic system.

It is a current opinion that the biggest advantage of homogeneous organometallic catalysts with respect to the heterogeneous ones lies in the fact that in many cases the structure of the catalytic center (or more correctly the precursor) is well defined. This allows the understanding of the catalytic events at the molecular level and the tuning of the catalyst structure to obtain the best performance in term of selectivity. It is thus worth opening a brief digression on selectivity and selective catalysts, since a clear description of this non-usual term often facilitates the comprehension of problems that seem very complex only because of a technical terminology. A selective catalyst can be defined as an organized assembly of atoms and molecules, governing not only the kinetics of the system by lowering the activation barrier but also the paths of the reacting molecules. A portion of matter having these properties, with a metallic atom or metallic cluster at the center, has at least supra- or super-supramolecular character and has shape definitely usually larger than that of reactants and hence not smaller than 1 nm (nanometer) (Zecchina *et al.*, 2007).

Selective nanocatalysis can be thus defined as the science of synthesis and of *in situ* characterization of ensembles of atoms with supramolecular tailored size, displaying catalytic properties to tune with precision their activity and selectivity.

In the following we will show that these concepts are valid for the most common reactions catalyzed by homogeneous organometallic complexes.

Following a historical ordering, they are hydroformylation, partial oxidation of ethane into acetaldehyde and conversion of hydrocarbons to aldehydes, carbonylation, asymmetric hydrogenation and epoxidations, olefin oligomerization, olefin methatesis, and C–C coupling reactions.

The first significant homogeneous catalyst is the hydroformylation catalyst $HCo(CO)_4$. This catalyst was discovered by the German chemist Otto Roelen in 1938 (Figure 5.3). Otto Roelen (1897–1993), after his graduation from the Technische Hochschule of Stuttgart in 1922, worked in Mülheim (Ruhr region) with Franz Fischer and Hans Tropsch where he became expert in heterogeneous catalysis, in Fisher–Tropsch synthesis, and in the chemistry of CO at high pressure. While in Mülheim he discovered the hydroformylation (oxo) reaction. During the World War II, he was chief chemist at the Ruhrchemie. At the end of

(a)   (b)

Figure 5.3 (a) Otto Roelen (1897–1973) and (b) Walter Julius Reppe (1892–1969), the first protagonists of organometallic catalysis. Roelen discovered the first homogeneous $HCo(CO)_4$(hydroformylation) catalyst, while Reppe developed the acetylene chemistry with homogeneous catalysts. The first image refers to the plaque delivered by Gesellschaft Deutscher Chemiker (GDCh) for the discovery of oxo synthesis of Otto Roelen at Ruhrchemie (image in the public domain). *Source:* (b) Courtesy of BASF Company.

the war he was guest at a farm belonging to friends in the zone overrun by the Russian armies. However, he was soon released by the Russians. At the request of the English Ministry of Fuel and Power, he was brought to London for interrogation, where he described the hydroformylation (oxo) synthesis plants in detail to interrogators of British Scientific and Industrial Research Department. After the war, in 1948, a commercial plant was put on the stream by Exxon. Roelen was awarded the Adolf von Baeyer prize in 1963. In his honor, DECHEMA (Society for Chemical Engineering and Biotechnology) named the *Otto Roelen Prize* after him.

In the hydroformylation (oxo) process, patented in 1938, alkenes react with the synthesis gas in the presence of the homogeneous catalyst $HCo(CO)_4$ to give a mixture of branched and linear aldehydes:

$$RHC{=}CH_2 + CO + H_2 \rightarrow a\ RCH_2CH_2CHO + (1-a)\ RCH(CHO)CH_3$$

in which, formally, the addition of formaldehyde across the double bond has taken place. The reaction occurs under high CO pressure and is associated with formation of colloidal cobalt. As it occurs for the other homogeneous catalysts containing metal atoms in low valence state, $HCo(CO)_4$ is only the precursor, the active catalyst being most likely the coordinatively unsaturated $HCo(CO)_3$ complex formed by loss of one CO ligand. The presence of this labile intermediate has never been proved experimentally, even if kinetic experiments are in favor of this existence. The reaction mechanism has been intensively studied by the Italian organic chemist Piero Pino (Piacenti *et al.*, 1966).

The only further advancement on $HCo(CO)_4$-based hydroformylation catalysis occurred in 1960 when the researchers Lynn H. Slaugh and Richard D. Mullineaux (Mullineaux and Slaugh, 1966) working at the Shell chemical company modified the original compound by substituting one CO ligand with a bulky trialkylphosphine ligand. This modification caused a dramatic change in rate and selectivity. The electronic effect of substituting an electron-donating alkylated phosphine for one of the carbonyl ligands to produce $HCo(CO)_3$ (PR_3) resulted in stronger Co–CO bonding and a dramatic reduction in the CO partial pressures and temperatures without any decomposition of catalyst to cobalt metal. The bulky (PR_3) ligand influenced not only the electrons, population at cobalt center (so affecting the reaction rate) but also, owing to its bulky character, the selectivity. The $HCo(CO)_4$-based catalysts have today only historical interest because they are the first homogeneous precursor catalyst with a well-defined structure. Today their use as hydroformylation catalysts has been largely superseded by rhodium-based catalysts.

The second example of a process based on a homogeneous catalyst is the Wacker process, involving the transformation of ethane into acetaldehyde. The development of this process started in 1956 at Wacker Chemie, a worldwide operating company located in Bavaria.

In the early times acetaldehyde was obtained by hydration of acetylene to acetaldehyde, following a process discovered in 1881 by the Russian chemist Kutscheroff (1881), industrially performed using mercury(II) bromide as catalyst. The availability of ethylene as cheaper raw material prompted the researchers at Wacker Chemie to investigate the reaction of ethylene and oxygen over palladium supported on carbon. Unexpectedly, the experiment gave evidence for the formation of small amount of acetaldehyde, a valuable intermediate for chemical synthesis. New studies highlighted that, due to catalyst inactivation, the heterogeneous process initiated by Wacker was not economically feasible. The researchers so moved toward homogeneous Pd-based catalysts, and a pilot plant was in operation in 1958. The catalyst is $(PdCl_4)^{2-}$, and the Wacker catalytic reaction can be described as follows:

$$[PdCl_4]^{2-} + C_2H_4 + H_2O \rightarrow CH_3CHO + Pd + 2HCl + 2Cl-$$

As this reaction leads to the formation of Pd, the Pd(II) is restored with aid of $CuCl_2$ following the scheme

$$Pd + 2CuCl_2 + 2Cl^- \rightarrow [PdCl_4]^{2-} + 2CuCl$$

The so formed CuCl is then transformed into the original $CuCl_2$ by oxidation following the reaction

$$CuCl + \frac{1}{2}O_2 + 2HCl \rightarrow 2CuCl_2 + H_2O$$

In the reaction all catalysts are recovered, and only the alkene and oxygen are consumed. The function of copper(II) chloride is to act as an oxidizing agent, so avoiding the precipitation of Pd(0) metal and the stopping of the reaction. Notice that this reaction was first reported by Phillips (1894a, b) and that the Wacker reaction was first published by Smidt *et al.* (1959). On this basis it can be thus inferred that, as in the case of many other industrial processes, the realization of the Wacker process was due to the contribution of several excellent industrial researchers. The mechanism has attracted and is still attracting experimental (Baeckvall, Akermark, and Ljunggren *et al.*, 1979; Anderson, Keith, and Sigman, 2010) and computational studies (Comas-Vives *et al.*, 2010; Keith, Oxgaard, and Goddard, 2006). Without entering into the details of the whole mechanism, it can be concluded that the original $[PdCl_4]^{2-}$ complex is only the precursor of the really active species, which is thought to be formed by loss of a chlorine ligand with formation of a coordination vacancy where ethene becomes coordinated.

The reactions of acetylene using homogeneous catalysts leading to more complex molecules were studied at the beginning of the twentieth century by the German chemist Walter Julius Reppe (1892–1969), a very important researcher, whose personal biography is intimately associated with the development of German chemistry between the two world wars.

Walter Reppe studied the chemistry of acetylene and of its derivatives in great detail (Reppe and Schweckendieck, 1948) and published in 1981 an exhaustive treaty on the subject (Reppe, 1981). Reppe is undoubtedly the chemist who has developed the chemistry of acetylene to the highest level of sophistication (Reppe and Schweckendieck, 1948; Reppe, Schlichting, and Meister, 1948; Reppe *et al.*, 1948) (Figure 5.3). He obtained his doctorate in Munich in 1920 and at the end of World War I assumed a position at the Badische Aniline und Soda Fabrik (BASF) Company.

At BASF Reppe began to synthesize vinyl ether from acetylene and alcohol. Reppe's technique known as "ethynylation" became soon suitable at the commercial scale, and in the following years Reppe focused on other reactions such as the "cyclic oligomerization" of acetylene (Reppe and Schweckendieck, 1948; Reppe, Schlichting, and Meister, 1948; Reppe *et al.*, 1948) and the carbonylation reactions (Reppe, 1953a, b; Reppe and Kröper, 1953; Reppe *et al.*, 1953a, b; Reppe and Vetter, 1953). The cyclic oligomerization of acetylene to cyclooctatetraene (Reppe *et al.*, 1948) is involving four acetylene molecules:

$$4 \ HC \equiv CH \xrightarrow{\text{Cat}} \text{(cyclooctatetraene)}$$

In the tetramerization process the catalyst precursor is $Ni(CN)_2$ in tetrahydrofuran. This precursor is thought to be transformed into a $Ni(C_2H_2)_2$ complex, which is the catalyst initiating the cycle of reactions from acetylene to cyclooctatetraene (Straub and Gollub, 2004).

In 1939 he patented one of the most significant applications of acetylene chemistry, the synthesis of polyvinylpyrrolidone (PVP), initially used as a blood plasma substitute and later in many applications in industrial production.

In 1939 Reppe also developed the carbonylation process using $Ni(CO_4)$ catalyst in which acetylene was reacting with carbon monoxide and water to give acrylic acid:

$$\equiv + \ CO \ + \ H_2O \longrightarrow \text{(acrylic acid)}$$

The process has some similarity with the Roelen hydroformylation process. $Ni(CO_4)$ is likely the third homogeneous catalyst ever employed in industry. In the period 1939–1950 (Reppe, 1949), Reppe succeeded also in producing vinyl ether by reacting acetylene with alcohols. Important reactions must be attributed to him and his group, including those involving hydrogen cyanide and chloridric acid to give vinyl compounds and aldehydes to give dyols.

At the end of the war, Reppe was detained by the US Army from June 1945 in a concentration camp and subjected to harsh interrogation about his research on acetylene.

He was upset because the Americans were suspecting him of collaboration with Nazi power. However, he was not implicated in any war crimes at the Nuremberg process, and he said to have joined the party in 1944 only to protect his group of young chemists. Reppe was released from internment in September 1947, and the valuable acetylene catalytic reactions were considered as war reparations. He was later named research director of BASF. A report about the Reppe interrogation can be found in the paper of Bigelow, who, as a lieutenant colonel in the Chemical Warfare Service, spent many months interviewing Walter Reppe (Bigelow, 1947). After the war, Reppe and Magin discovered in 1951 also the copolymerization of ethylene and CO (100–400 bar) to form polyketones by using $K_2[Ni(CN)_4]$ in water as catalyst precursor (Reppe and Magin, 1948).

Even today, BASF manufactures a large fraction of acrylic acids and their esters by hydro-carbonylation of acetylene following the Reppe process.

Later the same reaction scheme, where methyl alcohol and propyne are used instead of H_2O and ethyne in the presence of a palladium complex as catalyst, gave methyl methacrylate (MMA), which is an important monomer because its polymerization gives a polymeric material characterized by good optical and mechanical properties (Shell). Although most of the industrial processes developed by Reppe and coworkers have been superseded, largely because of the shift from coal to oil feedstocks, he was undoubtedly one of the leader scientists showing the utility of homogeneous metal catalysts in large-scale synthesis of organic compounds. There is no doubt that his research motivated the flowering of organometallic chemistry and its industrial applications.

Methanol carbonylation is another old process based on homogeneous catalysts. Carbonylation chemistry was pioneered by Otto Roelen (Ruhrchemie) and Walter Reppe (IG Farben, later BASF) in the late 1930s. They discovered that metal carbonyls, $HCo(CO)_4$, $Ni(CO)_4$, and $Fe(CO)_5$, can act as homogeneous carbonylation catalysts. Today we know that these complexes are only the precursors of the real active species. This lesson was not forgotten at BASF where the first methanol-to-acetic acid carbonylation industrial process was commercialized in 1960. It used an iodide-promoted cobalt catalyst and required very high pressures as well as high temperatures but gave acetic acid with high selectivity. It is thought that under high CO pressure, $HCoCO_4/CoCO_4^-$ equilibrium is existing in solution and that these species (together with CH_3I acting as co-catalysts) are the real catalysts' precursors. The situation changed in the mid-1960s with the discovery that organophosphine-substituted rhodium and palladium complexes are active under milder reaction conditions. The major advance came in 1966 with the discovery by Monsanto of rhodium–iodide catalysts for the carbonylation of methanol, which led to the first commercial unit in 1970. The species participating in this reaction are $[RhI_2(CO)_2]^-$ and CH_3I. The advantages over the cobalt-catalyzed BASF process consist in significantly milder conditions allowing substantial costs savings and in higher selectivity for acetic acid (Thomas and Suss-Fink, 2003). The only disadvantage is the use of rhodium, a costly precious metal.

Another homogeneous catalyst (or more correctly precursor) coming in chronological order is the Wilkinson–Osborn hydrogenation catalyst; it is the father of a multitude of important asymmetric hydrogenation catalysts discovered few years later. The discovery of this catalyst, together with that of sandwich compounds (which are at the origin of the rich family of metallocene olefin polymerization catalysts discovered by Kaminsky and Britzinger as documented in Chapter 4), brought international renown to Wilkinson (1973 Nobel Prize). Geoffrey Wilkinson (1921–1996), graduated at Imperial College London in 1941. During the war he was recruited by Professor Friedrich Paneth as chemist for the nuclear energy project and sent to Canada, where he stayed until 1946. For the next four years he was at the University of California,

Berkeley, and was always involved in nuclear energy projects. He then moved to MIT and began to be interested in the ruthenium, rhodium, and rhenium transition metal complexes containing carbon monoxide, olefins, and hydrides as ligands. Back to England in 1955, he was appointed the chair of Inorganic Chemistry at Imperial College in the University of London. The interest in transition metal complexes led him to work on homogeneous catalytic reactions such as olefin hydrogenation and hydroformylation.

The Wilkinson catalyst is the Rh(I) complex Cl Rh (I)[P(Ph)$_3$]$_3$ containing three bulky P(PH$_3$)$_3$ ligands, which are shielding completely the Rh(I) cationic center:

$$\text{Ph}_3\text{P} - \underset{\underset{\text{Ph}_3\text{P}}{|}}{\overset{\overset{\text{PPh}_3}{|}}{\text{Rh}}} - \text{Cl}$$

This complex is acting as fairly efficient olefin hydrogenation catalyst (Osborn *et al.*, 1966).

To become the "active species," at least one P(Ph)$_3$ ligand must be lost in order to make the metal atom coordinatively unsaturated with an "openside" where substrates (hydrogen and olefin) can easily enter and chemically interact. After the loss of one P(Ph)$_3$ ligand and interaction with H$_2$ and ethene, the putative coordination state of Rh passes from 4 to 6 because of formation of a dihydride and simultaneous coordination of an olefin molecule (Figure 5.4). From these considerations it is inferred that Cl Rh (I)[P(Ph)$_3$]$_3$ is only the precursor of the real active species and that the statement that the Wilkinson catalysts have a *defined structure* is valid only for the precursor species, since the structure of the really active and coordinatively unsaturated species is more the result of indirect consideration than of real characterization with physical methods.

In relation with homogeneous hydrogenation catalysis, the contribution of John Osborn (1939–2000) must be emphasized (Figure 5.5). John Osborn began studying for his PhD degree with Geoffrey Wilkinson at Imperial College in 1963 when he made his earliest contributions (Osborn–Wilkinson catalyst). In 1967 he was appointed as an assistant professor at Harvard University and was promoted to associate professor in 1972. In 1975 he moved at the Université Louis Pasteur in Strasbourg. He also showed that RhH(CO)(PPh$_3$)$_3$ was an effective and selective catalyst for the hydroformylation of olefins. In his commemoration of Osborn, Richard R. Schrock (2005 Nobel Prize) wrote:

> Over a period of thirty years, through a combination of synthesis and elaborate mechanistic studies, he and his coworkers elucidated basic reactions in organometallic chemistry (Schrock, 2001).

The discovery of the catalytic activity of the Wilkinson–Osborn compound stimulated in the 1970–2000 period the synthesis of a large variety of analogues

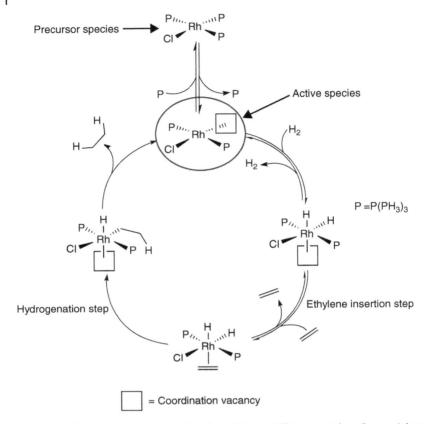

Figure 5.4 Olefin hydrogenation cycle of the Osborn–Wilkinson catalyst. *Source:* Adapted from Luo, Oliver, and McIndoe (2013).

containing different phosphine ligands with fluorine instead of hydrogen and was expected to have enhanced hydrogenation activity (Richter *et al.*, 2000).

One of the innovations associated with this research is that, by employing perfluorinated ligands, the catalyst can be contained in a perfluorocarbon phase while the substrate and product reside in a second organic phase. This elegant concept, applied, for example, to the rhodium-catalyzed hydroformylation (Horvath and Rabai, 1994) allows separation of the catalyst from the product, a fact not common in homogeneous catalysis.

The Crabtree catalyst and its modifications synthesized in the 1975–1985 period is another example of hydrogenation catalysts containing Ir(I). Like Wilkinson and Osborn, Robert H. Crabtree (1948–), professor of chemistry at Yale University, is undoubtedly another important contributor to the development of hydrogenation catalysts. He is known for the iridium-based

Figure 5.5 (a) Sir Geoffrey Wilkinson (1921–1996) and (b) John Osborn (1939–2000). Wilkinson in collaboration with Osborn synthesized the homogeneous hydrogenation Rh(I) complex known as Osborn–Wilkinson catalyst. *Source*: (a) The Wilkinson image is in the public domain: author smoke foot. (b) Schrock, 2001. Reproduced with permission of John Wiley & Sons.

Crabtree's catalyst for hydrogenations and for his famous textbook on organometallic chemistry.

Crabtres's catalyst

As in the case of the Wilkinson catalysts, the precursor structure is always fourfold square coordinated. However, the structure of ligands is different, and the Cl^- counter-anion is substituted by counter-anions $(PF_6)^-$ of more complex structures, with smaller coordination ability (Crabtree and Morris, 1977; Crabtree *et al.*, 1982). These compounds are more active than the parent Osborn–Wilkinson complex.

The enhancement in activity comes from the combination of two factors: the low coordinating character of the counter-anion (which allows a more facile olefin insertion) and the catalyst's longevity. In conclusion, there is no doubt that the Osborn–Wilkinson compound has been an ideal playground for the synthesis of new catalytic compounds, mainly obtained by modification of the structure of the ligands and of the counter-anion.

Soon after the Osborn–Wilkinson and Crabtree hydrogenation catalysts discoveries (which are not enantioselective), the need of chiral hydrogenation catalyst started to be felt. As the problem of chiral catalysis is complex, before starting to describe the advancements in this area, a short diversion about chirality and chiral compounds may be appropriate.

Already in 1815 Jean-Baptiste Biot (a French physicist) showed that certain chemical compounds are able to rotate the plane of a beam of polarized light, a property named "optical activity." Only in 1848, Louis Pasteur proposed that this property had a molecular basis originating from some form of molecular *dissymmetry*, while the term *chirality* was coined by Lord Kelvin a year later. The real origin of chirality was only determined in 1874 by van 't Hoff and Joseph Le Bel who independently proposed the tetrahedral geometry of carbon compounds and understood that the arrangement of groups around this tetrahedron was the real source of the optical activity and chirality of the resulting compound. Emil Fischer performed what can now be regarded as the first example of chiral synthesis, that is, the enantioselective elongation of sugars via a process, which is now known as the Kiliani–Fischer synthesis (Fischer and Hirschberger, 1889). A substance is defined *homochiral* if all the constituent molecules are of the same chiral form (enantiomer). Homochirality is a common property of natural amino acids and sugars. The origin of this property, not fully understood, will be discussed in the next chapter. It is unclear if the homochiral character of natural compounds has a purpose, and it has been even proposed that it can be a form of information storage (Julian, Myung, and Clemmer, 2005).

It is interesting that different enantiomers of a given compound often react differently with the various reagents. For instance, the aspartame, an artificial, non-saccharide sweetener used as a sugar substitute in some foods and beverages, has two enantiomers and only the L-aspartame is responsible for the sweet taste. Also it has been shown that in some antidepressant drugs, which are sold as racemic mixtures, only one enantiomer has beneficial effects (Lepola, Wade, and Andersen, 2004; Hyttel *et al.*, 1992). In the same way in penicillamine (that exists in two optical forms) one isomer is toxic (Jaffe, Altman, and Merryman, 1964). The most known case of the occurrence of two enantiomers, one almost lethal and the other safe, is that of thalidomide, a medicine that in the postwar era was marketed as a tranquilizer and, afterward, used against nausea in pregnant women. In Germany shortly after the drug was in the market, between 5000 and 7000 infants were born with malformation of limbs (phocomelia), and worldwide, about 10,000 cases were reported. Only 50% of the infants survived. The negative effects of thalidomide stimulated the introduction of more stringent drug regulations and control (Mcbride, 1961). In relation to the necessity to synthesize pure enantiomeric drugs, it is worth noting that in 2001 Ryōji Noyori in his Nobel lecture was complaining that

> ... *even in the early 1990s, about 90% of synthetic chiral drugs were still racemic.*

It can be stated with confidence that following the thalidomide disaster (Cushny, 1903; Cushny and Peebles, 1905), the development and licensing of chiral drugs changed dramatically in more positive way.

However, enantioselective synthesis is difficult to achieve. In fact enantiomers possess identical energies, and hence, in the absence of specific catalysts, the enantiomers are produced in equal quantities so leading to a racemic mixture. A common solution to obtain a single enantiomer is to introduce a "chiral" feature able to promote its formation over the other, via interactions at the transition state level (asymmetric induction). Shortly we will describe in detail the methods utilized in the asymmetric hydrogenations and asymmetric oxidation.

One important question is the origin of the chiral molecules that can be used to promote the chiral synthesis. The question is how, in the absence of some form of a homochiral template, the reactions occurring in prebiotic conditions have produced enantiomeric products instead of racemic mixtures. Experimental investigations have yielded several viable models invoking both chemical and physical mechanisms for amplification of enantiomeric excess (Hein and Blackmond, 2012; Jaakkola, Sharma, and Annila, 2008). The chirality question will be more extensively tackled in the context of the branch of science named prebiotic chemistry, which is treated in the last chapter.

Coming back to the Wilkinson and Crabtree's catalyst, we recall that, like the heterogeneous counterparts, they catalyze the hydrogenation reaction

$$CRR'{=}CH_2 + H_2 \rightarrow HCRR'{-}CH_3$$

However, also in this instance only racemic mixtures are formed. When the growing importance of L-enantiomers in pharmaceutical chemistry became evident, the important task now was to create enantioselective catalysts. The chemist community and homogeneous catalysts, owing to their highly tunable structure, appeared as the candidates able to take on the challenge.

It is precisely in this area that Knowles, Noyori, and Sharpless (2001 Nobel Laureates) brought substantial innovations. In particular, in 1968 Knowles, being informed of the results of Shiro Akabori and coworkers on asymmetric hydrogenation on Ni (Akabori and Izumi, 1962) and having realized that key to asymmetric synthesis is the structure of the chiral ligand, substituted the achiral $P(Ph)_3$ ligand in the Osborn–Wilkinson catalyst with the ancillary "chiral" ligand $P(Ph)(Me)(Propyl)$ (Figure 5.6). In this way he obtained a new hydrogenation catalyst with chiral activity.

Figure 5.6 (a) William Knowles (1917–2012), (b) Ryōji Noyori (1938–), and (c) Henry Boris Kagan (1930–). They are the main protagonists of the chiral hydrogenation catalysis. Knowles and Noyori were awarded the 2001 Nobel Prize. *Source*: (a) The Knowles image is in the public domain. (b) Courtesy of Archivio Storico University of Bologna. (c) Reproduced with permission of Kagan.

Although the initially obtained asymmetric yields were only about 15%, this modest result was extremely encouraging because it demonstrated that the strategy was correct. The biography of William Standish Knowles (1917–2012) is interesting because it shows that the career of this outstanding researcher developed completely at the Monsanto industry where he began his asymmetric project in the mid-1960s. That project developed a catalyst that synthesized individual enantiomers of chiral compounds, removing the need for costly and time-consuming separations of racemic mixtures (Knowles, 1983).

Knowles was also the first to use enantioselective metal catalysis from laboratory scale to industrial-scale synthesis; in fact by using the DIPAMP ligand (a chelating diphosphine, whose structure is illustrated in the following), he developed a chiral hydrogenation step in the L-DOPA production process.

DIPAMP ligand

In his Nobel lecture (2001), Knowles (2002) summarized the relevance of his discoveries saying that

> *these soluble hydrogenation catalysts have started a new era in catalytic processes. Since we are now dealing with pure complexes we can design something to do just the job you want.*

Knowles retired in 1986, although he continued to consult for several more years.

Following Akabori and Knowles, it is evident that the success in the synthesis of an asymmetric catalyst is largely determined not only by the metal center but also by the structure of the ancillary chiral ligands. These key points were tackled by Ryōji Noyori (2001 Nobel Laureate) (Noyori, 2002), who developed novel asymmetric hydrogenation catalysts based on Ru and on the new efficient ligand BINAP (Figure 5.6).

The research in this area was strongly encouraged in 1992 by the Food and Drug Administration in the United States, which introduced a guideline regarding "racemic switches," in order to favor the diffusion of clinical drugs consisting of single enantiomers. Following Noyori (2002), the chiral ancillary ligands

> ... *must possess suitable and sophisticated three-dimensional structures and functionality to generate sufficient reactivity and the desired stereoselectivity for the reaction of reactant and substrate* ...

To perform this function the design of the ligand is the key step. The ligand that allowed a substantial improvement in the synthesis of chiral molecules in hydrogenation of functionalized ketones is BINAP, which is a tridimensional chelating ligand that can be produced in R and S forms.

The BINAP chelates the Ru Cl_2 L_2 (L = solvent molecules) with formation of a complex structurally similar to the Osborn–Wilkinson and Knowles catalysts' base on Rh. As usual, the complex is only the precursor of the active species because in the presence of H_2, one Cl atom is substituted by one hydride groups. Instead of the two original L ligands, the complex can coordinate two further groups, for instance, a diketone, which is then hydrogenated:

R and S BINAP ligands

In his biographical notes, Ryōji Noyori describes how his love for chemistry was stimulated at the age of 12 by a conference on nylon where the lecturer from DuPont company was explaining, in an advertizing way, that this new fiber could be synthesized from coal, air, and water. Following this narration, he was very much impressed by the power of chemistry because it can create important products from almost nothing! He became student at Kyoto University, then instructor in the research group of Hitoshi Nozaki, and later associate professor at Nagoya University. He spent a postdoctoral period at Harvard where he interacted with John Anthony Osborn who at that time was assistant professor at Harvard. This interaction greatly enhanced the interest of Noyori in asymmetric hydrogenation catalysis. In the same period he became acquainted with Barry Sharpless (2001 Nobel Prize). After the Harvard period he returned to Nagoya, becoming a full professor in 1972. He is still based at Nagoya, though he is also now president of RIKEN, a multisite national research initiative.

Before abandoning the asymmetric hydrogenation catalysis, it is appropriate to illustrate also the contribution of Henry Boris Kagan, since he published the first example of a chiral bidentate diphosphine ligand (named DIOP) in 1971 (Dang and Kagan, 1971) (Figure 5.6). This achievement led to the discovery of many related ligands in a variety of practical applications.

DIOP ligand

Kagan's example of an efficient C_2 symmetry ligand stimulated the early development of asymmetric catalysis (Kagan and Dang, 1972; Kagan, 1975). He described also in 1973 the first example of a chiral "homogeneous" supported catalyst, by the synthesis and use of a DIOP fragment covalently bounded to the "Merrifield resin" (a polystyrene resin based on a copolymer of styrene and chloromethyl-styrene, which has been invented by Robert Bruce Merrifield, 1984 winner of the Nobel Prize in Chemistry). Today Kagan (1930–) is emeritus professor at the Université Paris-Sud in France. There are few doubts that he was one of the pioneers of asymmetric catalysis, and for this contribution he has received many important awards. Unlike Knowles and Noyori, he was not awarded the Nobel Prize in 2001. On that occasion, the science minister of France wrote to the Nobel Foundation that Henri Kagan was the pioneer of catalytic asymmetric synthesis, citing his 1971 paper on the catalytic asymmetric hydrogenation of olefins. Kagan stated that he was not behind the science minister's initiative and would not comment on the matter (Goodman, 2001).

In conclusion, we can say that after the contributions of Knowles, Nojori, and Kagan, enantioselective catalysis (known traditionally as asymmetric catalysis) through the use of coordination complexes as catalysts rendered chiral by the insertion of chiral ligands (chiral auxialiaries) in the coordination sphere (Glorius and Gnas, 2006; Evans, Helmchen, and Rüping, 2007) became a very popular branch of homogeneous catalysis. Today most enantioselective catalysts are effective at low concentrations (Heitbaum, Glorius, and Escher, 2006; Blaser and Schmidt 2004) and are well suited to industrial-scale synthesis.

5.4 Asymmetric Epoxidation Catalysts

In the field of asymmetric catalysis, epoxidation of olefinic molecules occupy a central position. Barry Sharpless demonstrated that in the presence of the ancillary chiral ligand, the complex [Ti(OPr)$_4$] is able to perform enantioselective partial oxidations of prochiral unsaturated substrates (for instance, allylic alcohols) (Katsuki and Sharpless, 1980) (Figure 5.7). The asymmetric induction is achieved by adding an enantiomerically enriched tartrate (diethyl tartrate (DET)) derivative using an organic peroxide (for instance, t-Bu-OOH) as oxidant following the scheme (Sharpless, 2002)

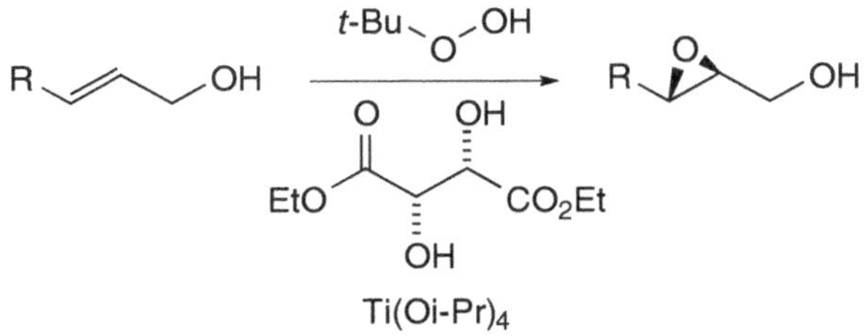

Enantioselective epoxidation reaction

Depending on the chiral character of the DET, enantiomeric epoxides are generated with high yield. Interesting, the addition of molecular sieves (zeolite 3A) enhances the reactivity. The Sharpless epoxidation catalyst's success is due to two major features. First, the products frequently have enantiomeric excesses above 90%, and second, epoxides can be easily converted into diols, aminoalcohols, or ethers. For this reason chiral epoxides play a very important role in the synthesis of natural products.

The breakthrough came when M. G. Finn's thesis results (Woodard, Finn, and Sharpless, 1991; Finn and Sharpless, 1991) reported that the compound [Ti(DIPT) (O-i-Pr)$_2$]$_2$ (diisopropyl tartrate: DIPT) is from 5 to 10 times more active than Ti(OR)$_4$ as a catalyst for the formation of racemic epoxy alcohol. The tartrate-induced acceleration of the titanium-catalyzed epoxidation reaction opened the way toward the use of DET as ancillary ligand and toward many applications for chiral synthesis of different compounds (Pedersen $et\ al.$, 1987). The structure of the catalyst is still uncertain, although all studies point toward a dimeric [Ti(tartrate)(OR)$_2$] form. The putative structure is as follows:

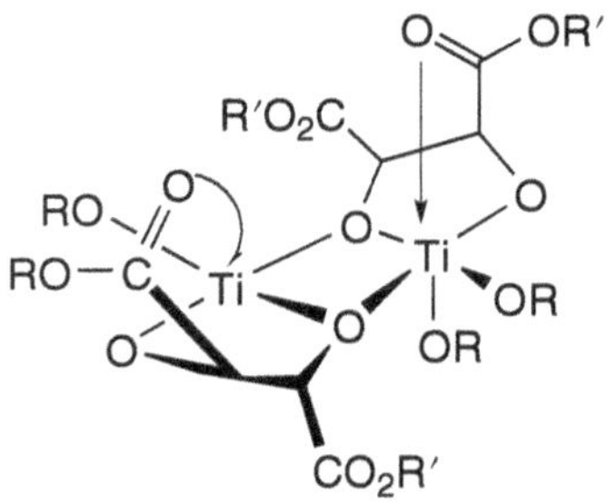

**Putative structure of the
Sharpless epoxidation catalyst**

Figure 5.7 (a) Karl Barry Sharpless (1941–), (b) Eric Jacobsen (1960–), and (c) Maurice Brookhart (1943–). They were the protagonists of chiral epoxidation and post-metallocene catalysis. *Source*: (a) Reproduced with permission of Sharpless. (b) Reproduced with permission of Jacobsen. (c) https://commons.wikimedia.org/wiki/File :Brookhart_Maurice.jpg. Used under CC BY-SA 3.0 https://creativecommons.org/licenses /by-sa/3.0/deed.en.

Sharpless, born in Philadelphia (1941), earned his PhD from Stanford University in 1968 and continued postdoctoral work at Stanford and Harvard universities. He was appointed professor at MIT in the 1970–1977 and 1980–1990 periods as well as at Stanford University in 1977–1980. He currently holds the W. M. Keck professorship in chemistry at the Scripps Research Institute. In the 1990–2000 period he also developed the OsO_4 catalyzed asymmetric dihydroxylation using dihydroquinine (DHQ) as ancillary ligand. He was awarded the Nobel Prize (with Noyori and Knowles) in 2001. While in MIT Sharpless collaborated with Eric N. Jacobsen (1960–), who has later developed new enantioselective epoxidation catalysts (Jacobsen catalysts) (Jacobsen *et al.*, 1991) whose usefulness is demonstrated by aiding the synthesis of biologically important active compounds like the important anticancer agent Taxol (Deng and Jacobsen, 1992). Jacobsen, after gaining his PhD in 1986 at the University of California, Berkeley, joined the laboratory of Barry Sharpless at MIT as a postdoctoral fellow. He spent a period at the University of Illinois at Urbana–Champaign before relocating to Harvard in 1993 where he was Sheldon Emery Professor of Chemistry (Figure 5.7). He is a prominent contributor to the field of organic chemistry and is best known for the development of the famous Jacobsen epoxidation catalyst whose structure is shown in the following:

Jacobsen epoxidation catalyst

It is generally accepted that after the addition of the simple oxidant (NaOCl) to the Mn(III) system, O=Mn(V) is formed and is thought to be the active oxidant species.

It is thought that the substrate is approaching the metal-oxo O=Mn(V) bond over the diamine bridge, where the bulky tert-butyl groups on the periphery of the ligand do not inhibit the alkene's approach. This mechanism, which was originally proposed by John Groves (Groves and Nemo, 1973), is commonly referred to as a "side-on perpendicular approach." Although the overall mechanism and the pathway of alkene approach are still debated, there is no doubt about the supramolecular character of the catalysts. In other words, it can be once more inferred that, in order to perform complex and highly selective pathways, the catalysts must possess a complex tridimensional structure with open windows to allow the interaction with the substrates.

5.5 Olefin Oligomerization Catalysts

In the previous chapter we reported on the metallocene polymerization catalysts whose primogenitor is the Cp_2ZrCl_2 compound discovered by Sinn and Kaminsky. This catalyst is homogeneous in origin, but due to the fast polymerization reaction, it becomes quickly supported on the polymer and thus becomes heterogeneous.

This compound when treated with MAO is able to efficiently polymerize ethylene and propylene with formation of long polymeric chains. In other words, this catalyst is favoring ethylene insertion into the growing chain more than chain termination via β-*elimination*. Recall from Chapter 3 that this property is also shared by Ti in the Ziegler–Natta catalysts.

The catalytic oligomerization of ethylene is nevertheless a topic of considerable and growing interest from both the industrial and academic point of view. In fact the demand of linear α-olefins up to C_{10} is growing very fast. For these reasons the search of efficient catalysts for ethylene oligomerization has been and is still representing an important research line. The research has been is based on two concepts: (1) change of the metal center with special interest toward late transition metals like Co, Ni, Pd, and Rh and (2) modification of the ligand sphere. The state of the research in this area is well documented of the review by Speiser, Braunstein, and Saussine (2005).

The main contribution in this area comes from M. Brookhart (1943–) (Svejda, Johnson, and Brookhart, 1999; Svejda and Brookhart, 1999), professor of chemistry (1990–) at the University of North Carolina (Figure 5.7). His research is well known for the development of the so-called post-metallocene catalysts:

Classical Brookhart catalyst:
M = Ni; X = halogen

A second major focus of Brookhart's group concerns the C–H and C–C bond activation by transition metal complexes and their incorporation into catalytic cycles. For his contributions in the aforementioned areas, Brookhart has received many awards.

An interesting property of these catalysts is the ubiquitous role of Ni(II) whose ability to favor ethylene dimerization is known since the Karl Ziegler's time. As we have already reported in Chapter 3, during the investigation on the catalytic properties of aluminum alkyls in 1953, Erhard Holzkamp, a Karl Ziegler student, observed that when ethylene reacted with triethylaluminum, a smooth dimerization to butene occurred instead. It was found that traces of nickel in the autoclave and traces of acetylene were responsible for the formation of soluble dimerization catalysts of unknown structure (Wilke, 2003). Due to properties of Ni(II) and to the structure and electron-donating properties of the ligand sphere, the β-elimination is favored (Ittel, Johnson, and Brookhart, 2000). From this it can be safely inferred that both the electronic structure of the metal center (preferably a late transition metal in low oxidation state like Ni^{2+}) and the fine-tuning of the coordination sphere are important.

5.6 Organometallic Metathesis

In 2005 Robert Howard Grubbs (1942–) and Richard R. Schrock (1945–) were awarded with the Nobel Prize in Chemistry together with the French chemist Yves Chauvin (1930–2015) for their fundamental contributions to the understanding of olefin metathesis, a class of important organic reactions that has completely changed the organic synthetic chemistry in the second half of the last century. Olefin metathesis reaction using both homogeneous and heterogeneous catalysts is now employed in more than 300 patented applications in the petroleum, rubber, and chemical industries. It is used to make all kinds of products – from tire rubber to detergents and plastic auto parts to perfumes.

Chauvin first proposed the widely accepted mechanism of alkene metathesis using transition metal complexes as catalysts, and then Schrock and Grubbs developed new organometallic catalysts that carried out the reactions more efficiently.

The term "metathesis" is widely used in chemistry to describe reactions in which two small units of a molecule are interchanged between pairs of such molecules. The observation on metathesis reactions goes back to 1955 when a homogeneous Ti(II)-catalyzed polymerization of norbornene was observed by Anderson and Merckling (1955). In the 1950–1970 period, many observations concerning metathesis reactions coming from petrochemical companies utilizing heterogeneous catalysts were appearing. In Chapter 4 we reported on the early observations by Eleuterio (DUPont), Banks, and Bailey (Phillips Petroleum). In 1966 the chemist Nissim Calderon, working at the Goodyear Tire & Rubber Company, discovered another strange reaction (Calderon *et al.*, 1967, 1968) that lead to the redistribution of olefin fragments by scission and regeneration of carbon–carbon double bonds.

The Goodyear researchers named the reaction "olefin metathesis" from the Greek word μετάθεσις, meaning put in a different order. At about the same time, Johannes C. Mol and others at the University of Amsterdam, in the Netherlands, independently reached the same conclusion with propylene and carbon-14-labeled propylene in the presence of a heterogeneous catalyst (Mol, Moulijn, and Bollhouwer, 1968). The race to explain the reaction was on. According to a once-popular hypothesis, which was favored by Calderon and is sometimes referred to as the conventional mechanism, a cyclobutane intermediate complex with the metal could be formed (Calderon *et al.*, 1968). In 1971, Roland Pettit, chemistry professor at the University of Texas, proposed a different explanation, that is, the formation of a tetramethylene complex, in which four methylene units are bonded to a central metal atom (Lewandos and Pettit, 1971). The time was thus mature for further advancements.

In 1971 Yves Chauvin in collaboration with his student Hérisson (1970) explained for the first time metathesis in detail, even though he did not realize immediately the importance of his explanation (Hérisson and Chauvin, 1971). Chauvin showed also that the reaction involves two double bonds (Figure 5.10).

In particular, Chauvin and Hérisson suggested that olefin metathesis is initiated by a metal carbene (indicated as CM or M=). The proposed reaction scheme is reported in Figure 5.8.

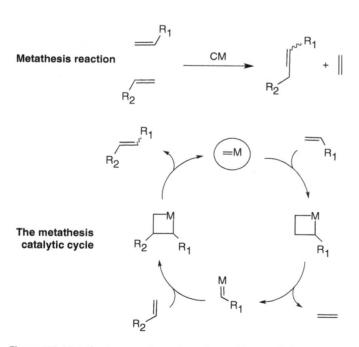

Figure 5.8 Metathesis general reaction scheme. M= stands for the metal carbene.

The catalysts used by Chauvin and Hérisson were $WCl_6/Sn(CH_3)_4$, $WOCl_4/Sn(C_4H_9)_4$, and $WOCl_4/Al(C_2H_5)_2Cl$. Chauvin assumed that these precursors are transformed into W=CHR upon contact with the reactants and that this metal carbene is the real catalytic species (Soufflet, Commereuc, and Chauvin, 1973). Chauvin's description of metathesis led Grubbs and Schrock to develop catalysts that carried out the reaction more efficiently.

Chauvin has always claimed that three papers published in 1964 led him to formulate his hypothesis on the presence of a metal-carbene intermediate in the olefin metathesis reaction. In relation to this, the first example was a compound with a metal–carbon double bond discovered in 1964 by the 1973 Nobel Laureate Emil Otto Fischer (Fischer and Maasböl, 1964). The same as for coordinated CO in carbonylic complexes, competition for the d-electrons of the metal takes place between π bonding ligands and the empty p-orbital on the carbon of the $H_2C=$ group. The valence state of the metal center and the electron-donating character of the ligands determine the extent of the d-π overlap in $M=CH_2$. The second came from Giulio Natta's description of the ring-opening polymerization of cyclopentene with triethylaluminum and hexachlorotungsten (Natta, Dall'Asta, and Mazzanti, 1964), and the third was the paper on the disproportionation of propylene from Banks and Bailey (1964) at the Phillip Petroleum Company.

The papers of Natta and of Banks and Bailey indicated that cyclopentene polymerization and propylene disproportionation are the same reaction. Therefore Chauvin assumed that they must involve the same type of intermediate species and hypothesized that the intermediate products could be metal carbenes. Slowly, support to Chauvin's hypothesis started to come from various laboratories. It is not possible to discuss all contributions in this field. We only mention here the papers of Michael Franz Lappert (professor of chemistry at the University of Sussex) and coworkers concerning rhodium-based catalysts (Cardin, Doyle, and Lappert, 1972; Cardin *et al.*, 1973) and the results obtained by Charles P. Casey and Terry J. Burkhardt at the University of Wisconsin, thanks to the discovery that the (diphenylcarbene)-pentacarbonyl-tungsten reacts with isobutene to form as major product a new olefin, 1,1-diphenylethene (Casey and Burkhardt, 1974). Further confirmation of Chauvin's hypothesis followed from several other laboratories, for instance, from that of Samuel J. Danishefsky, professor at Columbia University and at the Memorial Sloan Kettering Institute (Danishefsky and Kitahara, 1974) and from that of the Frederick D. Lewis group at Northwestern University (Lewis, Hoyle, and Johnson, 1975). Thomas J. Katz, professor of chemistry at Columbia University, and his graduate student James McGinnis (Katz and McGinnis, 1975) were the first, however, to unambiguously substantiate in 1975 the Chauvin carbene mechanism for the olefin metathesis, suggesting that initiators for the reactions be sought among isolable metal-carbene complexes.

Katz reported the first use of an isolable metal-carbene complex (diphenyl-carbene)-pentacarbonyl-tungsten to initiate the metathesis of unsymmetrically substituted ethylenes (Katz, Lee, and Acton, 1976; Katz and Lee, 1980; Katz *et al.*, 1980; Katz, 2006; Katz and Sivavec, 1985).

Despite Katz's seminal papers, the personage who gave real concrete support to Chauvin's theoretical ideas about metathesis was Robert Howard Grubbs, who, after gaining his PhD in 1968 at Columbia University, under the supervision of Ronald Breslow (known for his outstanding contributions to biomimetic chemistry), started to be interested in metathesis when appointed to the University of Michigan (Grubbs, 1972) (Figure 5.10). In 1978 he moved to the California Institute of Technology, where his interests became mainly dedicated to the group of catalyst carrying his name for olefin metathesis. These catalysts perform a complete new series of reactions that entail the redistribution of fragments of alkenes by opening and regenerating carbon–carbon double bonds. Grubbs also contributed to the development of the so-called living polymerization catalysts (Bielawskia and Grubbs, 2007).

The Grubbs's catalysts are ruthenium(II) carbenoid complexes (Grubbs, 1972, 2004; Scholl *et al.*, 1999), which, in addition to their broad usage in academic research, are now used commercially to prepare new pharmaceuticals, composites for structural applications, and for the conversion of biorenewable carbon sources into fuels and commodity chemicals (Figure 5.9).

Catalysts for other useful transformations were also developed and studied by Grubbs's group at Caltech. One postdoctoral fellow, Dr. Akira Miyashita, had a major role in collaborating with Grubbs in his metathesis activity. Miyashita prematurely died in 2004, but he had in fact moved with Grubbs from Michigan to Caltech in 1978 and contributed a number of new processes as member of Grubbs's group (Grubbs and Miyashita, 1977; Grubbs *et al.*, 1978).

In 1978, when Grubbs arrived at Caltech, another high level DuPont chemist, Fred Tebbe (1935–1995), reported on the structure and reactivity of a complex

Coordination vacancies

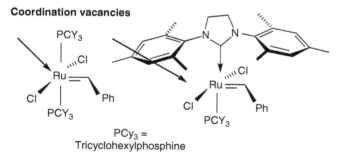

PCy$_3$ =
Tricyclohexylphosphine

Figure 5.9 Typical Grubbs carbenoid complexes. The metal center is passing from fivefold to sixfold coordination during the reaction cycle. In the figure PCy$_3$ stands for tricyclohexylphosphine.

later named "Tebbe reagent," an organometallic complex that contains both a titanium and an aluminum metallic centers, linked by a pair of bridging ligands (Tebbe, Parshall, and Reddy, 1978). In his Nobel lecture the contribution of Tebbe was recognized by Grubbs' who showed that the Tebbe reagent provided the basis for the synthesis of the metallacyclobutane intermediate, a derivative of cyclobutane in which a methylene group was replaced by a metal atom.

Other studies of Grubbs's group explored the mechanism of the Ziegler–Natta catalysts in the ethylene and propylene polymerization and thus determined the critical steps in this reaction (Tritto and Grubbs, 1990; Clawson *et al.*, 1985).

More bound to the industrial organization (DuPont) was the second putative father of the metathesis synthesis, Richard Royce Schrock (1945–), an American chemist who received his PhD in inorganic chemistry from Harvard University in 1971 under the supervision of Professor John Anthony Osborn, to whom he was deeply dedicated and wrote a very moving obituary upon his death (Shapley, Schrock, and Osborn, 1969; Schrock and Osborn, 1971; Vengrovius, Sancho, and Schrock, 1981) (Figure 5.10).

Tebbe reagent

In 1975 Schrock joined MIT and became a full professor in 1980.

At DuPont Schrock investigated the synthesis of tantalum alkylidenes, a class of compounds that turned out to be a crucial for understanding the catalytic cycle of olefin metathesis. His later work at MIT was devoted to the study of a group of molybdenum alkylidenes and alkylidyne catalysts, which are active in olefin and alkyne metathesis, respectively. Schrock succeeded in demonstrating that metallacyclobutanes and metallacyclobutadienes are the key intermediates in olefin and in alkyne metathesis, respectively.

The prototypical Schrock catalyst is $(R''O)_2(R'N)Mo(CHR)$, where R = tert-butyl, $R' = 2,6$-diisopropylphenyl, and $R'' = C(Me)(CF_3)_2$.

Prototypical Schrock catalyst structure

Figure 5.10 (a) Yves Chauvin (1930–2014), (b) Robert Grubbs (1942–), and (c) Richard Schrock (1945–). Protagonists of metathesis catalysis, they were awarded the 2005 Nobel Prize in Chemistry. Chauvin image by courtesy of CPE Lyon. The Grubbs and Schrock images are in the public domain (authors Saibo and A. mela, respectively).

Such catalysts are now commercially available from several chemical industries and are normally used in synthetic applications of olefin metathesis. Schrock's work is still ongoing in regard to understanding the metathesis selectivity and to developing new catalyst architectures. Besides metathesis, his work is today oriented toward elucidating the mechanism of dinitrogen fixation in order to develop homogeneous single-molecule catalysts able to form ammonia from dinitrogen and hydrogen, mimicking the activity of nitrogenase enzymes in biology.

In 1990 Schrock and his associates synthesized a class of metal-containing compounds, in particular titanium, called Schrock carbenes, which Schrock had been developing since the 1970s. An example of the titanium-based carbenic structure is shown in the following scheme:

Typical Schrock carbenic structure

Last, the extension of metathesis catalysis to biological systems must be briefly mentioned. One of the first to realize the importance of extending metathesis research to biological systems was the German chemist Sabine Laschat, professor of organic chemistry at the University of Stuttgart. Laschat has performed important research on the chemo-enzymatic syntheses of natural products, investigating their mode of action for the development of novel ligands in asymmetric catalysis (Rabe *et al.*, 2009) as well as for the synthesis of biocompatible hydrogels and thermotropic liquid crystals (Sauer *et al.*, 2009).

An important coworker of Sabine Laschat has been Nicolai Cramer whose main research program encompasses enantioselective metal-catalyzed transformations and their implementation for the synthesis of biologically active molecules (Cramer *et al.*, 2005).

5.7 Cross-Coupling Reactions

Another group of catalytic reactions of outstanding importance in organic synthesis concerns cross-coupling reactions that lead to the formation of C—C bonds. These include the Heck, Negishi, Suzuki, Sonogashira, Stille, and Hiyama reactions as we describe below:

Heck	$R_1 - X + R_2C=C-H \rightarrow R_2C=C-R_1 + HX$
Negishi	$R_1 - ZnX + R_2 - X \rightarrow R_1 - R_2 + ZnX_2$
Suzuki	$R_1 - B(OH)2 + R_2 - X \rightarrow R_1 - R_2 + XB(OH)_2$
Sonogashira	$R_1 - X + HC \equiv CR \rightarrow R_1 - C \equiv C - R + HX$
Stille	$R_1 - Sn(R)3 + R_2X \rightarrow R_1 - R_2 + Sn - (R)3X$
Hiyama	$R-1-X + R_2 - Si(R)3 \rightarrow R_1 - R_2 + X - Sn(R)3$

In these reactions, R_1 and R_2 represent unlike alkyl, alkenyl, or aryl groups and X is an halogen. All are reactions of the C–C coupling type. They all occur in the presence of a Pd(0) homogeneous catalyst. The common compound is $Pd(PPh_3)_4$ tetrakis(triphenylphosphine) palladium(0), first prepared by Lamberto Malatesta and his group in Milan in the 1960s (Malatesta and Angoletta, 1957). This complex reversibly dissociates PPh_3 ligands in solution, so reactions attributed to $Pd(PPh_3)_4$ arise from $Pd(PPh_3)_3$ or even $Pd(PPh_3)_2$. This is the reason why in the most widely accepted mechanisms the catalyst is assumed to be $Pd(PPh_3)_2$.

Richard F. Heck (1931–) after obtaining his PhD in 1954 at UCLA spent some time at the Swiss Federal Institute of Technology in Zurich, where he worked with Professor Vladimir Prelog, who later became a 1975 Nobel Laureate (Figure 5.11). In 1955 Heck returned to UCLA. His first project involved Ziegler–Natta catalysts, a topic where he gained experience in transition metal chemistry. He then began (1958) an investigation of the hydroformylation reaction, which resulted in a rationalization of the mechanism of this reaction. This research was an overwhelming success and was the first demonstration of C—C cross coupling. In 1971 Heck accepted a position at the University of Delaware. In the following years Heck worked toward developing more "user friendly" catalysts. From these studies palladium phosphines emerged as potential catalysts. This discovery, reported in 1972 (Heck and Nolley, 1972), formed the basis of the reaction known as Heck reaction (Heck, 1977). Today the Heck reaction is an important tool for all organic chemists. Heck retired from the University of Delaware in 1989, where he is now emeritus professor. Besides the Nobel Prize in 2010, he has been the recipient of many other awards.

The second protagonist in C—C cross coupling is the Japanese chemist Ei-ichi Negishi born in 1935 in Changchun, China (Figure 5.11). He returned to Tokyo after World War II (1945), where he would receive his Bachelor of Engineering in 1958. A turning point in his professional career was a Fulbright Scholarship in 1960 and the associated PhD degree in synthetic organic chemistry at the University of Pennsylvania. In 1966 he became a postdoctoral researcher at Purdue University and later assistant professor in 1968, working with Nobel

(a)

(b)

(c)

Figure 5.11 (a) Richard F. Heck (1931–), (b) Ei-ichi Negishi (1935–), and (c) Akira Suzuki (1930–). They were awarded the 2010 Nobel Prize in Chemistry for their discoveries *of palladium-catalyzed cross couplings in organic synthesis.* These reactions are known as Heck, Negishi, and Suzuki reactions. Other cross-coupling reactions discovered in the same period are the Sonogashira, Stille, and Hiyama cross-coupling reactions. *Source*: (a)(b)(C) Courtesy of Nobel Media AB.

Laureate Herbert C. Brown. In 1972, he went on to become associate professor at Syracuse University where, in 1979, he was promoted to full professor. In the same year, he returned to Purdue University where he spent the next four years investigating the possible uses of d-group transition metals as catalysts. From these studies the cross-coupling reaction today known as Negishi reaction was developed (Negishi, 1982, 2007). The catalyst is a palladium complex $Pd(0)L_n$ whereby the ligand L can be typically triphenylphosphine or BINAP. In 1999 Negishi was named the Herbert C. Brown Distinguished Professor of Chemistry at Purdue University. Among the many awards the 2010 Nobel Prize must be mentioned.

The third protagonist in C—C coupling is Akira Suzuki (1930–) (Figure 5.11). Born in Hokkaido, Japan, he entered Hokkaido University in Sapporo where he obtained his PhD. During the investigations he conducted for his thesis, he realized that organometallic compounds were versatile intermediates for organic synthesis. After completing his PhD he became research assistant and, in 1961, assistant professor at the newly founded Synthetic Chemical Engineering Department at Hokkaido University. In 1973, he succeeded Professor H. Otsuka in the Applied Chemistry Department. In total he spent 35 years at Hokkaido University. After retirement from the Hokkaido University in 1994, he served at two private universities. He finally retired from university work in 2002.

For Suzuki, a very important and inspiring experience was a two-year stay at Purdue University in Indiana (1963–1965) where he became familiar with the organoboron compounds and their chemistry. It is from this experience that the C–C coupling chemistry performed with organic compounds started and became important (Miyaura, Yamada, and Suzuki, 1979; Miyaura and Suzuki, 1995; Suzuki and Brown, 2003). As for the previous examples of C–C coupling reactions, the catalysts are preferentially Pd(0) complexes.

Other important contributions to the C–C cross coupling are due to Kenkichi Sonogashira (1931–) who discovered the coupling reaction carrying his name in 1975 when he was professor of chemistry at Osaka University (Sonogashira, Tohda, and Hagihara, 1975). Sonogashira was later professor at Osaka City University and retired in 2004. Another protagonist was John Kenneth Stille (1930–1989) professor at Colorado State University from 1977. Stille was killed at age 59 in the United Airlines Flight 232 crash at Sioux City, Iowa. To him we owe the "Stille reaction" (Milstein and Stille, 1978; Stille, 1986). Yet another contribution to C–C coupling reactions was in 1988 by T. Hiyama (Hatanaka and Hiyama, 1988). The Hiyama coupling is a palladium-catalyzed cross-coupling reaction of organo-silanes with organic halides. Today cross-coupling reactions are still a rapidly developing area.

We mentioned palladium-catalyzed arylation of carbonyl compounds and nitriles, amination, etheration, and thiolation, but these developments are only a few examples of the evolution of cross-coupling reactions. Some further topics under examination and awaiting innovation concern the use of new ligands substituting the tetrakis ligand. Other innovations are expected in Pd recovery, which would also depend on anchoring the catalysts on supports.

Before ending our discussion of the Pd-catalyzed cross-coupling reactions, it may be appropriate to add that a few scientists think that in the reaction mixture small colloidal particles of metal are formed and that the reaction is occurring on the surface. This could imply that the catalyst is only partially homogeneous and that ligand-stabilized metal nanoclusters have an active role. Of course the problem of potential formation of colloidal particles is not limited to the cross-coupling catalyst and can be a common situation for all catalysts based on noble metals like Pt, Rh, and Ir (Widegren and Finke, 2003).

5.8 Pd(II)-Based Complexes and Oxidation of Methane to Methanol

Besides Pd(0)-based catalysts, there are Pd(II) complexes that can act as catalysts as well. One of the main involved reactions is the oxidation of methane to methanol. We have already noted that partial oxidation of methane to methanol is a reaction that nature solves with enzymatic catalysis (methane monooxygenase) and that this is still a very challenging problem (Labinger, 2004; Holmen, 2009). The challenge has also stimulated a rich number of studies in the area of homogeneous catalysis.

Among the various catalytic paths leading to methane functionalization is the so-called Periana-Catalytica method. This method merits a brief mention because it can cause methane to partially oxidize to methanol in good yield (Labinger, 2004; Holmen, 2009; Mironov *et al.*, 2013). The reaction is catalyzed by a Pt(II) complex in concentrated sulfuric acid following the scheme

$$CH_4 + H_2SO_4 \xrightarrow[\text{H}_2\text{SO}_4, \ 220\ ^\circ\text{C}]{\text{(bpym)PtCl}_2} CH_3OH + SO_2 + H_2O$$

The catalyst is assumed to be the homogeneous complex

Pt^{II}-X

derived from the (bpym) PtCl$_2$ precursor

by addition of concentrated H$_2$SO$_4$. The mechanism is still the object of intense debate, so it will not be discussed in detail. Unfortunately, this interesting catalytic path, which combines the action of a precious metal complex and a strong Brønsted acid, is not economically practical owing in large part to the excessive quantities of sulfuric acid that would have to be handled and recycled. Its relevance is consequently confined only to the fundamental aspects, which must be faced in order to achieve highly selective alkane oxidation.

5.9 Non-transition Metal Catalysis, Organocatalysis, and Organo-Organometallic Catalysis Combination

In the previous paragraphs we have illustrated the terrific success of catalysis based on transition metal complexes in many reactions, including hydrogenation, metathesis, and cross couplings. As consequence of this success, it was general belief that transition metals are the essential components of homogeneous catalysis.

Transition metal catalysts are often air and moisture sensitive or are present as contaminants in products. In particular, due to the drawbacks associated with the noble metal separation, there is space for alternative approaches.

Therefore a number of groups have focused on non-transition metal catalysis and organocatalysis in which molecules not containing transition metals or fully organic are the catalytically active species. As consequence the development of alternative methods and concepts for performing highly selective reactions is becoming a major task in modern organic chemistry and indeed non-transition metal catalysis and organocatalysis have seen a rise in popularity in the 2000–2010 period when measured in terms of publications.

These catalyses interact with the substrate in a unique way. Whereas transition metal catalysts interact with substrates by coordination at vacant sites of the transition metal exploiting the d-electron manifold, non-transition metal and organocatalysts interact with the substrate either by forming reactive covalently bonded intermediates or by forming hydrogen-bonded or ion-pair complexes.

The definition of organocatalysis refers to a form of catalysis where the rate of a chemical reaction is increased by an organic compound consisting of carbon, hydrogen, sulfur, and other nonmetal elements like P and B (Berkessel and Groeger, 2005, 2007; Notz, Tanaka, and Barbas, 2004). The definition of non-transition metal catalysis is less restrictive because it includes purely organic catalysts and catalysis with molecular species containing alkaline, alkaline earths metals, and Al. A clean-cut division between the two types of catalysis is difficult to make, and two definitions are often used for the same catalysts.

In this section we attempt to concisely show that non-transition metal catalysis and organocatalysis are lively examples of transition metal-free catalysis and represent branches of catalysis science that are rapidly developing today.

To keep the discussion brief, only two cases will be illustrated, metal-free H_2 activation catalysis and amino catalysis.

5.9.1 Metal-Free Hydrogen Activation and Hydrogenation

The most important example of hydrogen activation on metal-free molecular species is the interaction of H_2 with frustrated Lewis pairs (FLPs). FLP is a combination of a Lewis acid and a Lewis base in which steric demands preclude the classical adduct formation.

Phosphine/borane combinations

$$R_3P \cdots B(C_6F_5)_3$$

where R are bulky ligands (typically $C_6H_2Me_3$) that are the simplest examples of FLPs (Stephan, 2010). In fact, because of the steric hindrance associated with bulky R_3 and $(C_6F_5)_3$ ligands, P and B cannot form a normal Lewis acid–Lewis pair adduct and can split hydrogen in a heterolytic way at very moderate temperature and pressure following the scheme

$$R_3P + B(C_6F_5)_3 + H_2 \rightarrow [R_3PH]^+[HB(C_6F_5)_3]^-$$

resulting in a phosphonium borate salt. The process is not reversible and hence cannot be utilized for hydrogenation reactions where activated hydrogen is hydrogenating a gas substrate. Yet it has been demonstrated that by simply changing the boron ligands, the process become reversible.

This observation has opened the way toward hydrogenation reactions. Indeed, if proton hydride transfer could occur from the phosphonium borate to a substrate, the resulting phosphine/borane would react again with H_2 and regenerate the salt and thus drive the catalytic hydrogenation of the substrate.

Before describing the hydrogenation reactions realized so far, it is appropriate to mention that fueled by this exciting discovery, a large variety of FLPs has been synthesized that cannot be described in detail in this book.

Many substrates have also been tested. The main success has been obtained using imines as substrates that are hydrogenated to amines. N-heterocycles like quinoline have also been reduced.

Despite the limited number of reduced substrates, it is a widespread conviction that the obtained results are opening a promising field and that today we are still in the infancy of nonmetal hydrogenation catalytic reactions. The reactivity of FLPs is not restricted to hydrogen activation. They also activate and CO_2 (Menard and Stephan, 2010). While the capture of N_2O results in a quite robust product, the capture of CO2 is reversible. FLPs containing AL instead of B are active in the reduction of CO_2 to methanol.

5.9.2 Amino Catalysis

In this section only one example of the very numerous reactions catalyzed by a purely organic compound is reported where the catalyst is represented by a simple molecule L-proline, which displays secondary amine functionality and carboxylic groups. A typical reaction is shown in the following scheme, where the methyl vinyl ketone reacts with 2-methylcyclohexane-1,3-dione 2 in the presence of L-proline 3, to give a more complex ketone with high yield:

Example of amino catalysis

Because of the very recent origin of this catalysis, it is difficult to present an exhaustive review. For the interested reader, we recommend the recent review of Bertelsen and Jørgensen (2009). We only mention that carboxylic and amine groups are often present (as active species) also on the channels of many enzymes. This seems to indicate that the gap between the metal and enzyme catalysis is now being gradually filled by an organocatalysis. In addition aminocatalysis based on proline and amino acids as catalysts is attracting much attention because it may help explain the role of catalysis in prebiotic chemistry (as it will be shown in Chapter 6).

Today the development of organo-organometallic catalysts represents a growing area. An example is shown by the combination BINOL-phosphoric acid (BINOL:1,1'-Bi-2-naphthol) with transition metal centers. First we recall that BINOL-derived phosphoric acid esters provide a class of strong, sterically tunable acids and that they are highly enantioselective organocatalysts currently being used in many acid-catalyzed reactions. Recently the combination of asymmetric BINOL–phosphoric acid

BINOL–phosphoric acid

with inactive neutral Ir complexes (not described for sake of brevity) has evolved as a new strategy to carry out enantioselective hydrogenations that could not be performed in a traditional way by employing separately one of the two catalysts. This approach represents an example of organo-organometallic catalysis combination with dual character (Rueping, Koenigs, and Atodiresei, 2010) and is reported here only to illustrate the versatilities and potentialities of the new methods.

5.10 Bio-inspired Homogeneous Catalysts

As we will see in Chapter 7, in nature metal porphyrin complexes are widely present in the active sites of enzymes (cofactors) responsible for catalytic aerobic oxidations, reduction and transport of dioxygen, and destruction of peroxides.

The structure of porphyrin complexes is similar to that of phthalocyanine metal complexes, which are purely synthetic. The similarity with the cofactors of enzymes has stimulated the study of the catalytic properties of synthetic metal porphyrin and phthalocyanine complexes (mainly of Fe(II), Cu(II), Co(II), Ni(II), and Zn(II)). These studies can be considered as forming the core of the chapter devoted to bio-inspired catalysis (Figure 5.12).

From the scheme of BINOL-phosphoric acid it is clearly inferred that the metal centers in the basic structures are only fourfold coordinated and hence possessing unsaturated coordinative positions. Furthermore the majority of metal ions can easily change the oxidation state. Taken together these facts make the metal centers potentially active in binding substrates (for instance, oxygen or hydrogen peroxide) and so show redox catalytic activity. This justifies why, although not industrially employed, the catalytic properties of metalloporphyrins have been intensively studied (Costas, 2011).

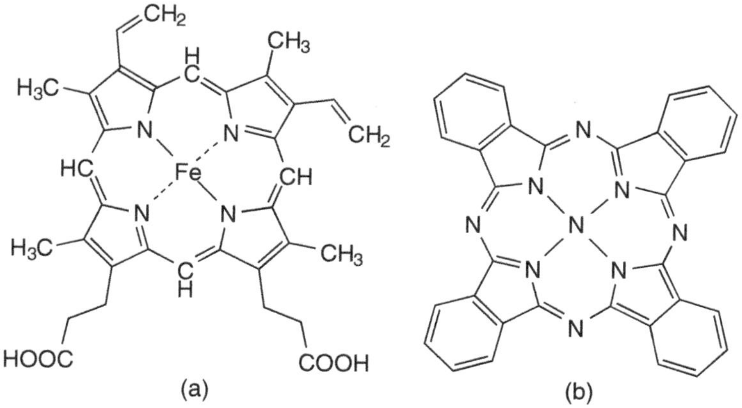

Figure 5.12 Basic structure of (a) metal porphyrin (heme group of hemoglobin) and (b) metal phthalocyanine.

The catalytic properties of metal phthalocyanines have also been widely investigated. These purely synthetic compounds are very attractive as catalysts not only because of structural analogy with porphyrin complexes but also because of their accessibility in terms of the cost and straightforward preparation on a large scale as well as their chemical and thermal stability. Furthermore these complexes are widely used in different fields of material science. The large variety of redox reactions catalyzed by metal phthalocyanines cannot be reviewed in this book. For more information, the interested reader is referred to the Alexander Sorokin review (2013) where the methods for immobilization of these molecular catalysts on solid substrates and in the cavities of microporous solids like zeolites are also described. To conclude, we should mention that the described complexes are only a small part of the large variety of porphyrinoid macrocyclic structures synthesized and investigated so far.

References

Akabori S. and Izumi Y. (1962) Method of preparing a raney nickel optically active hydroxy acid hydrogenation catalyst. Hyogo-ken, US 3,203,905.

Anderson A. W. Merckling (1955) Polymeric bicycle[2.2.1]-2-heptene. M. G. US patent. 2721.189, 1954; *Chemical Abstracts*, **50**, 3008.

Anderson, B.J., Keith, J.A., and Sigman, M.S. (2010) Experimental and computational study of a direct O_2-coupled Wacker oxidation: water dependence in the absence of Cu salts. *Journal of the American Chemical Society*, **132**, 11872–11874.

Baeckvall, J.E., Akermark, B., and Ljunggren, S.O. (1979) Stereochemistry and mechanism for the palladium(II)-catalyzed oxidation of ethene in water (the Wacker process). *Journal of the American Chemical Society*, **101**, 2411–24.

Banks, R.L. and Bailey, G.C. (1964) Olefin disproportionation: a new catalytic process. *Industrial & Engineering Chemistry Product Research and Development*, **3**, 170–173.

Berkessel, A. and Groeger, H. (2005) *Asymmetric Organocatalysis*. Wiley-VCH, Weinheim.

Berkessel, A. and Groeger, H. (2007) Asymmetric organocatalysis, special issue, organocatalysis. *Chemical Reviews*, **107**, 5413–5883.

Bertelsen, S. and Jørgensen, K.A. (2009) Organocatalysis – after the gold rush. *Chemical Society Reviews*, **38**, 2178–2189.

Bielawskia, C.W. and Grubbs, R.H. (2007) Living ring-opening metathesis polymerization. *Progress in Polymer Science*, **32**, 1–29.

Bigelow, M.H. (1947) Reppe's acetylene chemistry. *Chemical and Engineering News*, **25**, 1038–1042.

Blaser, H.U. and Schmidt, E. (eds) (2004) *Asymmetric Catalysis on Industrial Scale, Challenges, Approaches and Solutions*. Wiley-VCH Verlag GmbH & Co.

Calderon, N., Chen, H.Y., and Scott, K.W. (1967) Olefin metathesis - a novel reaction for skeletal transformations of unsaturated hydrocarbons. *Tetrahedron Letters*, **8**, 3327–3329.

Calderon, N., Ofstead, E.A., Ward, J.P. *et al.* (1968) Olefin metathesis I. Acyclic vinylenic hydrocarbons. *Journal of the American Chemical Society*, **90**, 4133–4140.

Cardin, D.J., Doyle, M.J., and Lappert, M.F. (1972) Rhodium(I)-catalysed dismutation of electron- rich olefins: rhodium(I) carbene complexes as intermediates. *Journal of the Chemical Society, Chemical Communications*, **16**, 927–928.

Cardin, D.J., Çetinkaya, B., Doyle, M.J., and Lappert, M.F. (1973) The chemistry of transition-metal carbene complexes and their role as reaction intermediates. *Chemical Society Reviews*, **2**, 99–144.

Casey, C.P. and Burkhardt, T.J. (1974) Role of metal-carbene complexes in cyclopropanation and olefin metathesis reactions. *Journal of the American Chemical Society*, **96**, 7808–7809.

Clawson, L., Soto, J., Buchwald, S.L. *et al.* (1985) Olefin insertion in a metal alkyl in a Ziegler polymerization system. *Journal of the American Chemical Society*, **107**, 3377–3378.

Comas-Vives, A., Stirling, A., Lledós, A., and Ujaque, G. (2010) The Wacker process: inner- or outer-sphere nucleophilic addition? New insights from ab initio molecular dynamics. *Chemistry - A European Journal*, **16**, 8738–8747.

Costas, M. (2011) Selective C–H oxidation catalyzed by metalloporphyrins. *Coordination Chemistry Reviews*, **255**, 2912–2932.

Crabtree, R.H. and Morris, G.E. (1977) Some diolefin complexes of iridium(I) and a trans-influence series for the complexes [IrCl(cod)L,]. *Journal of Organometallic Chemistry*, **135**, 395–403.

Crabtree, R.H., Demou, P.C., Eden, D. *et al.* (1982) Dihydrido olefin and solvento complexes of iridium and the mechanisms of olefin hydrogenation and alkane dehydrogenation. *Journal of the American Chemical Society*, **104**, 6994–7001.

Cramer, N., Laschat, S., Baro, A. *et al.* (2005) Enantioselective total synthesis of cylindramide. *Angewandte Chemie*, **44**, 820–822.

Cushny, A.R. (1903) Atropine and the hyoscyamines-a study of the action of optical isomers. *Journal of Physiology*, **30**, 176–94.

Cushny, A.R. and Peebles, A.R. (1905) The action of optical isomers: II. Hyoscines. *Journal of Physiology*, **32**, 501–510.

Dang, T.P. and Kagan, H.B. (1971) The asymmetric synthesis of hydratropic acid and amino-acids by homogeneous catalytic hydrogenation. *Journal of the Chemical Society D: Chemical Communications*, 481.

Danishefsky, S.J. and Kitahara, T. (1974) Useful diene for the Diels–Alder reaction. *Journal of the American Chemical Society*, **96**, 7807–7808.

Deng, L. and Jacobsen, E.N. (1992) A practical, highly enantioselective synthesis of the taxol side chain via asymmetric catalysis. *Journal of Organic Chemistry*, **57**, 4320–4323.

Evans, D.A., Helmchen, G., and Rüping, M. (2007) Chiral auxiliaries in asymmetric synthesis, in *Asymmetric Synthesis – The Essentials* (ed. M. Christmann). Wiley-VCH Verlag GmbH & Co, pp. 3–9.

Finn, M.G. and Sharpless, K.B. (1991) Mechanism of asymmetric epoxidation. 2. Catalyst structure. *Journal of the American Chemical Society*, **113**, 113–126.

Fischer, E. and Hirschberger, J. (1889) Über MannoseII. *Berichte der Deutschen Chemischen Gesellschaft*, **22**, 365–376.

Fischer, E.O. and Maasböl, A. (1964) On the existence of a tungsten carbonyl carbene complex. *Angewandte Chemie, International Edition*, **3**, 580–581.

Glorius, F. and Gnas, Y. (2006) Chiral auxiliaries – principles and recent applications. *Synthesis*, **12**, 1899–1930.

Goodman, S. (2001) French Nobel protest makes chemist a cause célèbre. *Nature*, **414**, 239.

Groves, J.T. and Nemo, T.E. (1973) Epoxidation reactions catalyzed by iron porphyrins - oxygen transfer from iodosylbenzene. *Journal of the American Chemical Society*, **105**, 5781–5786.

Grubbs, R.H. (1972) Olefin metathesis. I. Acyclic vinylenic hydrocarbons. *Journal of the American Chemical Society*, **94**, 2538–2540.

Grubbs, R.H. (2004) Olefin metathesis. *Tetrahedron*, **60**, 7117–714.

Grubbs, R.H. and Miyashita, A. (1977) The metallacyclopentane–olefin interchange reaction. *Journal of the Chemical Society, Chemical Communications*, 864–865.

Grubbs, R.H., Miyashita, A., Liu, M., and Burk, P. (1978) Preparation and reactions of phosphine nickelocyclopentanes. *Journal of the American Chemical Society*, **100**, 2418–2425.

Hatanaka, Y. and Hiyama, T. (1988) Cross-coupling of organosilanes with organic halides mediated by a palladium catalyst and tris(diethylamino)sulfonium difluorotrimethylsilicate. *Journal of Organic Chemistry*, **53**, 918–920.

Heck, R.F. (1977) Transition metal-catalyzed reaction of organic halides with CO, olefins and acetylenes. *Advances in Catalysis*, **26**, 323–349.

Heck, R.F. Jr., and Nolley, J.P. (1972) Palladium-catalyzed vinylic hydrogen substitution reactions with aryl, benzyl, and styryl halides. *Journal of Organic Chemistry*, **37**, 2320–2322.

Hein, J.E. and Blackmond, D.G. (2012) On the origin of single chirality of amino acids and sugars in biogenesis. *Accounts of Chemical Research*, **45**, 2045.

Heitbaum, M., Glorius, F., and Escher, I. (2006) Asymmetric heterogeneous catalysis. *Angewandte Chemie, International Edition*, **45**, 4732–4762.

Hérisson J.-L. (1970) Catalyse de transformation des oléfines par les complexes du tungstène. II. Télomérisation des oléfines cycliques en présence d'oléfines acycliques. PhD thesis. Faculté des Sciences de l'Université de Paris.

Hérisson, J.-L. and Chauvin, Y. (1971) Catalysis of olefin transformations by tungsten complexes. II. Telomerization of cyclic olefins in the presence of acyclic olefins. *Makromolekulare Chemie*, **141**, 161–176.

Holmen, A. (2009) Direct conversion of methane to fuels and chemical. *Catalysis Today*, **142**, 2–8.

Horvath, I.T. and Rabai, J. (1994) Facile catalyst separation without water: fluorous biphase hydroformylation of olefins. *Science*, **266**, 72–76.

Hyttel, J., Bøgesø, K.P., Perregaard, J., and Sánchez, C. (1992) The pharmacological effect of citalopram resides in the (S)-(+)-enantiomer. *Journal of Neural Transmission*, **88**, 157–160.

Ipatieff, V.N., Pines, H., and Schmerling, L. (1940) Isomerization accompanying alkylation II. The alkylation of benzene with olefins, naphthenes, alcohols and alkyl halides. *Journal of Organic Chemistry*, **5**, 253–263.

Ittel, S.D., Johnson, L.K., and Brookhart, M. (2000) Late-metal catalysts for ethylene homo- and copolymerization. *Chemical Reviews*, **100**, 1169–1204.

Jaakkola, S., Sharma, V., and Annila, A. (2008) Cause of chirality consensus. *Current Chemical Biology*, **2**, 153–158.

Jacobsen, E.N., Zhang, W., Muci, A.R. *et al.* (1991) Highly enantioselective epoxidation catalysts derived from 1,2-diaminocyclohexane. *Journal of the American Chemical Society*, **113** (18), 7063–7064.

Jaffe, I.A., Altman, K., and Merryman, P. (1964) The antipyridoxine effect of penicillamine in man. *Journal of Clinical Investigation*, **43**, 1869–1873.

Julian, R.R., Myung, S., and Clemmer, D.E. (2005) Do homochiral aggregates have an entropic advantage? *Journal of Physical Chemistry B*, **109**, 440–444.

Kagan, H.B. (1975) Asymmetric catalysis by chiral rhodium complexes in hydrogenation and hydrosilylation reactions. *Pure and Applied Chemistry*, **43**, 401–421.

Kagan, H.B. and Dang, T.P. (1972) Asymmetric catalytic reduction with transition metal complexes. I. A catalytic system of rhodium(I) with (−)-2,3-O-isopropylidene-2,3-dihydroxy-1,4-bis(diphenylphosphino) butane, a new chiral diphosphine. *Journal of the American Chemical Society*, **94**, 6429–6433.

Katsuki, T. and Sharpless, K.B. (1980) The first practical method for asymmetric epoxidation. *Journal of the American Chemical Society*, **102**, 5974–5976.

Katz, T.J. (2006) Corrections to a history of olefin metathesis. *New Journal of Chemistry*, **30**, 1844–1847.

Katz, J. and Lee, S.J. (1980) Initiation of acetylene polymerization by metal carbenes. *Journal of the American Chemical Society*, **102**, 422–424.

Katz, T.J. and McGinnis, J. (1975) Mechanism of the olefin metathesis reaction. *Journal of the American Chemical Society*, **97**, 1592–1594.

Katz, T.J. and Sivavec, T.M. (1985) Metal-catalyzed rearrangement of alkene alkynes and the stereochemistry of metallacyclobutene ring opening. *Journal of the American Chemical Society*, **107**, 737–738.

Katz, T.J., Lee, S.J., and Acton, N. (1976) Stereospecific polymerizations of cycloalkenes induced by a metal-carbene. *Tetrahedron Letters*, **47**, 4231–4247.

Katz, T.J., Savage, E.B., Lee, S.J., and Nair, M. (1980) Induction of olefin metathesis by acetylenes. *Journal of the American Chemical Society*, **102**, 7940–7942.

Keith, J.A., Oxgaard, J., and Goddard, W.A. (2006) Inaccessibility of β-hydride elimination from —OH functional groups in Wacker-type oxidation, III. *Journal of the American Chemical Society*, **128**, 3132–3133.

Knowles, W.S. (1983) Asymmetric hydrogenation. *Accounts of Chemical Research*, **16**, 106–112.

Knowles, W.S. (2002) Asymmetric hydrogenations. *Angewandte Chemie, International Edition*, **41**, 1998–2007.

Kutscheroff, M. (1881) Über eine neue Methode direkter Addition von Wasser (Hydratation) an die Kohlenwasserstoffe der Acetylenreihe. *Berichte der Deutschen Chemischen Gesellschaft*, **14**, 1540–1542.

Labinger, J.A. (2004) Selective alkane oxidation: hot and cold approaches to a hot problem. *Journal of Molecular Catalysis A: Chemical*, **220**, 27–35.

Lepola, U., Wade, A., and Andersen, H.F. (2004) Do equivalent doses of escitalopram and citalopram have similar efficacy? A pooled analysis of two positive placebo-controlled studies in major depressive disorder. *International Clinical Psychopharmacology*, **19**, 149–55.

Lewandos, G.S. and Pettit, R. (1971) Mechanism of the metal-catalyzed disproportionation of olefins. *Journal of the American Chemical Society*, **93**, 7087–7088.

Lewis, F.D., Hoyle, C.E., and Johnson, D.E. (1975) Abnormal regioselectivity in the photochemical cycloaddition of singlet trans-stilbene with conjugated dienes. *Journal of the American Chemical Society*, **97**, 3267–3268.

Luo, J., Oliver, A.G., and McIndoe, J.S. (2013) A detailed kinetic analysis of rhodium-catalyzed alkyne hydrogenation. *Dalton Transactions*, **42**, 11312–11316.

Malatesta, L. and Angoletta, M. (1957) Palladium(0) compounds. Part II. Compounds with triarylphosphines, triaryl phosphites, and triarylarsines. *Journal of the Chemical Society*, 1186–1188.

Mcbride, W.G. (1961) Thalidomide and congenital abnormalities. *Lancet*, **278**, 1358–1362.

Menard, G. and Stepham, D.W. (2010) Room temperature reduction of CO_2 to methanol by Al-based frustrated Lewis pairs ald ammonia borane. *Journal of the American Chemical Society*, **132**, 1796–1797.

Milstein, D. and Stille, J.K. (1978) A general, selective, and facile method for ketone synthesis from acid chlorides and organotin compounds catalyzed by palladium. *Journal of the American Chemical Society*, **100**, 3636–3638.

Mironov, O.A., Bischof, S.M., Konnick, M.M. *et al.* (2013) Using reduced catalysts for oxidation reactions: mechanistic studies of the "Periana-Catalytica" system for CH_4 oxidation. *Journal of the American Chemical Society*, **135**, 14644–14658.

Miyaura, N. and Suzuki, A. (1995) Palladium-catalyzed cross-coupling reactions of organoboron compounds. *Chemical Reviews*, **95**, 2457–2483.

Miyaura, N., Yamada, K., and Suzuki, A. (1979) A new stereospecific cross-coupling by the palladium-catalyzed reaction of 1-alkenylboranes with 1-alkenyl or 1-alkynyl halides. *Tetrahedron Letters*, **20**, 3437–3440.

Mol, J.C., Moulijn, J.A., and Bollhouwer, C. (1968) Carbon C14 studies on the mechanism of disproportionation of propene. *Journal of the Chemical Society, Chemical Communications*, **11**, 633.

Mullineaux, R.D. and Slaugh, L.H. (1966) Patent US3239570 A assigned to Shell Oil Co, 8 Mar.

Natta, G., Dall'Asta, G., and Mazzanti, G. (1964) Stereospecific homopolymerization of cyclopentene. *Angewandte Chemie, International Edition*, **3**, 723–729.

Negishi, E. (1982) Palladium-or nickel-catalyzed cross-coupling. A new selective method for carbon–carbon bond formation. *Accounts of Chemical Research*, **15**, 340–348.

Negishi, E. (2007) Transition metal-catalyzed organometallic reactions that have revolutionized organic synthesis. *Bulletin of the Chemical Society of Japan*, **80**, 233–257.

Notz, W., Tanaka, F., and Barbas, C.F. (2004) Enamine-based organocatalysis with proline and diamines: the development of direct catalytic asymmetric aldol, Mannich, Michael, and Diels–Alder reactions. *Accounts of Chemical Research*, **37**, 580–591.

Noyori, R. (2002) Asymmetric catalysis: science and opportunities. Nobel Lecture. *Angewandte Chemie, International Edition*, **41**, 2008–2022.

Olah, G.A. (2005) Beyond oil and gas: the methanol economy. *Angewandte Chemie, International Edition*, **44** (**18**), 2636–2639.

Olah, G.A. and Schlossberg, R.H. (1968) Method for measuring acid strength on solid superacid catalyst. *Journal of the American Chemical Society*, **90**, 2726–2731.

Olah, G.A., Kuhn, S.J., Tolgyesi, W.S., and Baker, E.B.J. (1962) Stable carbonium ions II. *American Chemical Society*, **84**, 2733–2740.

Olah, G.A., Bollinger, J.M., Cupas, C.A., and Lukas, J. (1967) Stable carbonium ions. XXXIV. 1-Methylcyclopentyl cation. *Journal of the American Chemical Society*, **89**, 2692–2694.

Osborn, J.A., Jardine, F.H., Young, J.F., and Wilkinson, G. (1966) The preparation and properties of tris(triphenylphosphine)halogenorhodium(I) and some reactions thereof including catalytic homogeneous hydrogenation of olefins and acetylenes and their derivatives. *Journal of the Chemical Society A*, 1711–1732.

Pedersen, S.F., Dewan, J.C., Eckman, R.R., and Sharpless, K.B. (1987) Unexpected diversity in the coordination chemistry of tartrate esters with titanium(IV). *Journal of the American Chemical Society*, **109**, 1279–1282.

Phillips, F.C. (1894a) Palladium induced oligomerization of acetylene. *American Chemical Journal*, **16**, 255–259.

Phillips, F.C. (1894b) Untersuchungen über die chemischen Eigenschaften von Gasen. I. Mitteilung. Erscheinungen bei der Oxydation von Wasserstoff und Kohlenwasserstoffen. *Zeitschrift für Anorganische Chemie*, **6**, 213–228.

Piacenti, F., Pino, P., Lazzaroni, R., and Bianchi, M. (1966) Influence of carbon monoxide partial pressure on the isomeric distribution of the hydroformylation products of olefins. *Journal of the Chemical Society C: Organic*, 488–492.

Pines, H. (1981) *The Chemistry of Catalytic Hydrocarbons Conversion*. Academic Press, New York.

Pines, H. and Stalik, W.M. (1977) *Base Catalyzed Reactions of Hydrocarbons and Related Compounds*, Academic Press, New York.

Rabe, V., Frey, W., Baro, A. *et al.* (2009) Syntheses, crystal structures, spectroscopic properties, and catalytic aerobic oxidations of novel trinuclear non-heme iron complexes. *European Journal of Inorganic Chemistry*, **31**, 4660–4674.

Reppe, W.J. (1949) *Acetylene Chemistry*, vol. **17**. Charles A. Meyer & Cie. Inc., New York.

Reppe, W. (1953a) Carbonylierung I. Über die Umsetzung von Acetylen mit Kohlenoxyd und Verbindungen mit reaktionsfähigen Wasserstoffatomen. *Justus Liebigs Annalen der Chemie*, **582**, 1–37.

Reppe, W. (1953b) Carbonylierung V. Zur Kenntnis der Metallcarbonyle und Metallcarbonylwasserstoffe. *Justus Liebigs Annalen der Chemie*, **582**, 116–132.

Reppe, W.J. (1981) *Acetylene Chemistry*. US Department of Commerce.

Reppe, W. and Kröper, H. (1953) Carbonylierung II. Carbonsäuren und ihre Derivate aus olefinischen Verbindungen und Kohlenoxyd. *Justus Liebigs Annalen der Chemie*, **582**, 38–71.

Reppe W., Magin A. (1948) US Patent 2, 577. 228.

Reppe, W. and Schweckendieck, W.J. (1948) Cyclisierende Polymerisation von Acetylen.III Benzol, Benzolderivate und hydroaromatische Verbindungen. *Justus Liebigs Annalen der Chemie*, **560**, 104–116.

Reppe, W. and Vetter, H. (1953) Carbonylierung VI. Synthesen mit Metallcarbonylwasserstoffen. *Justus Liebigs Annalen der Chemie*, **582**, 133–161.

Reppe, W., Schlichting, O., and Meister, H. (1948) Cyclisierende Polymerisation von Acetylen. II. Über die Kohlenwasserstoffe $C_{10}H_{10}$, $C_{12}H_{12}$ und Azulen. *Justus Liebigs Annalen der Chemie*, **560**, 93–104.

Reppe, W., Schlichting, O., Klager, K., and Toepel, T. (1948) Cyclisierende polymerisation von acetylen I Über cyclooctatetraen. *Justus Liebigs Annalen der Chemie*, **560**, 1–92.

Reppe, W., Kröper, H., Pistor, H.J., and Kutepow, N.V. (1953a) Carbonylierung III. Umsetzung von Alkoholen und offenen Äthern mit Kohlenoxyd zu Carbonsäuren. *Justus Liebigs Annalen der Chemie*, **582**, 72–86.

Reppe, W., Kröper, H., Pistor, H.J., and Weissbarth, O. (1953b) Carbonylierung IV. Einwirkungvon Kohlenoxyd und Wasser auf cyclische Äther. *Justus Liebigs Annalen der Chemie*, **582**, 87–116.

Richter, B., Spek, A.L., van Koten, G., and Deelman, B.J. (2000) Fluorous versions of Wilkinson's catalyst. Activity in fluorous hydrogenation of 1-alkenes and recycling by fluorous biphasic separation. *Journal of the American Chemical Society*, **122**, 3945–3951.

Rueping, M., Koenigs, R.M., and Atodiresei, I. (2010) Unifying metal and brønsted acid catalysis – concepts, mechanisms, and classifications. *Chemistry – A European Journal*, **16**, 9350–9365.

Sauer, S., Saliba, S., Tussetschläger, S. *et al.* (2009) p-Alkoxybiphenyls with guanidinium head groups displaying smectic mesophases. *Liquid Crystals*, **36**, 275–299.

Schmidt-Rohr, K. and Chen, Q. (2008) Parallel cylindrical water nanochannels in Nafion fuel-cell membranes. *Nature Materials*, **7**, 75–83.

Scholl, M., Ding, S., Lee, W., and Grubbs, R.H. (1999) Synthesis and activity of a new generation of ruthenium-based olefin metathesis catalysts coordinated with 1,3-dimesityl-4,5-dihydroimidazol-2-ylidene ligands. *Organic Letters*, **1**, 953–956.

Schrock, R.R. (2001) In memory of John Anthony Osborn. *Advanced Synthesis & Catalysis*, **343**, 3–4.

Schrock, R.R. and Osborn, J.A. (1971) Preparation and properties of some cationic complexes of rhodium(I) and rhodium(III). *Journal of the American Chemical Society*, **93**, 2397–2407.

Shapley, R.R., Schrock, R.R., and Osborn, J.A. (1969) Rhodium catalysts for the homogeneous hydrogenation of ketones. *Journal of the American Chemical Society*, **91**, 2816–2819.

Sharpless, B. (2002) Searching for new reactivity (Nobel lecture). *Angewandte Chemie International Edition In English*, **41**, 2024.

Sheldon, R.A. (2000) Atom efficiency and catalysis in organic synthesis. *Pure and Applied Chemistry*, **72**, 1233–1246.

Smidt, J., Hafner, W., Jira, R. *et al.* (1959) Katalytische Umsetzungen von Olefinen an Platinmetall-Verbindungen Das Consortium-Verfahren zur Herstellung von Acetaldehyd. *Angewandte Chemie*, **71**, 176–182.

Sonogashira, K., Tohda, Y., and Hagihara, N.A. (1975) Convenient synthesis of acetylenes: catalytic substitutions of acetylenic hydrogen with bromoalkenes, idoarenes, and bromopyridines. *Tetrahedron Letters*, **50**, 4467–4470.

Sorokin, A.B. (2013) Phthalocyanine metal complexes in catalysis. *Chemical Reviews*, **113**, 8152.

Soufflet, J.P., Commereuc, D., and Chauvin, Y. (1973) Catalyse de transformation des oléfines par les complexes du tungstène. Forme possible des intermédiaires. *Comptes Rendus de l'Académie des Sciences Paris*, **276**, 169–171.

Speiser, F., Braunstein, P., and Saussine, L. (2005) Catalytic ethylene dimerization and oligomerization: recent developments with nickel complexes containing P,N-chelating ligands. *Accounts of Chemical Research*, **38**, 784–793.

Stephan, D.W. (2010) Activation of dihydrogen by non metal systems. *Chemical Communications*, **46**, 8526–8533.

Stille, J.K. (1986) The palladium catalyzed cross coupling reaction of organotin reagents with organic electrophiles. *Angewandte Chemie International Edition in English*, **25**, 508–524.

Straub, F. and Gollub, C. (2004) Mechanism of Reppe's nickel-catalyzed ethyne tetramerization to cyclooctatetraene: a DFT study. *Chemistry – A European Journal*, **10**, 3081–3090.

Suzuki, A. and Brown, H.C. (2003) Organic syntheses via boranes, in *Suzuki Coupling*, vol. **3**. Aldrich Chemical Co., Inc., Milwaukee, WI.

Svejda, S.A. and Brookhart, M. (1999) Ethylene oligomerization and propylene dimerization using cationic (R-diimine)nickel(II) catalysts. *Organometallics*, **18**, 65–74.

Svejda, S.A., Johnson, L.K., and Brookhart, M. (1999) Low-temperature spectroscopic observation of chain growth and migratory insertion barriers in (R-diimine)NiII olefin polymerization catalysts. *Journal of the American Chemical Society*, **121**, 1034–1035.

Tebbe, F.N., Parshall, G.W., and Reddy, G.S. (1978) Olefin homologation with titanium methylene compounds. *Journal of the American Chemical Society*, **100**, 3611–3613.

Thomas, C.M. and Süss-Fink, G. (2003) Ligand effects in the rhodium-catalyzed carbonylation of methanol. *Coordination Chemistry Reviews*, **243**, 125–142.

Tritto, I. and Grubbs, R.H. (1990) Conversion of titanacyclobutane complexes for ring opening metathesis polymerization into Ziegler–Natta catalysts. *Studies in Surface Science and Catalysis*, **56**, 301–312.

Vengrovius, J.H., Sancho, J., and Schrock, R.R. (1981) Alkyne metathesis. *Journal of the American Chemical Society*, **103**, 3932–3934.

Widegren, J.A. and Finke, R.G. (2003) A review of the problem of distinguishing true homogeneous catalysis from soluble or other metal-particle heterogeneous catalysis under reducing conditions. *Journal of Molecular Catalysis A: Chemical*, **198**, 317–341.

Wilke, G. (2003) Fifty years of Ziegler catalysts: consequences and development of an invention. *Angewandte Chemie, International Edition*, **42**, 5000–5008.

Woodard, S.S., Finn, M.G., and Sharpless, K.B. (1991) Mechanism of asymmetric epoxidation.1.Kinetics. *Journal of the American Chemical Society*, **113**, 106–113.

Zecchina, A., Elena, G.E., and Bordiga, S. (2007) Selective catalysis and nanoscience: an inseparable pair. *Chemistry – A European Journal*, **13**, 2440–2460.

6

Material Science and Catalysis Design

6.1 Metallic Catalysts

In the previous chapters we showed that many metals like Pt, Pd, Ru, and Ni are active catalysts in many reactions, including hydrogenation, oxidation, and Fischer–Tropsch reactions. Real heterogeneous catalysts are constituted by small (typically nanometer-sized) metal particles anchored on a support with a large fraction of the metal atoms exposed to reactants. The beneficial role of high dispersion of metal particles was clearly recognized as early as 1909 and was the motivation for awarding Paul Sabatier his Nobel Prize *"for his method of hydrogenating organic compounds in presence of finely disintegrated metals."* So far in this book we also saw that poisoning of the surface of a metal active in hydrogenation reactions with ancillary ligands with chiral character induces selective hydrogenation properties of olefin substrates (Akabori, Ohno, and Narita, 1952) via formation of reaction pockets formed by the chiral molecules adsorbed on the surface. Similarly, when alloyed with an inactive one (Pb, Au, etc.), metallic particles constituted by an active metal (Pt, Pd, Ni, etc.) become active in partial hydrogenation of triple C≡C to double C=C bonds (Lindlar and Dubuis, 1973) and in hydrogenation of C=O groups, leaving intact the C=C bonds present in the same molecule (Riedl and Pfleiderer, 1936). In a fully analogous way, supported Pt particle poisoned with $ZnSO_4$ presents selective activity in partial hydrogenation of benzene to cyclohexene (Asaki catalysts) (Nagahara *et al.*, 1997). Taken together these facts indicate that properly designed surface poisoning can help determine selectivity in many reactions catalyzed by finely dispersed metals. It is therefore useful to consider more fully the few general concepts concerning the effect of catalysts design on the surface properties of metal particles and especially regarding the role of particle shape and size, metal-support interaction, surface poisoning, and alloying. This discussion is an essential part of material science approach to catalysis by metals.

The Development of Catalysis: A History of Key Processes and Personas in Catalytic Science and Technology,
First Edition. Adriano Zecchina and Salvatore Califano.
© 2017 John Wiley & Sons, Inc. Published 2017 by John Wiley & Sons, Inc.

Our first consideration is the particle size effect. About the role of particle size, it must be said that many reactions, such as hydrocarbon hydrogenations on Pt, Pd, and Ni, are thought to be *"structure insensitive"* because they are progressing at approximately the same rate on metal particles of sizes larger than about 1 nm that show a bulk-like metallic behavior (Xu *et al.*, 1994). However, when particle sizes lower than 1 nm are involved (for instance, metal particles encapsulated into the cavities of microporous materials like zeolites), the situation becomes more complex because the average coordination number decreases and because the lattice parameter contracts (quantum effect). This fact has been demonstrated experimentally (Schauermann *et al.*, 2013). It is also known that in small particles the electronic energy levels are not continuous as in bulk materials; they are discrete, and the spacing between electronic energy levels is dependent on the surface-to-volume ratio and the shape of the particle. As a consequence electron donation or acceptance to and from reacting molecules, and hence the ability to act as catalysts, is influenced by the particle size (Che and Bennett, 1989). This is well documented for Au, which was considered inactive from a catalytic point of view for a long time but which has revealed a completely different behavior when synthesized with nanometric dimensions (Haruta, 1997). Similarly, Ag nanocrystals have the capability to dissociate O_2 to atomic oxygen species, while on bulk Ag surfaces the adsorbed oxygen species at 80 K is predominantly O_2^-. Small Cu and Pd nanoparticles are able to retain adsorbed CO up to much higher temperatures if compared with the bulk metals; very small Ni nanoparticles are even able to dissociate CO at 300 K, forming carbidic species on the particle surface. The activation energy for CO dissociation changes with the increasing size of the Ni particles, a pattern that affects the performance of Ni nanoparticles in Fischer–Tropsch synthesis of hydrocarbons from synthesis gas.

Second is the role of particle shape, since different facets of the metal particles can be associated with distinctly different catalytic activities. As the shape of the supported particle is largely determined by the interaction with the support (amorphous, crystalline, or microporous), the control of particles' shapes is another consideration of the material science approach to catalysis. This has motivated explorations of physical methods that can be used to control particle morphology. In particular, different sizes and shapes of metal particles have been preferentially studied by high resolution transmission electron microscopy (HRTEM) or scanning tunneling microscopy (STM), and the literature is extensive on this subject.

Third is the metal-support interaction. This primarily concerns the nucleation of metal particles on flat surfaces and on defective sites of the support. The STM image in Figure 6.1a shows the preferential nucleation of metal (gold) particles with low coordinated atoms present at the edges of the

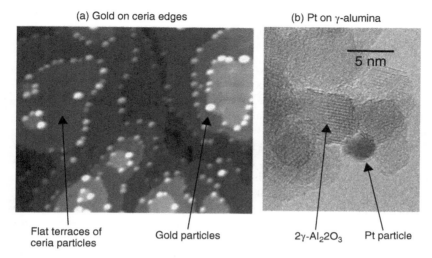

(a) Gold on ceria edges

(b) Pt on γ-alumina

5 nm

Flat terraces of ceria particles

Gold particles

2γ-$Al_2$2O_3

Pt particle

Figure 6.1 (a) STM image of gold particles preferentially nucleated at the edges of ceria terraces. (b) TEM image of a Pt nanoparticle interacting with two γ-Al_2O_3 nanocrystals. *Source*: (a) Adapted from Lu *et al.*, 2007. (b) Courtesy of E. Groppo and M. Manzoli.

crystallites of the ceria support. In the same image notice that the gold particles do not nucleate on the flat, defect-free portions of the ceria surface. Figure 6.1b provides another example of the propensity of metal particles (Pt) to interact with the metal support. Here the simultaneous interaction of the Pt nanoparticle is with two γ-Al_2O_3 nanocrystals.

In some instances the metal-support interaction is strong enough to encapsulate the metal particle by a thin film of the supporting oxide. This phenomenon (strong metal-support interaction (SMSI)) particularly occurs when reducible supports like TiO_2, Fe_2O_3, and Fe_3O_4 are present as these can inhibit the catalytic activities, though there are known to have been few cases of amplification (Wang, Kaden, and Dowler, 2012).

Fourth is surface poisoning via alloying with inactive elements as we described earlier. It must be added that a less documented but very important aspect is poisoning by carbon, which acts as a common surface modifier. This phenomenon is particularly relevant when supported metal particles with size smaller than 1 nm are used because they easily react with carbon residues with the formation of carbides.

In summary, metal catalysts dispersed on a support are highly complex multicomponent materials that throughout the twentieth century have been optimized to produce *millions of turnovers, with high reaction rates and high selectivity* (Schauermann *et al.*, 2013). These achievements were due to the efforts of a great many researchers within different industrial laboratories, usually proceeding via empirical methods. Only in recent years have advanced physical methods started to be applied to resolve the remaining problems.

6.2 Oxides and Mixed Oxides

Oxides are combination of metals with oxygen, with M_xO_y as the general formula.

Depending on the nature of the metal, they can develop acid–basic or redox properties, or both. For instance, Na, Mg, Al, Si, Ti, and Zr oxides (which do not contain easily reducible metals) will show dominant basic and acid properties. However, in V, Cr, Mo, Fe, Cu, and Ce oxides, the redox properties prevail. Mixed oxides with spinel and perovskite structure can also form a more rich and important family of oxidic materials in which redox and acid–base properties can coexist.

The acid and basic properties of oxides and the associated selectivity can occur in many reactions because their surfaces have coordinatively unsaturated positively (metal cations) and negatively (O^{2-} anions) charged centers potentially behaving as Lewis acid and Lewis bases working as isolated entities or in combination. The coordinative unsaturation of the individual centers of this distribution depends on the type (index) of exposed faces, while the effective charge carried by each surface center depends primarily on the ionicity of the oxide under consideration. From this it is evident that the faces of ionic oxides are regions where complex acid–base reactions can occur. Another consideration is that cation–anion pairs can behave in a concerted way favoring the heterolytic dissociation of adsorbed molecules. This is the case of ZnO, where the concerted action of surface Zn^{2+}–O^{2-} pairs leads to heterolytic H_2 dissociation with a simultaneous formation of ZnH hydrides and OH groups (Scarano *et al.*, 2006).

When the positive center has high charge and small radius (like Al^{3+}, Ti^{4+}, Zr^{4+}), the Lewis acid property predominates and the oxide develops Lewis acid properties. This empirical rule cannot be applied to Si^{4+}. In fact in this instance the Si^{4+} radius is so small that the resulting Si–O bond is covalent, and the SiO_2 solid does not show any distinct acid and basic properties.

The other extreme is when the charge of the cation is low and the radius high. This occurs when the basic property of O^{2-} dominates, and the oxide is behaving like a Lewis solid base (as Na_2O, MgO, CaO, etc.).

Native oxide surfaces (including SiO_2) are covered by hydroxyl groups. As these groups are potential proton donors or acceptors, it is evident that Brønsted acidity and basicity must be considered as well. Yet, depending on the hydroxylation state, Lewis and Brønsted groups are often simultaneously present on the same surfaces.

In oxides with redox properties, the reducibility and tendency of oxygen to leave the surface and to participate to oxidation reactions following the Mars–van Krevelen mechanism must be taken into account. One example is V_2O_5, which is the active catalyst for SO_2 to SO_3 oxidation, as we documented in Chapter 5. Another example is MoO_3 and Re_2O_7, whose reducible character can lead to formation of low valence Mo^{x+} and Re^{x+} surface species active in metathesis reactions. An important property of few oxides (namely TiO_2 and ZnO) is their semiconductor character, a fact that finds application in photocatalysis as it will be documented in Chapter 7.

In summary, it is difficult to formulate a comprehensive and general theory of the acid–base, redox, and semiconducting properties of oxide surfaces. Each oxidic system can only be treated separately as a case study (Zecchina, Lamberti, and Bordiga, 1998). However, there is no doubt that oxidic systems represent a category of versatile materials whose catalytic properties can be designed with unprecedented accuracy.

Kōzō Tanabe and Wolfgang F. Hölderich (Tanabe and Holderich, 1999) have estimated that in the 1960–2000 period, more than 300 solids with acid and base properties have been developed. Today solid acid–base catalysis is one of the economically and ecologically important fields in catalysis and in petrochemistry. The same authors have tried to evaluate how many oxides and mixed oxides found applications as catalysts in industrial processes. The resulting number is around 60 (Tanabe, 1970).

The list of most commonly used oxides and mixed oxides with a variable acidic character includes SiO_2, $SiO_2–TiO_2$, $SiO_2–Al_2O_3$, Al_2O_3 (also as carrier of Lewis and Brønsted acids like BF_3 or HF), TiO_2, and ZrO_2 (both pure and $-SO_3H$ doped), while MgO and $Al_2O_3–Na_2O$ and $Al_2O_3–K_2O$ are prototypes of basic solids. Oxides and mixed oxides with redox properties are V_2O_5, Cr_2O_3, Fe_2O_3, CeO_2, $SiO_2–V_2O_5$, $SiO_2–CrO_3$, $SiO_2–Re_2O_7$, and so on. Here we will mainly consider the oxides with acidic or basic character together with some of the associated catalyzed reactions developed in the 1970–2000 period. The catalytic properties and applications of several oxides with a redox character were already discussed in Chapters 4 and 5.

6.2.1 SiO₂ and SiO₂-Based Catalysts and Processes

Silica is a highly covalent solid, and the particles covered by Si–OH groups have a weak Brønsted acid character. A fully dehydroxylated silica surface cannot be obtained without extensive particle sintering and a decrease of the surface area. The surface structure of this amorphous material has been widely investigated via many techniques (in particular, IR spectroscopy) and by quantum modeling (Ugliengo *et al.*, 2008).

In the twentieth century high surface area amorphous silica found dozens of industrial applications as adsorbent, filler, and support for catalysts, though it does not seem to have, in its pure form, many catalytic applications. The few catalytic applications of pure silica are associated with the surface Si–OH groups that are weakly acidic. This group has been found to catalyze the cyclohexanone oxime-caprolactam isomerization reaction (Lasier and Rigby, 1941)

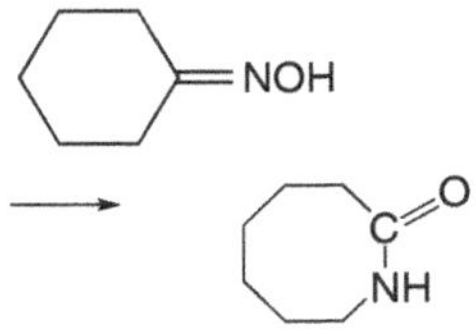

Before this discovery, such an isomerization reaction was performed in the presence of strong acids. As caprolactam is an intermediate for the synthesis of nylon-6, the importance of the observation concerning the catalytic activity of $\equiv$Si—OH groups might be easily understood. However, the amorphous nature of SiO_2 caused its yield and selectivity to be unsatisfactory.

As it will be illustrated later in our discussion of microporous materials, the process became selective and industrially relevant when the catalytic properties of $\equiv$Si—OH groups located in the channels of silicalite (a crystalline zeolitic material discovered the last decade of the twentieth century) were discovered and exploited.

Beside their use as weak acid centers, silanols can be utilized to anchor species like chromates (Phillips catalysts for ethylene polymerization), vanadium pentoxide (which form surface vanadates active in oxidation reactions), and molybdenum and rhenium oxides (metathesis catalyst). It has been found that silanols can be used to anchor finely dispersed TiO_2 clusters or single Ti^{4+} species, which are active in the ammoximation reaction that transform cyclohexanone (produced by the oxidation of cyclohexane in air or alternatively by the partial hydrogenation of phenol) into the oxime:

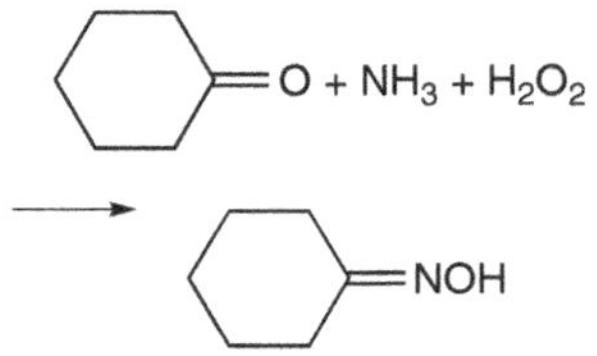

However, also in this application, due to the amorphous nature of silica, the resulting catalysts showed poor selectivity and were abandoned.

Only more recently much better results in term of selectivity were obtained with titanium silicalite (Roffia *et al.*, 1987), a crystalline microporous material whose structure will be documented in Section 6.5 in which we discuss zeolitic materials.

In the last decade of the twentieth century, Silica has found further applications as a support of acidic species like $Al_2(SO_4)_3$, CF_3SO_3H, and SbF_5 for industrialized hydrocarbon isomerization reactions. The heterogeneous catalysis in the studies made by George A. Olah (1994 Nobel Laureate) using liquid superacids CF_3SO_3H and $HF-SbF_5$ influenced this development. Silica is presently finding more application as the support of finely divided metals in hydrogenation reactions. The complete list of the processes where SiO_2 is involved can be found in Tanabe and Hölderich (1999).

6.2.2 Al_2O_3 and Al_2O_3-Based Catalysts and Processes

Alumina (in its different crystallographic forms) is more ionic than silica and has uses as adsorbent, support, catalyst binder, and additive to acid zeolites. Alumina exposes on the surface coordinatively unsaturated $Al^{3+}-O^{2-}$ ions acid–base pairs and defect sites (Wischert *et al.*, 2012). When basic molecules like CO or pyridine are put into contact with the surface, $Al^{3+}-CO$ and adducts $Al^{3+}-Py$ are preferentially formed (Gribov *et al.*, 2010).

However, interaction with alcohols, ROH (which contains the polar $O^{\delta-}H^{\delta+}$ groups), requires the complementary $Al^{3+}-O^{2-}$ pairs to be present on the surface for the formation of surface –OH and –OR species. Indeed aluminas have proved to be very active in the dehydration of alcohols to olefins and to ethers, as studied by Herman Pines (processes developed since the 1980–1985 period) and described in Chapter 5. In particular, aluminas are applied to produce dimethyl ether from methanol, which is the first step of methanol-to-olefin (MTO) process as we will illustrate Section 6.6 in our discussion of zeolites. Aluminas are also used as catalysts for the synthesis of chloroform from methyl alcohol and HCl (1978). Chlorinated alumina can behave as a strong Lewis acid, and when supporting finely divided platinum, it is as we saw in Chapter 5 an efficient reforming catalyst. Al_2O_3 can serve as the support of both acid and basic species. In fact, when combined with BF_3, which is a Lewis acid, the resulting $BF_3-Al_2O_3$ system has been used in alkane isomerization (1993), and $Al_2O_3-Na_2O$ and $Al_2O_3-_-K_2O$ systems were used as basic catalysts for reactions involving the double bond isomerization (1980–1990) (Hattori, 2001).

6.2.3 $SiO_2-Al_2O_3_-$ and $SiO_2-Al_2O_3$-Based Catalysts and Processes

Recall from Chapter 5 that the Houdry's hydrocarbon cracking catalyst is essentially constituted by impure $SiO_2-Al_2O_3$. The ascertained negative impact of the impurities (Fe) on selectivity of the Houdry's catalysts moved the researchers to prepare pure amorphous silica–alumina. Then, not much time after the preparation of pure silica–alumina, the discovery of synthetic zeolites came in the 1960–1970 period (a fact that will be described in a separate paragraph) and displaced pure high surface area silica–aluminas as the cracking catalysts. $SiO_2-Al_2O_3$ and acid catalysts were progressively abandoned. To understand the complex structure and the acid properties of $SiO_2-Al_2O_3$, it is useful to know that the surface of amorphous silica–alumina consists of (1) isolated aluminum grafted to the silica surface (isolated Lewis sites), (2) a very small amount of aluminum in the silica network that, like in zeolites, brings about the strong Brønsted acidity due to the presence of a proton compensating the charge defect, and (3) small patches of aluminum oxides (Chizallet and Raybaud, 2009). Today high surface area silica–aluminas are often present as the support in modern composite hydrocracking catalysts, which contain an acid zeolite fraction as the shape-selective catalyst in order to produce middle distillates from heavy oil fractions (Tanabe and Holderich, 1999).

6.2.4 MgO⁻ and MgO-Based Catalysts and Processes

MgO is an ionic solid whose surface properties are dominated by O^{2-} Lewis basicity. For this reason it is a prototype of solid base, a fact that is associated with interesting catalytic properties. A list of processes catalyzed by MgO can be found in the review of Hattori (2001). We only mention that the discovery of its activity in alkylation of aromatic rings (1976) has attracted much interest because this reaction was supposed to be uniquely catalyzed by acids. A representative reaction is the industrialized alkylation of phenol with methanol to give 2,6-xylenol. The Li_2O/MgO system has also been studied because of its interesting activity in the oxidative coupling of methane (Lunsford, 2000; Holmen, 2009).

MgO can form high surface area solid solutions with NiO and CoO, so affording a method to obtain isolated and fivefold coordinated Ni^{2+} and Co^{2+} ions on the low index surfaces of the MgO matrix (Cimino *et al.*, 1999; Cimino and Stone, 2002).

6.2.5 ZrO₂ and ZrO₂-Based Catalysts and Processes

ZrO_2 in the pure form is rarely used as an acid catalyst, while sulfated ZrO_2 (which has strongly acidic groups on the surface) is a strong solid acid catalyst that has found industrial application for isomerization of alkanes (1992). On the basis of this property, this solid has been often classified as a solid superacid. Similarly ZrO_2 supporting triflic acid is active in the same reactions (1994). When doped with Na_2O, ZrO_2 is active in alcohol dehydration with the formation of olefin. When doped with Cr_2O_3 (which, because to its reducible character, is a mild hydrogenation catalyst), $ZrO_2-Cr_2O_3$ is applied to hydrogenate benzoic acid (which is prepared by oxidation of toluene) to the corresponding aldehyde (process industrialized in 1986). A complete list of processes involving ZrO_2-based systems can be found in the Tanabe and Hölderich review (Tanabe and Holderich, 1999).

6.3 Design of Catalysts with Shape and Transition-State Selectivity

Toward the twentieth century, in parallel with the unprecedented development of petrochemistry with regard to energy production and synthesis of chemical compounds, the problems associated with the formation of undesired products and pollution (both associated with poor selectivity of industrial catalysts) could not be further procrastinated (Anastas, Kirchhoff, and Williamson, 2001). Emphasis thus shifted to the improvement of catalytic selectivity. Catalytic selectivity is a problem particularly relevant in the production of fine chemicals and pharmaceutical compounds by heterogeneous catalysts (Sheldon and Downing, 1999).

The need to control catalyst selectivity was also fueled by the new emerging processes dictated by environmental problems like efficient water splitting to produce hydrogen by photocatalysts (Balzani and Juris, 2001; Kalyanasundaran and Graetzel, 2010), efficient conversion of fuels to electricity through electrochemical fuel cells (Steele and Heinzel, 2001), selective conversion of biomass to fuels and bulk chemicals (Hara, 2010), and greenhouse gases (e.g., CO_2 and CH_4) reduction and utilization (Arakawa *et al.*, 2001). The pollution that could be controlled and even abated by the appropriate use of new catalysts was, *inter alia*, demonstrated by a three-way catalytic converter introduced in 1975 in California to reduce the pollutant emission from automobiles (Bagot, 2004).

All together these studies have dictated the research direction toward shape and transition-state selectivity.

The shape-selectivity concept, which originates from the field of enzymatic catalysts, was first proposed by Paul Weisz and Vincent J. Frilette in 1960 (Weisz and Frilette, 1960). Today at least 17 commercial processes are based on shape selectivity (Degnan, 2003).

The widely accepted and applicable principles concerning shape and transition-state selectivity can be summarized as follows:

1. *Reactant shape selectivity.* Discriminates among competing reactants on the basis of their size (size exclusion at the pore mouth of catalysts micropores).
2. *Product shape selectivity.* Discriminates between products based on their size in the pore diameter where the active center is located.
3. *Transition-state selectivity.* Imposes steric constraints on the transition state at the geometry of the cage walls surrounding the active sites.

In many instances it is difficult to discriminate among the different selectivity types that act in a specific catalyst, since the actual behavior is based on a combination of several factors.

From the preceding definitions it should be evident that catalysts showing shape and transition-state selectivity would have pockets, pores, channels, and cages of molecular size and active sites located at their walls. In fact catalysts showing such properties are often microporous and have an elaborate design.

6.4 Zeolites and Zeolitic Materials: Historical Details

Zeolites are the most famous microporous material family with shape-selective properties; many other microporous systems, including aluminum phosphates (ALPO), heteropolyacids, metal–organic frameworks (MOFs), and porous polymers, should be mentioned as well. The zeolites popularity is measured by about 50,000 papers published in the 1995–2013 period, one-tenth of them being specifically devoted to catalysis.

Zeolites occur in nature and were known for almost 250 years as aluminosilicate minerals. It was only with the advent of synthetic zeolites (Milton, 1959; Breck, 1964, 1975) that the exceptional properties of these solids started to be appreciated, even though the molecular sieving property of a few natural zeolites was known since 1930.

However, we should emphasize that only after the seminal work of Barrer (Barrer, Peterson, and Schoenborn, 1966; Barrer, 1979) and Milton, Occelli, and Robson (1989) has this class of porous materials begun to be applied in catalysis. In particular it was the Barrer's descriptions in the 1940s of gas separations with zeolite minerals that demonstrated potential commercial applications. Because zeolite minerals were rare and unavailable in practical quantities, Robert M. Milton at the Union Carbide laboratory started in mid-1949 an exploratory program to synthesize zeolites in the laboratory. By year-end, he had discovered a new and practical method to make several novel synthetic zeolites, including two that were going to gain commercial prominence, called zeolite A and zeolite X. In 1952, Milton was joined at Union Carbide laboratories by a young inorganic chemist, Donald W. Breck (1921–1980), who continued the discovery and study of more synthetic zeolites including one called zeolite Y (Flanigen, 2005). Breck became quickly recognized in the initial development of synthetic molecular sieves and one of the founders of the International Zeolite Association. His landmark book published in 1974, *Zeolite Molecular Sieves: Structure, Chemistry, and Use*, summarized the first 25 years of zeolite science and technology and occupies a unique place in the library of any scientist involved in the science of the molecular sieve.

In his memory the Breck award, administered by the International Zeolite Association, was established and sponsored by Union Carbide Corporation.

In 1954, Union Carbide commercialized synthetic zeolites as a new class of industrial materials for separation and purification. From 1940, the number of synthesized structures exploded and the same happened for their applications. Today, about 200 different structures are known (Baerlocher, McCusker, and Olson, 2007) and about 80 are microporous.

The definition and use of the term "zeolite" has evolved and changed, especially over the past decade, to include non-aluminosilicate compositions and structures containing open cavities in the form of channels and cages, reversible hydration–dehydration characteristics. For this reason we will often use the term zeolitic materials. A concise chronological list of the discoveries of the principal zeolitic and microporous materials is reported below. The list is not fully exhaustive.

Time of initial discovery	Composition
Late 1940s to early 1950s	Low Si/Al ratio zeolites (A, X, Y)
Mid-1950s to late 1960s	High Si/Al ratio zeolites (ZSM-5, BEA, MORD, etc.)
Early 1970s	SiO_2 molecular sieves (silicalite)
Late 1970s	$AlPO_4$ molecular sieves (ALPO)
Late 1970s to early 1980s	SAPO and MeAPO molecular sieves
Late 1970s	Metallo–silicates
Early to mid-1980s	$AlPO_4$-based mol. sieves, titanium silicalite
Early to mid-1990s	Mesoporous molecular sieves (MCM-41)
	Octahedral–tetrahedral frameworks (ETS-10)
Late 1990s	MOFs
2000s	Germanosilicate zeolites

Most data concerning the history of high silica zeolites and silicalite can be found in Flanigen's review (Flanigen, 2001). As for aluminophosphate molecular sieves, they were first described by Wilson *et al.* (1982) at Union Carbide. These new generations of molecular sieve materials, designated $AlPO_4$-based molecular sieves, include more than 24 structures and 200 compositions (Flanigen *et al.*, 1987). Additional members of the aluminophosphate-based molecular sieve family, for example, SAPO and MeAPO were discovered in 1986 (Flanigen *et al.*, 1986). Mesoporous molecular sieves were first synthesized in 1990 by Japanese researchers (Yanagisawa, Harayama, and Kuroda, 1990). They were later produced also at Mobil Corporation laboratories (Beck *et al.*, 1992) and named Mobil Crystalline Materials (MCM). In 1969, Grace described the first modification chemistry based on steaming zeolite Y to form an "ultra-stable" Y zeolite. In 1967–1969, Mobil Oil reported the synthesis of the high silica zeolites beta and ZSM-5.

Information about the structures of some zeolites will be given in the next section where we describe separately the catalytic properties of the zeolitic materials.

In the list the mesoporous structure with entirely organic composition, like the great variety of porous polymers, are omitted because they will be described separately. The same will be done for MOFs.

As we briefly mentioned earlier, the history of zeolites is dominated by the figure of Richard Barrer (1910–1996), who is considered as the father of zeolite chemistry, and indeed he is an outstanding example of an academic scientist whose activity had a profound influence on industrial chemistry of second half of twentieth century (Figure 6.2). Richard Barrer was born in Wellington, New Zeeland. His first degree was from Canterbury College, the same college where he had studied under Ernest Rutherford, the father of modern nuclear

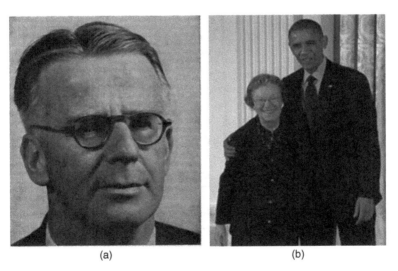

(a)   (b)

Figure 6.2 (a) Richard Barrer (1910–1996) and (b) Edith Flanigen (1929–) shown with US President Barack Obama in 2014. They can be considered as the main protagonists of zeolite chemistry and synthesis. Barrer, father of zeolite chemistry, discovered the potentialities of zeolites in gas separations. Flanigen synthesized a large variety of zeolites, which have found application in catalysis. For this she can be considered as one of the most inventive chemists of all time. *Source*: (a) Reproduced with permission of Publishing Contracts & Copyright Executive of RSC. (b) https://commons.wikimedia.org/wiki/File:Edith-flanigen-barack-obama.png.

science. In 1932, thanks to a fellowship, he immigrated to Britain where he joined the renowned Sir Eric Rideal's Colloid Science Laboratory in Cambridge. Rideal encouraged his colleagues to produce their own ideas, and after reading McBain's *Sorption of Gases by Solids* (McBain, 1932), Richard Barrer focused on the adsorption of gases in zeolites, especially chabazite, whose crystals possess strictly regular pore and channel structures of molecular dimensions. In this way one of the most significant new fields of research in the twentieth century initiated its growth. After obtaining his PhD at Cambridge, in 1948 he became professor at the University of Aberdeen for the period 1948–1954; in 1954–1976 he was professor of Chemical Physics at the Imperial College, London, a position that he held until retirement. During his department headship at Imperial College, he hired Sir Geoffrey Wilkinson (Nobel laureate) as professor of Organic and Inorganic Chemistry. Barrer's work led to the synthesis of zeolites by the Union Carbide and Mobil Corporations and their use in the conversion of crude oil to high octane fuels. His research also introduced zeolite A (Breck, Eversole, and Milton, 1956) as an environmentally safe component of detergents, used worldwide from the 1970s. He was nominated for the Nobel Prize in 1996 but died from cancer before a decision had been made. The zeolite mineral barrerite is named in his honor. Barrerite is one of the rarest zeolites existing in nature. It was first described in 1974 for an occurrence in Sardinia, Italy, at the Sant' Efisio Tower on Cape Pula in Cagliari Province.

Another chemist active in the chemistry of zeolites in the 1960s is the American industrial chemist Edith Marie Flanigen, a classic high quality industrial researcher, though not rare in industrial laboratories (Figure 6.2).

Born in Buffalo, New York, after completing the university studies at Syracuse University, Edith joined the Linde Division at Union Carbide in 1952, where she was first assigned to silicone chemistry, and it was only in 1956 that she joined the molecular sieve or zeolites group. For a time in the 1960s, however, Edith was engaged in a program unrelated to zeolites, the synthesis of emeralds. At that time, when masers were a hot research area, a reliable supply of single crystals of emeralds was important. Flanigen and a coworker developed an emerald synthesis process that grew emeralds to a high level of perfection able to deliver powerful laser beams from microwave lasers.

However, it was primarily Flanigen's work on zeolite materials that in 1992 earned her the first Perkin Medal ever awarded to a woman; the Perkin Medal is the highest honor for outstanding work in applied chemistry in the United States. In the late 1970s, the team led by Flanigen filed more than 30 patents (Skeels and Flanigen, 1999) and had succeeded in developing a whole new generation of synthetic zeolites.

The ingredients for this success can be inferred from Flanigen's acceptance speech for the Perkin Medal, where she described her fellow researchers this way:

Creative minds stretched and emboldened by excellence in their educational training. Dreamers, visionaries, free spirits. At home with concepts. Thinkers with uncanny chemical intuition. Persistent, almost stubborn, in their resolve. With a childlike impetus at play- and just a little bit of luck.

Edith Flanigen's work made the production of zeolite Y (an aluminosilicate) commercially viable (Wilson *et al.*, 1982). Her molecular sieves have made gasoline production more efficient, cleaner, and safer worldwide. Edith Flanigen's more recent work on aluminophosphates has applications for lubricating oils (Flanigen, Patton, and Wilson, 1988). She was the first to synthesize the fully siliceous crystalline zeolitic material silicalite (Flanigen, 1976).

6.5 Zeolites and Zeolitic Materials Structure

Zeolites are aluminosilicates that are built from an infinitely extending three-dimensional network of $[SiO_4]$ and $[AlO_4]^{-1}$ tetrahedra (Breck, 1974). They have actually porous crystal structures characterized by intersecting channels of molecular dimension that allow a large surface area and thus a large number of catalytic sites. At the channels' intersection small cages are present, and these cages can behave as nanoreactors. In addition they have exchangeable cations located in well-defined crystallographic positions of the channels and cages. The framework structure of the first synthesized zeolites (faujasite in the Y and X versions) is shown Figure 6.3.

Depending on the Si/Al ratio of their framework, synthetic zeolites with a faujasite structure are divided into X and Y zeolites. In X zeolites the Si/Al ratio is comprised between 2 and 3, while in Y zeolites it is higher than 3. Due to the presence of Al in the structure, the framework is carrying a negative

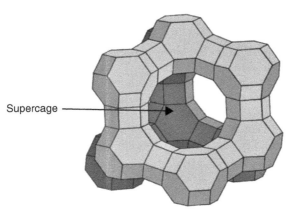

Supercage

Faujasite

Figure 6.3 Faujasite structure. Each segment stays for a SiOSi or SiOAl group. The cations counterbalancing the negative charge of the framework (Na^+, NH_4^+, and H^+) are omitted.

charge, which is balanced by the positive charges of cations in extraframework positions (Tosheva, 2001). The counterions relevant in catalysis are located in the supercage, which is accessible to molecules of appropriate shape. The ions can be monovalent like Na^+, K^+, and NH_4^+ or bivalent like Zn^{2+} and Cu^{2+}, or even trivalent (rare earth cations). Cu-exchanged zeolites are solid with relevant catalytic properties for NO_x abatement exploiting the reaction of NO_x with ammonia (Braundberger *et al.*, 2008).

When the charge compensating cation associated with the framework aluminum is NH_4^+, heating the zeolite between $300°–400°C$ leads to elimination of NH_3 with formation of H^+, which acts as a strong Brønsted acid (Bhatia, 2000). If the zeolite is heated to a temperature higher than $450°C$, the Al^{3+} ions can go, at least partially, into extraframework position, and Lewis acid sites are formed (Weitkamp, 1999). The thermal stability of the zeolites increases with the Si/Al ratio of the framework (Y is more stable than X). For catalytic cracking, the Y zeolite is used in the protonic form and can also contain rare earth cations.

Besides Y zeolite other very stable synthetic zeolites characterized by lower Al content in the structure and smaller channels diameter have been synthesized. Among all we choose to mention ZSM-5, chabazite, mordenite, and beta-zeolites, which in the protonic form, have found important applications in the petroleum refinery industry. Flanigen succeeded in synthesizing a zeolitic material with a ZSM-5 structure but without Al in the structure (silicalite) (Figure 6.4). The successful synthesis of this structure (Flanigen *et al.*, 1978) has broken the conviction that the stability of MFI and other zeolite structures is correlated with the presence of Al in the structure (Flanigen, 1991). The discovery of silicalite also opened the way toward the synthesis of titanium

Silicalite and HZSM-5

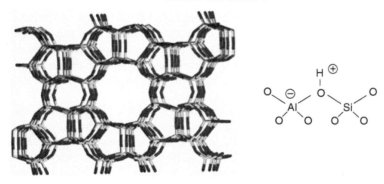

Figure 6.4 Silicalite framework. Black sticks stay for oxygen, while gray tetrahedral sticks stay for silicon. When a silicon atom is substituted by Al (ZSM-5), a positive compensating charge (H^+) is present (right-hand side), which acts as a strong Brønsted acid.

silicalite (TS-1), where Si is substituted by Ti (Taramasso, Perego, and Notari, 1983). The catalytic properties of zeolites, silicalite, and TS-1 will be described in the following section.

A small number of the framework Al and Si atoms in silicalite or ZSM-5 can be partially or totally substituted by trivalent metal ions like Fe^{3+}. Depending on the thermal treatment and the presence of moisture, the Fe^{3+} present in the framework can migrate in extraframework position with formation of iron oxide clusters whose size is limited by the cage's diameter. Fe–ZSM-5 is active in the oxidation of benzene with N_2O as the oxidant (Panov *et al.*, 1992). In this catalyst, various iron species are present in the cavities, including highly dispersed Fe_2O_3 particles and clusters together with grafted Fe(II) and Fe(III) single sites (Zecchina *et al.*, 2007). The most widely accepted opinion is that the active sites are isolated or dimeric Fe(II) species. The dimeric species could be structurally similar to those present in methane monooxygenase enzyme, whose structure will be discussed in the next section.

Besides silicalite several other fully siliceous zeolites have been synthesized in more recent times, including the siliceous analogue of zeolite A and of beta zeolite (Fernandez *et al.*, 2006).

Nearly in the same period, Beck *et al.* (1992) synthesized nanotubes of amorphous silica–alumina. The most popular component of this group of materials is MCM-41. The acronym MCM stays for *Mobil Composition of Matter*. In 1994 the Mobil research group constituted by C.T. Kresge, M.E. Leonowicz, W.J. Roth, J.C. Vartuli, and J.S. Beck was awarded the prestigious Breck Award *"for preparing the first ordered mesoporous silicate and aluminosilicate materials containing pores in the range of 16 to 100 Å."*

The striking fact about the MCM-41 is that, although composed of amorphous silica walls, it has a long-range ordered framework with uniform mesopores. These materials also have large surface areas, some more than $1000 \, m^2/g$. The pore diameter of these materials can be nicely controlled within a mesoporous range between 1.5 and 20 nm by adjusting the synthesis conditions. The tunable diameter of the pores makes these compounds fully suitable for hosting or anchoring large molecules and catalysts of a complex structure, an achievement not possible with zeolites. A drawback is that, due to the amorphous character of the SiO_2 framework, the structural properties of the active centers located on the walls are not clearly defined.

In summary, it can be safely stated that the crystalline character of most zeolitic materials, the tunable size of the channels and cages, the presence of a variety of isolated and clustered catalytic sites in the channels and cages, and zeolites and zeolitic materials represent a family of tunable structures showing similarities with enzymes, on one hand, and homogeneous catalysts, on the other hand (Zecchina, Bordiga, and Groppo, 2011). This dichotomy has led the scientific community to make many studies of the catalytic sites and structures located inside the framework.

Yet the structure of the species present or formed on interaction with molecules inside the pores of these microporous and mesoporous materials could not be studied with the classical surface science methods illustrated in Chapter 5. Only IR, Raman, UV–vis (in the diffuse reflectance mode), and NMR spectroscopies have proved their utility in the characterization of the species buried inside the channels.

In particular, Brønsted and Lewis acidities of zeolites are easily examined by FTIR analysis to determine the vibrational properties of Brønsted sites either before or after interaction with probe molecules (Ward and Hansford, 1969; Lamberti *et al.*, 2010). UV–vis spectroscopy in the reflectance mode has proved its utility as well.

Important contributions to the study of zeolites and mesoporous materials (Sayle *et al.*, 2006; Majda *et al.*, 2008) were made by the group of Professor Jacek Klinowski at the Department of Chemistry at the University of Cambridge, thanks to fundamental research in the field of solid-state NMR using the magic angle spinning technique (Hennel and Klinowski, 2005).

The structure and properties of the species formed inside the channels upon interaction with small molecules have further been intensively investigated by means of *ab initio* methods. Among the numerous researchers who contributed to this research field, Van Santen (1997), Richard Catlow (Sinclair and Catlow, 1996), Joachim Sauer (Hansen, Kerber, and Sauer, 2010), Piero Ugliengo (Bonelli *et al.*, 2000; Coasne and Ugliengo, 2012), and Alexis Bell (Bell, 2012) should be mentioned.

6.6 Shape-Selective Reactions Catalyzed by Zeolites and Zeolitic Materials

The combination of acid functionalities (or, in general, of catalytic centers) inside the channels, comprising shape and transition-state selectivity and easy regeneration, is the key point of the success of zeolitic materials in catalysis. This has allowed zeolites and zeotypes to represent a real revolution in the field of catalysis. For instance, the applications of zeolites containing metal clusters as cracking and reforming catalysts have allowed large-scale production of high quality fuels and bulk chemicals in the petroleum industry (Sinfelt, 2002; Taniewski, 2006). The catalytic activity of acid sites is either Brønsted or Lewis in character (Corma, 1995; Kazansky and Serykh, 2004). Ward proposed that two Brønsted acidic sites are needed for the formation, upon water elimination at high temperature, of a single Lewis Al^{3+} acidic site (Ward, 1968, 1969).

Zeolite catalysis has been expanding also into the areas of specialty and fine chemical synthesis (Davis, 1998). From ISI Web of Knowledge data, we can infer that the interest in zeolites and zeolitic materials for industrial catalytic applications is still very high and even growing.

It is difficult to give a comprehensive list of all scientists that have made important contributions in the zeolite area and their catalytic activity. Weitkamp (2000, 2012), Ferdi Schüth (Lim, Schrader, and Schueth, 2013; Vietze *et al.*, 1998), and Busca (2007) should be mentioned for their studies on the synthesis and catalytic properties of acidic zeolites. Another prominent scientist in this field is Avelino Corma who has authored a very large scientific volume on the synthesis and catalytic properties of many new materials and on their characterization (Corma, 1995; Corma, Domine, and Valencia, 2003). For these contributions in 2004, Corma (Instituto de Tecnologia Quimica, UPV-CSIC, Valencia, Spain) has been recipient of the Breck Award.

However, it is not possible to give a detailed description of all industrial processes where zeolitic materials are involved. Consequently only few representative examples will be given.

It should also be kept in mind that the examples discussed in the succeeding text are only a selection of the numerous shape-selective reactions catalyzed by zeolites and that they are reported only for the sake of illustration of the shape-selective properties.

6.6.1 Alkanes- and Alkene-Cracking and Isomerization

Catalytic cracking of medium-range petroleum fractions over zeolite-based acidic catalysts is responsible for the manufacture of over 50% of gasoline in the world. Alkanes represent the largest constituent of gas oils, and the mechanism behind their cracking reactions is of special interest. All hydrocarbon-cracking catalysts contain protic acid centers (Haag and Chen, 1987; Busca, 2007; Corma, 1995) with the structure shown in Figure 6.3.

Y zeolites with a Si/Al ratio in the 2.5–2 range and ZSM-5 zeolites with a Si/Al ratio in the 15–100 interval are the main catalysts. In the reactions of Brønsted acids with alkenes, the first stage is the formation of a π-complex involving the double bond of an alkene as demonstrated by IR imaging (Spoto *et al.*, 1994). The next stage is the formation of the carbenium ion in an equilibrium reaction:

$$H^+ + H_2C = CR_1R_2 \leftrightarrow H_2C^+ - CHR_1R_2$$

It should be noted that while the carbenium ions are likely the active species (transition-state complexes), what can be really observed in ZSM-5 by IR (Kazansky, 1999; Zecchina *et al.*, 2002) is the precursor where the carbenium ion fragment is still bonded to the zeolite (i.e., the carbenium is substituting H^+). In alkene cracking, the carbenium ion mechanism (complicated by oligomerization reactions and cracking of the oligomers) is generally accepted.

In contrast, the chemical mechanism of alkane cracking reactions in zeolites is more debated. In the product distribution obtained from cracking alkanes, alkenes, cycloalkanes, and alkyl aromatics, it is quite clear that carbenium and carbonium ion-like intermediates are present in zeolites during catalytic cracking of hydrocarbons (Haag, Lago, and Weisz, 1981). Mota *et al.* (1993) suggested that this is due to the formation of a pentacoordinated carbonium ion that attacks the zeolite acid site during alkane cracking. The short-lived carbonium ion would then rearrange and crack. This mechanism is close to that which Olah proposed in 1962–1967.

Skeletal isomerization of paraffins with carbon numbers between 5 and 10 is used to improve the octane number of gasoline. This process also requires very strong acid sites that are present in the zeolite channels. Conversion and selectivity are strongly increased when Pt or Pd nanoparticles are incorporated into the acidic zeolite and the reaction is carried out in the presence of H_2 (reforming catalysts). The proposed mechanism involves the formation of an intermediate olefin, which is produced by dehydrogenation of the alkane on the metallic site. The olefin diffuses to an acid site, where it is protonated to form a carbenium ion-like transition state that can isomerize or even crack depending on the hydrocarbon chain and the lifetime of the branched carbenium ion-like species.

Isomerization of hydrocarbons with shorter chains is more difficult, and zeolites with strongest acidity, such as mordenite and ZSM-5, are more convenient. The isomerization of *n*-butane to isobutane is particularly demanding and requires the formation of carbonium ions intermediates. Oxide-supported SbF_5–HF (similar to the Olah magic acid) catalysts are used to carry out this reaction.

6.6.2 Aromatic Ring Positional Isomerizations

In commercial applications, zeolites have demonstrated their superiority as catalysts. For instance, in the xylene isomerization process, HZSM-5 zeolites are not only active but also select the para-isomer:

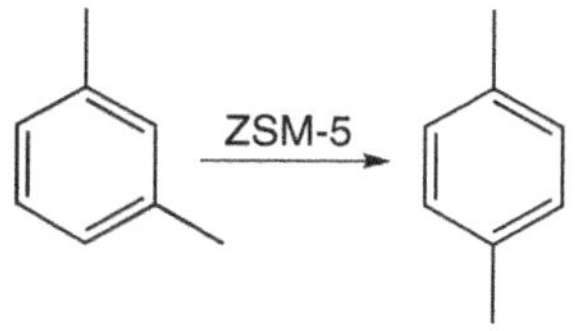

since the diffusion coefficient for *para*-xylene along the channels is more than two orders of magnitude higher than for *ortho*-xylene. This difference allows *para*-xylene to escape more rapidly outside the zeolite crystallite than the *ortho*-isomer. The role of protonic sites is similar to that discussed in Section 6.5.

6.6.3 Synthesis of Ethyl Benzene, Cumene, and Alkylation of Aromatic Molecules

The alkylation of aromatic hydrocarbons with olefins is a large-scale process in the chemical industry. In particular, the acid-catalyzed alkylation of benzene or toluene produces important intermediates, ethylbenzene, isopropylbenzene (cumene), *para*-diisopropylbenzene, C_{10}–C_{14} linear alkylbenzenes, and isopropyltoluenes (cymenes) (Bellussi *et al.*, 1995). In many petrochemical plants these alkylations are still performed using HF and $AlCl_3$–HCl as catalysts despite their several drawbacks, including corrosion. Corrosion can, of course, be partially deterred by supporting the acids. A catalyst of this type, not fully avoiding the corrosion problems, is the silica-supported phosphoric acid. Since 1976 ethylbenzene has been produced on a large scale using ZSM-5 catalysts by combining benzene (C_6H_6) and ethylene (C_2H_4) in an acid-catalyzed chemical reaction (Mobil process). Y-type zeolites have also been employed. A similar reaction leads to production of cumene from benzene and propylene (C_3H_6). Cumene is important because it is a chemical intermediate mainly used for the production of phenol and acetone. Mordenite and BEA zeolite are efficient and shape-selective catalysts for cumene synthesis. They are preferred to ZSM-5 because of their slightly larger channel diameters, which allow cumene to more easily diffuse out of the microcrystals.

In 2012 large fractions of ethylbenzene and cumene (which have different shapes):

Ethylbenzene   Cumene

were produced in this way.

Mesoporous silica–aluminas (MCM-41) are also active in the alkylation of toluene with propylene to form cymenes, which can easily diffuse in the mesopores. MCM-41 shows alkylation activities comparable to zeolites, to $AlCl_3$–HCl, and to supported phosphoric acid catalysts. Ethylene and propene

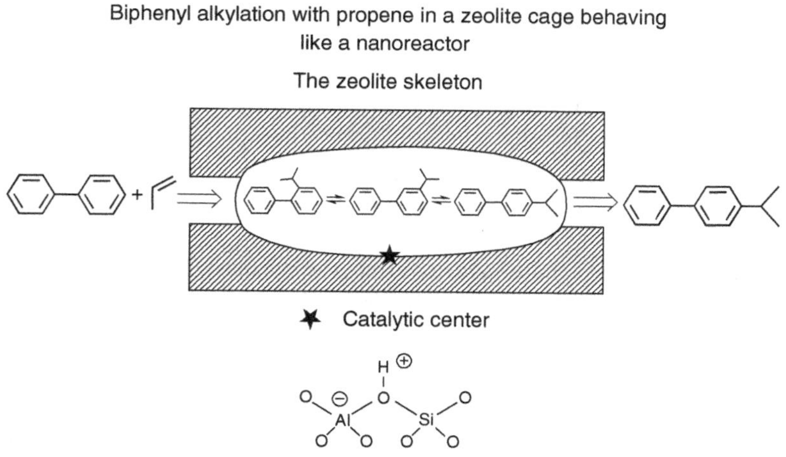

Figure 6.5 Shape selectivity in biphenyl alkylation with propene inside a zeolite cage behaving as a nanoreactor of molecular size. *Source*: Adapted from Pendyala *et al.*, 2013.

can be efficiently used also for the alkylation of more complex aromatics using high silica zeolites (ZSM-5) as selective catalysts. An example of shape-selective biphenyl alkylation reaction is shown in Figure 6.5.

Due to the small diameter of the channels, only the para-isomers are formed.

6.6.4   Friedel–Crafts Acylation of Aromatic Molecules

The Friedel–Crafts acylation reaction can be performed either in solution with $AlCl_3$ (homogeneous catalysis: old process) or with H-Beta zeolite (heterogeneous catalysis: new process). This reaction is shape selective because the acyl group is placed in the para position, so we will use it to compare homogeneous and heterogeneous processes and to illustrate the superior performance of zeolite catalyst in comparison with the homogeneous counterpart ($AlCl_3$).

In the homogeneous path, the yield is 85%, the solvent must be recycled, a phase separation is required, HCl is formed, and the catalyst suffers of degradation. In the heterogeneous path, the catalyst can be regenerated, no solvent is present, acetic acid is formed and can be more easily handled than HCl, and the yield is larger than 95% (because the catalyst is shape selective and formation of larger products is inhibited).

6.6.5   Toluene Alkylation with Methanol

The reaction of methanol with toluene to give paraxylene occurs in the HZSM-5 and mordenite channels following the scheme:

The reaction is highly shape selective, as the formation of *ortho-* and *meta*-isomers is inhibited because they do not diffuse efficiently through the channels' mouths.

6.6.6 Asaki Process for Cyclohexanol Synthesis

The industrial production of cyclohexanol by hydration of cyclohexene over HZSM-5 is a good example of a heterogeneous reaction where the zeolite catalyst (HZSM-5 or H-Beta) has led to significant advances with respect to former catalysts.

6.6.7 Methanol-to-Olefins (MTO) Process

As we mentioned in previous chapters, liquid hydrocarbon fuels are essential components in the global energy chain, owing to their high energy density and easy transportability. In a post-oil society (as envisioned by Nobel Laureate George A. Olah), fuel production would rely on alternative carbon sources, such as biomass and CO_2. The methanol-to-hydrocarbons (olefin) process is a key step in such a direction.

The reactions involved in this process are catalyzed by the zeolite acid sites and by compounds trapped within the voids forming the so-called hydrocarbon pool. Among the first species initially formed by proton-catalyzed dehydration of methanol are olefins following the reactions

$$2CH_3OH \rightarrow CH_2 = CH_2 + 2H_2O$$
$$3CH_3OH \rightarrow CH_3 - CH = CH_2 + 3H_2O$$

The produced olefins further interact with the protonic sites via carbenium ion intermediates, forming charged oligomeric species. These big compounds, called the "hydrocarbon pool," are trapped inside the pores. At steady state the hydrocarbon pool continually reacts with methanol to yield olefins (Dahl and Kolboe, 1994).

The possibility to convert methanol into hydrocarbons inspired Olah to advocate the methanol economy in 2006, and we will extensively report on this idea in Chapter 7 in the contest of the green economy.

6.6.8 Nitto Process

An industrially successful example of shape selectivity is the Nitto process developed in 1984 by researchers at Nitto Chemical, Japan (Fujita *et al.*, 1996), for the production of di- and monomethylamine from methanol and ammonia following the reaction

$$2MeOH + NH_3 \rightarrow Me_2NH/MeNH_2$$

These compounds are important intermediates for solvents, pharmaceuticals, and detergents. The catalyst is ion-exchanged mordenite. The appropriate diameter of the channels is preventing the formation of trimethylamine.

6.6.9 Butylamine Synthesis

The synthesis of butylamine from isobutene and ammonia in the pores of beta zeolite is another shape selective reaction discovered by BASF researchers, whose introduction in 1996 (Dingerdissen *et al.*, 1996) made obsolete the traditional processes based on the dangerous HCN molecule.

The amines synthesis via catalyzed gas-phase reactions is indicative of the utility of zeolites also for the intermediates in fine chemicals production.

6.6.10 Beckman Rearrangements on Silicalite Catalyst

Silicalite has an interesting application as a very weak acid catalyst in an industrially important reaction, the vapor-phase Beckmann rearrangement of cyclohexanone oxime to caprolactam (the Sumitomo process), occurring near 300°C.

The active sites for this reaction, which is also catalyzed by amorphous silica but less efficiently (Montedipe), are thought to be catalyzed by internal silanol nests. Hydroxyl nests are present at structural defects where, for instance, one silicon atom is missing in the periodic structure (indicated in the figure with a circle) and is replaced by a cluster of $\equiv$Si—OH groups (Bordiga *et al.*, 2000; Gu *et al.*, 2007) (Figure 6.6).

6.6.11 Partial Oxidation Reactions Using Titanium Silicalite

In the 1980s researchers at ENI, Italy, inserted titanium active sites in the framework of the silicalite molecular sieve and patented the first titanosilicate (Taramasso, Perego, and Notari, 1980, 1983), denoted as TS-1, where Ti is substituting Si in the structure. TS-1 is an important, environmentally benign selective oxidation catalyst that uses hydrogen peroxide as the oxidant, producing only water as a by-product (Bellussi and Rigguto, 1994) (Figure 6.7).

TS-1 is an outstanding example of a zeolitic material not containing aluminum, where both transition-state and product shape selectivities are

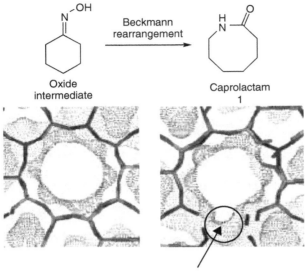

Figure 6.6 Beckmann rearrangement of cyclohexanone oxime to caprolactam in a silicalite cavity containing hydroxyl nests.

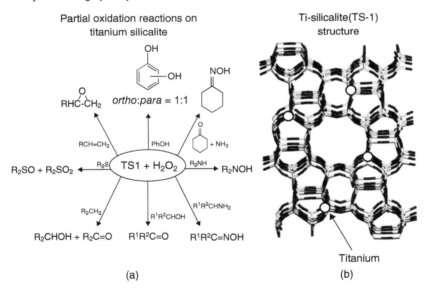

Figure 6.7 (a) Selective partial oxidation reactions catalyzed by TS-1 using H_2O_2 as oxidant. (b) TS-1 structure. The circles stay for Ti atoms substituting Si in the silicalite structure and acting as single-site catalytic centers. The TS-1 and silicalite frameworks are the same of ZSM-5, where a small fraction of silicon atoms is substituted by Al.

considered to be relevant. In this case the selectivity is dictated by the combination of isolated Ti(IV) species of a well-defined structure hosted in the MFI framework and acting as the hydrophilic "single site" and of channels of appropriate dimension and hydrophobic properties. The combination of these two factors results in an ensemble of nanometric dimensions, where the Ti centers can be reached only by reagents of the right size and appropriate hydrophilic/hydrophobic character. For this reason TS-1 is an election catalyst for the selective oxidation of organic substrate using hydrogen peroxide as oxidizing agent. In particular, we can cite phenol hydroxylation, olefin epoxidation, alkane oxidation, oxidation of ammonia to hydroxylamine, conversion of secondary amines to dialkyl hydroxylamines, conversion of secondary alcohols to ketones, and cyclohexanone ammoximation. In the phenol hydroxylation reaction, *para-* and *ortho-*isomers are exclusively obtained inside the channels of TS-1.

Of course, all these reactions involve only molecules, which can diffuse along the narrow channels of TS-1. A few years after the synthesis, it was discovered that the TS-1 is active in the alkane oxidation under mild conditions yielding ketones and alcohols. The results were reported in 1990 (Tatsumi *et al.*, 1990).

For the discovery of titanium silicalite, the ENI group, constituted by G. Bellussi, M. Clerici, V. Fattore, B. Notari, G. Perego, F. Buonomo, A. Esposito, F. Maspero, C. Neri, and U. Romano, was the recipient of the Breck award in 1992 with the following justification: *"for advancing our knowledge of the structures and properties of titanium MFI zeolites and for demonstrating both the potential and applications of these novel catalysts for partial oxidation reaction."* (Figure 6.8).

6.6.12 Nylon-6 Synthesis: The Role of Zeolitic Catalysts

The production of nylon-6 follows multistep processes that, due to space limitation, we cannot describe in detail. Among the industrially adopted processes, one path based on extensive application of zeolites as catalysts is highly environmentally clean.

The first step is benzene reduction to cyclohexene, as described in Chapter 5 in the section on heterogeneous hydrogenation catalysts. The second step is hydration with formation of cyclohexanol, using ZSM-5 as catalysts. The third step is heterogeneous dehydrogenation using a heterogeneous catalyst (chromium oxide). The fourth step is ammoxidation using TS-1 from ammonia and hydrogen peroxide. The fifth step is the Beckman rearrangement on silicalite with the formation of caprolactam (precursor of nylon 6).

6.6.13 Pharmaceutical Product Synthesis

The versatility of zeolitic material as a catalyst for the synthesis of chemicals of pharmaceutical interest can be illustrated by the preparation of paracetamol,

Figure 6.8 Picture from G. Bellussi of the ceremony for the attribution of the 1992 D. Breck Award to the Eni team: (1) A. Esposito, (2) C. Neri, (3) F. Buonomo, (6) G. Bellussi, (7) G. Perego, (8) U. Romano, (9) V. Fattore, (10) M. Clerici, (11) B. Notari. From IZA: (4) H. Karge, (5) P. Jacobs, (12) E.M. Flanigen.

a mild analgesic and a major ingredient in numerous cold and flu remedies, where HZSM-5, TS-1, and silicalite are involved in a three-step process. The details of the process are omitted for the sake of simplicity.

6.7 Organic–Inorganic Hybrid Zeolitic Materials and Inorganic Microporous Solids

6.7.1 Organic–Inorganic Hybrid Zeolitic Materials

A great achievement in the area zeolite synthesis was made by Tatsumi and his coworkers (Yamamoto *et al.*, 2003) who synthesized a new class of materials,

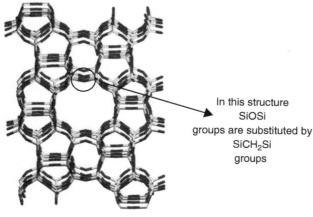

In this structure
SiOSi
groups are substituted by
SiCH₂Si
groups

MFI zeolite structure

Figure 6.9 Structure of an organic–inorganic hybrid zeolite. *Source*: Adapted from Tatsumi *et al.*, 2014.

coined ZOLs (zeolites with organic groups as a lattice), in which a methylene group is substituting oxygen in the framework of a MFI zeolite. ZOL materials were the first organic–inorganic hybrid composites in which an organic group was incorporated as part of the framework into a crystalline microporous material. The structure is shown in Figure 6.9.

The introduction of organic groups into the framework provided the material with enhanced hydrophobicity. However, to our knowledge, these interesting structures did not find application in catalysis.

6.7.2 ETS-10: A Microporous Material Containing Monodimensional TiO_2 Chains

The titanosilicate ETS-10 is a microporous material formed by the TiO_6 octahedra and the SiO_4 tetrahedra linked through bridging oxygen atoms. Each titanium octahedron incorporates a net negative charge of -2 that is compensated by extraframework charge-balancing cations (Na^+ and K^+ in the as synthesized material). The linked structural TiO_6 and SiO_4 units generate a three-dimensional interconnected system of channels. Besides its great exchange capacity toward divalent cations (which is the main characteristic of this material), one of the interesting features of ETS-10 is that the TiO_6 octahedra form linear chains isolated from one another by the siliceous matrix. This results in the generation of one-dimensional Ti–O–Ti–O–Ti wires that are responsible of the peculiar optical and electronic properties of the material (Damin *et al.*, 2004).

Due to the presence of the wires associated with collective properties similar to TiO_2, ETS-10 displays good selectivity in photocatalytic reactions, allowing the preferential degradation of only one given component from the mixture and overcoming the absence of selectivity, which is typical of heterogeneous photocatalysts (Llabres *et al.*, 2003).

6.7.3 Hydrotalcites: Microporous Solids with Exchangeable Anions

While zeolites exchange cationic species, other high surface area materials exchange anionic species. Hydrotalcite-like compounds with the formula $[Mg_{1-x}Al_x(OH)_2]^{x+} [A_{x/n}^{n-} \cdot H_2O]^{x-}$, where x is comprised in the 0–033 interval and A^{n-} is an exchangeable anion, are example of such solids.

The structure is similar to that of the mineral hydrotalcite. The importance of this family of compounds lies in the fact that, unlike zeolites, they can exchange anions such as F^-, CI^-, Br^-, SO_4^{2-}, and CO_3^{2-}. Upon heating, hydrotalcites are transformed into mixture of high surface area oxides.

The hydrotalcites have been used in this capacity or (mainly) after calcination. The most interesting properties of the oxides obtained by calcination are the following:

1. High surface area
2. Basic properties
3. Memory effect.
   The memory effect which allows the reconstruction, under mild conditions, of the original hydrotalcite structure when contacting the product of the thermal treatment with water solutions containing various anions.

The applications in catalysis are essentially as supports and as basic catalysts. For the scope of this book, we only mention their activity in polymerization of alkene oxides (a property shown also by MgO) and aldol condensation. An exhaustive review of the properties of this family of solids can be found in the Cavani *et al.* review (1991).

6.8 Microporous Polymers and Metal–Organic Frameworks (MOFs)

The research in the area of zeolites and zeolitic materials, although still very much limited to investigating zeolites as adsorbers and catalysts, can be considered as approaching maturity, even though the number of publications on catalytic applications is still growing (ISI Web of Science).

The same cannot be said for microporous polymers and MOFs. In this area the number of items published in the last few years shows a strong increment.

6.8.1 Microporous Polymers

It is commonly accepted that organic chemistry holds immense potential for the construction of new molecules. The same holds for the construction of

fully organic porous frameworks characterized by superior adsorptive properties and catalytic properties. Although the discovery of fully organic porous
frameworks is not recent, only in recent years has the interest been on the rise.
Among the explanations of this change is the increased ability of scientists to
connect small molecular fragments via coupling reactions as made possible by
the already described Heck, Negishi, Suzuki, Sonogashira, Stille, and Hiyama
processes (to mention only the most important).

The formidable possibilities of constructing extended tridimensional structures, for instance, by means, for instance, of Sonogashira coupling, as discussed in a preceding chapter and illustrated by the reactions shown below, are
representative of methodologies that obtain the extended structures (CPM-1
being the first example):

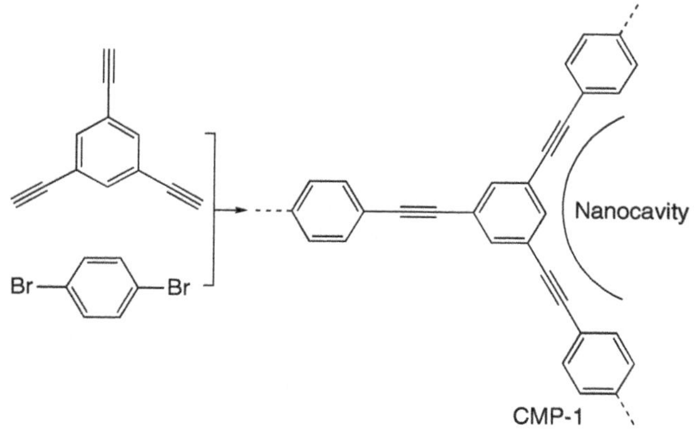

CMP-1

The so obtained robust microporous polymers belong to the so-called CMP
class, which can reach surface areas up to $1000 \, m^2/g$. The Sonogashira reaction
scheme also allows one to build polymeric material containing metallic functionalities with potential catalytic applications as shown schematically below
for the systems of CMP class containing Ir:

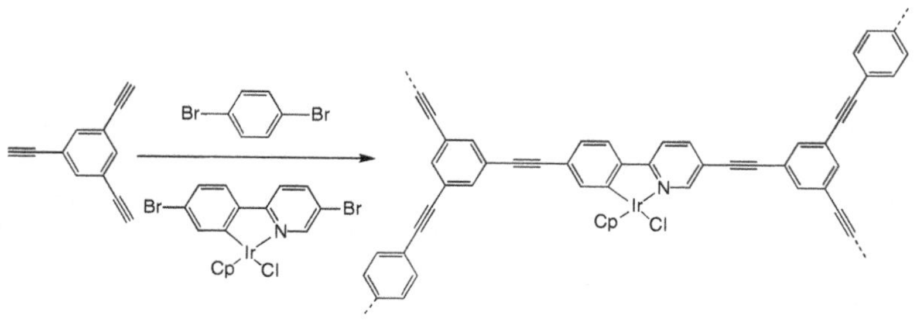

Alternatively, the Suzuki reaction can be used to obtain porphyrin-based polymers containing Fe (FeP-CMP).

If we additionally consider all the coupling possibilities ensured by the already described Suzuki, Negishi, Stille, and Hiyama reactions and by other cross-coupling reactions not described in this chapter, it is amazing how many microporous tridimensional structures can be produced (Xu *et al.*, 2013) with potential applications as adsorbers and catalysts. For instance, the FeP-CMP system described earlier (which can have surface area as high as $1300 \, \text{m}^2/\text{g}$) is a good olefin epoxidation catalyst that mimicks analogous homogeneous and enzymatic catalysts. It is outside the scope of this book to give a comprehensive description of this field as it has rapidly expanded in the 2000 to 2014 period, based on the data obtained from the ISI Web of Science. For the interested reader, the review by Dawson *et al.* (Dawson, Cooper, and Adams, 2012) can be useful.

6.8.2 Metal–organic Frameworks

Also research on the so-called MOFs has rapidly expanded. The last decade's to particularly large results obtained by consulting the ISI Web of Science shows nearly 700 papers published by 2013.

MOFs are compounds consisting of metal ions or clusters coordinated to often rigid organic molecules (linkers) to form one-, two-, or three-dimensional structures that can be microporous. Considering the variety of linkers and metal ions and clusters, the number of structures that can be synthesized is very huge. The flexibility with which the constituent's geometry, size, and functionality can be varied has led to the synthesis of a plethora of different MOFs over the past decade (Furukawa *et al.*, 2013).

The prototype structure is MOF-5, a scaffold constituted by (Zn_4O) clusters linked together by linkers. The simplest linker is

The diameter of the pores can be varied by changing the linker. The result is exceptionally lightweight scaffolding with an internal surface area up to $2000 \, \text{m}^2/\text{g}$, and the resulting structure is that shown in Figure 6.10. Internal cavities with larger diameter can be obtained by changing the linker. Furthermore the linker can be potentially designed to carry specific chemical functionalities, which has opened the way for catalytic applications.

Although the synthesis, structural, and adsorption aspects of these systems still dominate the rapidly growing literature, more and more members of the scientific community are starting to turn their attention to potential catalytic applications.

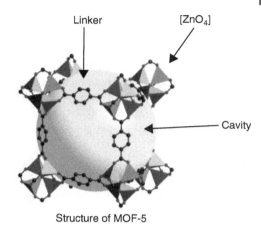

Figure 6.10 Structure of MOF-5. The internal cavities are represented by spheres. *Source*: Adapted from Rowsell and Yaghi, 2004.

Structure of MOF-5

It must be underlined that the application of MOFs to catalysis is still at the infancy.

The heterogeneous nature of MOFs, favors the separation of reactants and products, and their flexible design are definite elements of strength. However, so far, the essentially organic character of MOFs and sensitivity to moisture are limiting factors in their use at elevated temperatures and in presence of water.

References

Akabori, S., Ohno, K., and Narita, K. (1952) Akabori Reaction for Determining the C-terminus. *Bulletin of the Chemical Society of Japan*, **25**, 214–218.

Anastas, P.T., Kirchhoff, M.M., and Williamson, T.C. (2001) Catalysis as a foundational pillar of green chemistry. *Applied Catalysis A. General*, **221**, 3–13.

Arakawa, H., Aresta, M., Armor, J.N. *et al.* (2001) Catalysis research of relevance to carbon management: progress, challenges, and opportunities. *Chemical Reviews*, **101**, 953–996.

Baerlocher, C., McCusker, L.B., and Olson, D.H. (2007) *Atlas of Zeolites Framework Types*. Elsevier, Amsterdam.

Bagot, P.A.J. (2004) Fundamental surface science studies of automobile exhaust catalysis. *Materials Science and Technology*, **20**, 679–694.

Balzani, V. and Juris, A. (2001) Photochemistry and photophysics of Ru(II)-polypyridine complexes in the Bologna group. From early studies to recent developments. *Coordination Chemistry Reviews*, **211**, 97–115.

Barrer, R.M. (1979) Chemical nomenclature and formulation of compositions of synthetic and natural zeolites. *Pure and Applied Chemistry*, **51**, 1091–1100.

Barrer, R.M., Peterson, D.L., and Schoenborn, B.P. (1966) Separation of gases by zeolites. *Science*, **153**, 556.

Beck, J.S., Vartuli, J.C., Roth, W.J. *et al.* (1992) A new family of mesoporous molecular sieves prepared with liquid crystal templates. *Journal of the American Chemical Society*, **114**, 10834–10843.

Bell A. (2012) Applications of quantum theory and molecular dynamics to the analysis of zeolite-catalyzed reactions. Abstracts of papers of the American Chemical Society, 243, Meeting Abstract: 19-CATL.

Bellussi, G. and Rigguto, M.S. (1994) Metal ions associated to the molecular sieve framework, possible catalytic oxidation sites. *Studies in Surface Science and Catalysis*, **85**, 177–179.

Bellussi, G., Pazzuconi, G., Perego, C. *et al.* (1995) Liquid phase alkylation of benzene with light olefins catalyzed by β zeolites. *Journal of Catalysis*, **157**, 227.

Bhatia, S. (2000) *Zeolite Catalysis: Principles and Applications*. CRC Press, Boca Raton, FL.

Bonelli, B., Civalleri, B., Fubini, B. *et al.* (2000) Experimental and quantum chemical studies on the adsorption of carbon dioxide on alkali-metal-exchanged ZSM-5 zeolites. *Journal of Physical Chemistry B*, **104**, 10978–10988.

Bordiga, S., Roggero, I., Ugliengo, P. *et al.* (2000) Characterization of defective silicalites. *Journal of the Chemical Society, Dalton Transactions*, 3931–3929.

Braundberger, S., Krocher, O., Tissler, A., and Althoff, R. (2008) The state of the art in selective catalytic reduction of NO_x by ammonia using metal-exchanged zeolite catalysts. *Catalysis Reviews: Science and Engineering*, **50**, 492–531.

Breck, D.W. (1964) Crystalline molecular sieves. *Journal of Chemical Education*, **41**, 678.

Breck, D.W. (1974) *Zeolite Molecular Sieves – Structure, Chemistry and Uses*, pp. 313–320, 731–738. Wiley, New York.

Breck, D.W. (1975) Zeolites chemistry and reactions. *Material Science Research*, **10**, 211.

Breck, D.W., Eversole, W.G., and Milton, R.M. (1956) New synthetic crystalline zeolites. *Journal of the American Chemical Society*, **78**, 2338–2339.

Busca, G. (2007) Acid catalysts in industrial hydrocarbon chemistry. *Chemical Reviews*, **107**, 5366–5410.

Cavani, F., Clause, O., Trifiro, F., and Vaccari, A. (1991) Anionic clays with hydrotalcite-like structure as precursors of hydrogenation catalysts. *Advances in Catalysts Design*, 186–190.

Che, M. and Bennett, C.O. (1989) The influence of particle-size on the catalytic properties of supported metals. *Advances in Catalysis*, **36**, 55–72.

Chizallet, C. and Raybaud, P. (2009) Pseudo-bridging silanols as versatile brønsted acid sites of amorphous aluminosilicate surfaces. *Angewandte Chemie, International Edition*, **48**, 2891–2893.

Cimino, A. and Stone, F.S. (2002) Oxide solid solutions as catalysts. *Advances in Catalysis*, **47**, 141–306.

Cimino, A., Gazzoli, D., Indovina, V. *et al.* (1999) High and low surface area NiO–MgO and CoO–MgO solid solutions: a study of XPS surface composition and CO oxidation activity. *Topics in Catalysis*, **8**, 171–174.

Coasne, B. and Ugliengo, P. (2012) Atomistic model of micelle-templated mesoporous silicas: structural, morphological, and adsorption properties. *Langmuir*, **28**, 11131–11141.

Corma, A. (1995) Inorganic solid acids and their use in acid-catalyzed hydrocarbon reactions. *Chemical Reviews*, **95**, 559–614.

Corma, A., Domine, M., and Valencia, S. (2003) Water-resistant solid Lewis acid catalysts: Meerwein–Ponndorf–Verley and Oppenauer reactions catalyzed by tin-beta zeolite. *Journal of Catalysis*, **215**, 294–304.

Dahl, I.M. and Kolboe, S. (1994) On the reaction-mechanism for hydrocarbon formation from methanol over SAPO-34.1, isotopic labelling studies of the co-reaction of ethene and methanol. *Journal of Catalysis*, **149**, 458–464.

Damin, A., Llabres i Xamena, F., Lambert, C. *et al.* (2004) Ti quantum wires in the titanosilicate ETS-10. *Journal of Physical Chemistry*, **108**, 1328–1336.

Davis, M.E. (1998) Zeolite based catalysts for chemical synthesis. *Microporous and Mesoporous Materials*, **21**, 173–182.

Dawson, R., Cooper, A.I., and Adams, D.J. (2012) Nanoporous organic polymer networks. *Progress in Polymer Science*, **37**, 530–563.

Degnan, T.F. (2003) The implications of the fundamentals of shape selectivity for the petroleum and petrochemical industries. *Journal of Catalysis*, **216**, 32–46.

Dingerdissen, U., Kummer, R., Stops, P., Muller, U., Jurgen, H., and Eller, K. (1996) Preparation of amines from olefins over boron beta-zeolites. US Patent 6143934.

Fernandez, A.-B., Marinas, A., Blasco, T. *et al.* (2006) Insight into the active sites for the Beckmann rearrangement on porous solids by in situ infrared spectroscopy. *Journal of Catalysis*, **243**, 270–277.

Flanigen, E.M. (1976) in *Zeolite Chemistry and Catalysis*, vol. **171** (ed. J.A. Rabo), pp. 80–101. Elsevier, Amsterdam.

Flanigen, E.M. (1991) Zeolites and molecular sieves an historical perspective. *Studies in Surface Science and Catalysis*, **58**, 13–34.

Flanigen, E.M. (2001) Zeolites and molecular sieves: an historical perspective. *Studies in Surface Science and Catalysis*, **137**, 11–35.

Flanigen, E.M. (2005) Molecular sieve zeolites: an industrial research success story. *Research-Technology Management*, **48** (**4**), 29–33.

Flanigen, E.M., Bennet, J.M., Grose, R.W. *et al.* (1978) Silicalite, a new hydrophobic crystalline silica molecular sieve. *Nature*, **271**, 512–516.

Flanigen, E.M., Lok, B.M., Patton, R.L., Wilson, S.T. (1986) *Aluminophosphate Molecular Sieves and the Periodic Table*. Proceedings 7th International Zeolite Conference, Tokyo, p. 103 (eds. Murakami, Y., Ijima, A., and Ward, J.W.). Elsevier, Amsterdam.

Flanigen, E.M., Lok, B.M., Patton, R.L., and Wilson S.T. (1987) *Aluminophosphate Molecular Sieves and the Periodic Table*. New Developments in Zeolite Science and Technology. 1986 Proceedings 7th International Zeolite Conference, Tokyo (eds. Murakami Y., Ijima A., and Ward, J.W.), pp. 103–112. Elsevier, Amsterdam.

Flanigen, E.M., Patton, R.L., and Wilson, S.T. (1988) Structural, synthetic and physicochemical concepts in aluminophosphate-based molecular sieves. *Studies in Surface Science and Catalysis*, **37**, 13–27.

Fujita, T., Fukatsu, M., Niwa, K., and Ogura, K. (1996) Production of methylamines. US Patent 5688854.

Furukawa, H., Cordova, K.E., O'Keeffe, M., and Yaghi, O.M. (2013) The chemistry and applications of metal-organic frameworks. *Science*, **341**, 6149.

Gribov, E.N., Zavorotynska, O., Agostini, G. *et al.* (2010) FTIR spectroscopy and thermodynamics of CO and H_2 adsorbed on ɣ-, ϑ- and α-Al_2O_3. *Physical Chemistry Chemical Physics*, **12**, 6474–6482.

Gu, Z.Z., Chen, H.H., Zhang, S. *et al.* (2007) Rapid synthesis of monodisperse polymer spheres for self-assembled photonic crystals. *Colloids and Surfaces A: Physicochemical and Engineering Aspects*, **302**, 312.

Haag, W.O. and Chen, N.Y. (1987) in *Catalytic Design, Progress and Perspectives* (ed. L.L. Hegedus), pp. 163–212. Wiley, New York.

Haag, W.O., Lago, R.M., and Weisz, P.B. (1981) Transport and reactivity of hydrocarbon molecules in a shape-selective zeolite. *Faraday Discussions of the Chemical Society*, **72**, 317–330.

Hansen, N., Kerber, T., and Sauer, J. (2010) Quantum chemical modeling of benzene ethylation over H-ZSM-5 approaching chemical accuracy: a hybrid MP2:DFT study. *Journal of American Chemical Society*, **132**, 11525–11538.

Hara, M. (2010) Biomass conversion by a solid acid catalyst. *Energy Environmental Science*, **3**, 601–607.

Haruta, M. (1997) Size- and support-dependency in the catalysis of gold. *Catalysis Today*, **36**, 153–166.

Hattori, H. (2001) Solid base catalysts: generation of basic sites and application to organic synthesis. *Applied Catalysis A: General*, **222**, 247–259.

Hennel, J.W. and Klinowski, J. (2005) Magic angle spinning: a historical perspective, in *New techniques in solid-state NMR* (ed. J. Klinowski), pp. 1–14. Springer, Berlin.

Holmen, A. (2009) Direct conversion of methane to fuels and chemical. *Catalysis Today*, **142**, 2–8.

Kalyanasundaran, K. and Graetzel, M. (2010) Artificial photosynthesis: biomimetic approaches to solar energy conversion and storage. *Current Opinion in Biotechnology*, **21**, 298–310.

Kazansky, V.B. (1999) Adsorbed carbocations as transition states in heterogeneous acid catalyzed transformations of hydrocarbons. *Catalysis Today*, **51**, 419–434.

Kazansky, V.B. and Serykh, A.I. (2004) Unusual forms of molecular hydrogen adsorption by Cu+1 ions in the copper-modified ZSM-5 zeolite. *Catalysis Letters*, **98**, 77–79.

Lamberti, C., Zecchina, A., Groppo, E., and Bordiga, S. (2010) Probing the surfaces of heterogeneous catalysts by in situ IR spectroscopy. *Chemical Society Reviews*, **39**, 4951–5001.

Lasier W.A. and Rigby G.W. (1941) Catalytic process for the production of caprolactam, amino-capronitrile and hexamethylene diamine. US Patent 2,234,566.

Lim, I.H., Schrader, W., and Schueth, F. (2013) The formation of zeolites from solution – analysis by mass spectrometry. *Microporous and Mesoporous Materials*, **166**, 20–36.

Lindlar, H. and Dubuis, R. (1973) Palladium catalyst for partial reduction of acetylenes. *Organic Synthesis*, **5**, 88–89.

Llabrés i Xamena, F., Lamberti, C., Damin, A. *et al.* (2003) Enhancement of the ETS-10 titanosilicate activity in the shape-selective photocatalytic degradation of large aromatic molecules by controlled defect production. *Journal of the American Chemical Society*, **125**, 2264–2271.

Lu, J., Gao, H., Shaikhutdinov, S., and Freund, H. (2007) Gold supported on well-ordered ceria films: nucleation, growth and morphology in CO oxidation reaction. *Catalysis Letters*, **114**, 8–16.

Lunsford, J.H. (2000) Catalytic conversion of methane to more useful chemicals and fuels: a challenge for the 21st century. *Catalysis Today*, **63**, 165–174.

Majda, D., Paz, F.A.A., Friedrichs, O.D. *et al.* (2008) Hypothetical zeolitic frameworks: in search of potential heterogeneous catalysts. *Journal of Physical Chemistry C*, **112**, 1040–1047.

McBain, J.W. (1932) *The Sorption of Gases and Vapours by Solids.* G. Routledge & Sons, London.

Milton R.M. (1959) US Patent 2,882,243.

Milton, R.M., Occelli, L., and Robson, H.E. (1989) Zeolite synthesis. In ACS Symposium series, in *Molecular Sieve Science and Technology: A Historical Perspective*, vol. **398**. American Chemical Society, pp. 1–10.

Mota, C., Nogueira, L., Menzies, S. *et al.* (1993) H-D exchanged between zeolites and alkanes evidence the formation and rearrangement of pentacoordinated carbonium ions. *Studies in Surface Science and Catalysis*, **75**, 463–475.

Nagahara, H., Ono, M., Konishi, M., and Fukuoka, Y. (1997) Partial hydrogenation of benzene to cyclohexene. *Applied Surface Science*, **121/122**, 448–451.

Panov, G.I., Sheveleva, G.A., Kharitonov, A.S. *et al.* (1992) Oxidation of benzene to phenol by nitrous oxide over FeZSM-5 zeolites. *Applied Catalysis A*, **82**, 31–36.

Pendyala, V., Jacobs, G., Shafer, D. *et al.* (2013) Shape-selective alkylation of biphenyl with propylene using zeolite and amorphous silica–alumina catalysts. *Applied Catalysis A: General*, **453**, 195–203.

Riedl H. and Pfleiderer G. (1936) US Patent 2,158,525

Roffia, P., Padovan, M., Moretti, E., and De Alberti, G. (1987) European Patent 0208 311.

Rowsell, J. and Yaghi, O. (2004) Metal–organic frameworks: a new class of porous materials. *Microporous and Mesoporous Materials*, **73**, 3–14.

van Santen, R.A. (1997) Quantum-chemistry of zeolite acidity. *Catalysis Today*, **38**, 377–390.

Sayle, D.C., Mangili, B.C., Klinowski, J., and Sayle, T.X. (2006) Simulating self-assembly of ZnS nanoparticles into mesoporous materials. *Journal of the American Chemical Society*, **128**, 15283–15291.

Scarano, D., Bertarione, S., Cesano, F. *et al.* (2006) Plate-like zinc oxide microcrystals: synthesis and characterization of a material active toward hydrogen adsorption. *Catalysis Today*, **116**, 433–438.

Schauermann, S., Nilius, N., Shaikhutdinov, S., and Shamil, S. (2013) Nanoparticles for heterogeneous catalysis: new mechanistic insights. *Accounts of Chemical Research*, **46**, 1673–1681.

Sheldon, R.A. and Downing, R.S. (1999) Heterogeneous catalytic transformations for environmentally friendly production. *Applied Catalysis A*, **189**, 163–183.

Sinclair, P.E. and Catlow, C.R.A. (1996) Computational studies of the reaction of methanol at aluminosilicate Bronsted acid sites. *Journal of the Chemical Society – Faraday Transactions*, **92**, 2099–2105.

Sinfelt, J.H. (2002) Role of surface science in catalysis. *Surface Science*, **500**, 923–946.

Skeels, G.W. and Flanigen, E.M. (1999) Zeolite chemistry VII – framework substitution for aluminum in zeolites VIA secondary synthesis treatment. *Studies in Surface Science and Catalysis*, **49**, 331–341.

Spoto, G., Bordiga, S., Ricchiardi, G. *et al.* (1994) IR study of ethene and propene oligomerization on H-ZSM-5: hydrogen-bonded precursor formation, initiation and propagation mechanisms and structure of the entrapped oligomers. *Journal of the Chemical Society, Faraday Transactions*, **90**, 2827–2835.

Steele, B.C.H. and Heinzel, A. (2001) Materials for fuel-cell technologies. *Nature*, **414**, 345–352.

Tanabe, K. (1970) *Solid Acids and Bases: Their Catalytic Properties*. Academic Press, New York.

Tanabe, K. and Holderich, W.F. (1999) Industrial application of solid acid–base catalysts. *Applied Catalysis A: General*, **181**, 399–434.

Taniewski, M. (2006) Sustainable chemical technologies – development trends and tools. *Chemical Engineering & Technology*, **29**, 1397–1403.

Taramasso, M., Perego, G., and Notari, B. (1980) *Molecular Sieve Borosilicates*. Proceedings of 5th International Zeolite Conference, 40.

Taramasso, M., Perego, G., and Notari, B. (1983) Preparation of porous crystalline synthetic material comprised of silicon and titanium oxides. US Patent, US4410501 A.

Tatsumi, T., Nakamura, M., Negishi, S., and Tominaga, H. (1990) Selective oxidation of alkanes with. H2O2 catalysed by titanosilicate. *Journal of the Chemical Society, Chemical Communications*, **476**, 476–477.

Tatsumi, T., Liu, S.-B., Matsuhashi, H., Katada, N. (2014) *Acid-Base Catalysis Advanced Sciences and Spreading Applications to Solutions of Environmental, Resources and Energy Issues.* ABC-7, 7th international symposium on acid-base catalysis, Tokyo, May 12–15, 2013, Catalysis Today, 226, 1.

Tosheva, L. (2001) Molecular sieve macrostructures prepared by resin templating. PhD thesis. Lulea University of Technology.

Ugliengo, P., Sodupe, M., Musso, F. *et al.* (2008) Realistic models of hydroxylated amorphous silica surfaces and MCM-41 mesoporous material simulated by large-scale periodic B3LYP calculations. *Advanced Materials*, **20**, 4579–4585.

Vietze, U., Kraus, O., Laeri, F. *et al.* (1998) Zeolite-dye microlasers. *Physical Review Letters*, **81**, 4628.

Wang, H., Kaden, W., and Dowler, R. (2012) Model oxide-supported metal catalysts - comparison of ultrahigh vacuum and solution based preparation of Pd nanoparticles on a single-crystalline oxide substrate. *Physical Chemistry Chemical Physics*, **14**, 11525–11533.

Ward, J.W. (1968) The nature of active sites on zeolites. IV. The influence of water on the acidity of X and Y type zeolites. *Journal of Catalysis*, **11**, 238–250.

Ward, J.W. (1969) The nature of active sites on zeolites: VIII. Rare earth Y zeolite. *Journal of Catalysis*, **13**, 321–327.

Ward, J.W. and Hansford, R.C. (1969) The detection of acidity on silica-alumina catalysts by infrared spectroscopy – pyridine chemisorption: relationships between catalyst acidity and activity. *Journal of Catalysis*, **13**, 154–160.

Weisz, P.B. and Frilette, V.J. (1960) Intra-crystalline and molecular-shape selective catalysis by zeolite salts. *Journal of Physical Chemistry*, **64**, 382.

Weitkamp, L. (1999) *Puppe.* Springer Verlag, Berlin.

Weitkamp, J. (2000) Zeolites and catalysis. *Solid State Ionics*, **131**, 175–188.

Weitkamp, J. (2012) Catalytic hydrocracking mechanisms and versatility of the process. *ChemCatChem*, **4**, 292–306.

Wilson, S.T., Lok, B.M., Messina, C.A. *et al.* (1982) Aluminophosphate molecular sieves: a new class of microporous crystalline inorganic solids. *Journal of the American Chemical Society*, **104**, 1146–1147.

Wischert, R., Laurent, P., Copéret, C. *et al.* (2012) γ-Alumina: the essential and unexpected role of water for the structure, stability, and reactivity of "defect" sites. *Journal of American Chemical Society*, **134**, 14430–14439.

Xu, Z., Xiao, F., Purnell, S. *et al.* (1994) Size-dependent catalytic activity of supported metal clusters. *Nature*, **372**, 346–348.

Xu, Y., Jin, S., Xu, H. *et al.* (2013) Conjugated microporous polymers: design, synthesis and application. *Chemical Society Reviews*, **42**, 8012–8072.

Yamamoto, K., Sakata, Y., Nohara, Y. *et al.* (2003) Organic-inorganic hybrid zeolites containing organic frameworks. *Science*, **300**, 1470–1472.

Yanagisawa, T., Harayama, M., and Kuroda, K. (1990) Organic derivatives of layered polysilicates. III. Reaction of Magadiite and kenyaite with allydimethylchlorosilane. *Solid State Ionics*, **42**, 15.

Zecchina, A., Lamberti, C., and Bordiga, S. (1998) Surface acidity and basicity: general concepts. *Catalysis Today*, **41**, 169.

Zecchina, A., Spoto, G., Ricchiardi, G. *et al.* (2002) Zeolite characterization with spectroscopic methods. *Studies in Surface Science and Catalysis*, **142 A**, 3.

Zecchina, A., Rivallan, M., Berlier, G. *et al.* (2007) Structure and nuclearity of active sites in Fe-zeolites: comparison with iron sites in enzymes and homogeneous catalysts. *Physical Chemistry Chemical Physics*, **9**, 3483.

Zecchina, A., Bordiga, S., and Groppo, E. (eds) (2011) *Selective Nanocatalysts and Nanoscience: Concepts for Heterogeneous and Homogeneous Catalysis.* Wiley-VCH, Weinheim.

7

Photocatalysis

7.1 Photochemistry and Photocatalysis: Interwoven Branches of Science

Unlike normal reactions, in photochemical transformations, absorption of light (photons) causes a change in the electronic configuration of at least one reactant. The effect can be qualitatively described as absorption. During the absorption of light, a photon promotes one electron from a low energy to a high energy level (i.e., from ground to excited state), leaving a low-lying vacancy behind. The electron in the excited state is more loosely bound and hence can be easily transferred to a reactant with electron acceptor properties. Similarly, the empty state can behave as electron acceptor. It is quite evident that absorption of a photon confers both oxidizing and reducing character to the involved species. These facts often completely change the reactivity because excitation considerably extends the reaction spectrum of molecules by decreasing the energy barrier for electron transfer (Balzani, Ceroni, and Juris, 2014). Therefore photochemical reactions have often been applied to organic synthesis.

Upon absorption, photons do not leave anything behind, so they behave as reagents. Also catalysts (and hence the photocatalysts), by definition, decrease the energy barrier. However, the two processes should not be confused. The difference between photochemistry and photocatalysis is the presence, in the second case, of the catalysts that (unlike photons) are fully recovered after the reaction. Photochemistry and photocatalysis are often present in the same reaction system because, as suggested by Noyori (2010), the combination of both domains (photocatalysis and photochemistry) often enables synergy effects.

In photochemistry, light is absorbed by a molecule only when its energy (frequency) equals the difference between the ground and excited state. This is what restricts the number of cases where photoexcitation can occur. However, a way to circumvent this obstacle is to use photosensitizers because they allow

The Development of Catalysis: A History of Key Processes and Personas in Catalytic Science and Technology,
First Edition. Adriano Zecchina and Salvatore Califano.
© 2017 John Wiley & Sons, Inc. Published 2017 by John Wiley & Sons, Inc.

the conversion of light into chemical energy also by species that are not able to absorb that light. Photosensitizers in fact are molecules characterized by appropriate energy differences between the ground state and the excited state and whose function is to absorb photons and to transfer excitation to other species. The addition of these components often plays a fundamental role in the reaction rate. Though, in principle the role of sensitizer is only the transfer of excitation, the terms "sensitizer" and "photocatalyst" are often used and considered equivalent. However, the equivalence between sensitizer and photocatalyst is only verified when there is an efficient regeneration of the sensitizer at the end of the reaction. Only in this case can this typical photochemical component be considered a photocatalyst (Hoffmann, 2012). Sensitizer and photocatalyst often are contributing in a single photoactive system.

In regard to catalysis, photocatalysis merits special attention for several reasons. One reason is the externalities associated with fossil fuel consumption, and the development of renewable carbon-free energy using the sunlight is undoubtedly a major scientific and technological challenge characterizing the late twentieth century and the early decades of the twenty-first century. To illustrate the effort of the scientific community in the area of catalysis science, it is sufficient to say that since 1950 about 33,000 papers have been published, and most in the last 10 years. Conversion of solar light can generate electrical energy or chemical bonds where energy is stored. Silicon-based photovoltaic devices and TiO_2-based dye-sensitized solar cells are examples of cells that directly generate electricity. The silicon-based photovoltaic cells will not be described in this book because they are not photocatalytic. The most known dye-sensitized solar cells (DSSC) directly generating electricity are based on the TiO_2 semiconductor. The DSSC Gräetzel cells are partially based on photocatalytic processes. They were developed by B. O'Regan and M. Gräetzel at the École Polytechnique Fédérale de Lausanne (O'Regan and Graetzel, 1991). Michael Gräetzel was awarded the 2010 Millennium Technology Prize for this invention. These TiO_2-based cells are similar to those utilized for water splitting and CO_2 reduction. As DSSC cells cannot store energy in chemical bonds, they will not be described in detail.

Solar-driven water splitting and CO_2 photoreduction systems are the major ways of generating; the energy can be stored of new chemical bonds that in the form of fuel (H_2, alcohols). Energy storage in chemical bonds is the most efficient form of energy storage, and as a consequence this energy storage system has attracted research attention for the past three decades. It mimics photosynthesis, the extraordinary process that, about 2 billion years ago, nature developed for storing solar energy, by converting H_2O and CO_2 into carbohydrates and oxygen.

Photocatalysis is also important because it allows solar light to exploit and efficiently degrade pollutants using semiconductor oxides functioning as photocatalysts. Although chemical wastes can be primarily minimized by increasing the selectivity of industrial chemical processes via the introduction of innovative catalysts, highly industrialized societies producing high quantities of goods also produce large amounts of pollutants that are present in the liquid phase and air. Pollutant degradation by photocatalysis is increasingly becoming an important chapter of environmental science.

Although the recent development of photocatalysis was fueled by the energy and environmental needs of humanity, since the beginning of scientific chemistry, chemists have been interested in light as an energy source or reagent. Indeed photocatalysis is so intimately interwoven with photochemistry that it is conceptually difficult to separate these two branches of science.

7.2 Photochemistry Onset

The first paper exhibiting some scientific character in relation to photochemistry was that of Scheele (1772) on the effect of light on AgCl, and photography became established in several countries in the 1830s as a result (John and Field, 1963).

In 1817, the German chemist Theodor Grotthuss (1785–1822) formulated the first law of photochemistry. He already understood that a reaction could be produced by light and absorbed by molecules (Grotthuss, 1820). Grotthuss' idea remained unknown, until it was restarted in 1842 by the English, American-naturalized John W. Draper (1811–1882), a professor at New York University since 1837 and an expert of chemistry and of photographic processes. Draper developed the proposition in 1842 that only light rays that are absorbed can produce chemical change. For this reason Grotthuss' idea is better known today as the Grotthuss–Draper law.

In 1834 the chemist and apothecary Johann Bartholomäus Trommsdorff (1770–1837), founder of the first German pharmaceutical institute at Erfurt in 1795, observed a photochemical reaction of solid (−)-α-santonin (Trommsdorf, 1834). He discovered that crystals of α-santonin, when exposed to sunlight, turned yellow, giving rise to a burst, attributed later to a dimerization process. Only in 2007 was this reaction, occurring in a single crystal, fully described and understood (Natarajan *et al.*, 2007).

The first researcher to develop true photochemical concepts was Johannes Stark (1874–1957), a professor at the *Rheinisch-Westfälische Technische Hochschule* of Aachen, who in 1908 distinguished between primary and secondary processes taking place in a chemical system by light absorption (Stark, 1908). Stark defined the primary process as the immediate absorption of a photon by a molecule or an atom and the secondary process as the ensemble of "obscure" reactions started by the primary process.

In a famous 1912 paper, Einstein (1912) established the photochemical equivalence law.

This paper clearly shows how Einstein clearly realized that the basis of a photochemical process is built by the connection between Planck's law and the absorption process transforming radiant energy into molecular internal energy and then into kinetic energy. The chemical–physical processes consequent to the absorption or emission of electromagnetic radiations were connected to the structure of the molecular vibro-electronic levels by the Ukrainian physicist Aleksander Jablonski (1898–1980) who in 1935 represented schematically the absorption, fluorescence, and phosphorescence processes in a famous diagram (Jablonski, 1935) since then included in all spectroscopy and photochemistry books.

In the second half of the nineteenth century and at the beginning of the twentieth century, two Italian chemists, Emanuele Paternò and Giacomo Ciamician, both pupils of Cannizzaro at the University of Rome, built the true basis of organic photochemistry.

Giacomo Ciamician (1857–1922), born in a family of Armenian origin who immigrated to Italy around 1850, studied first at Vienna and then at Giessen where he received his PhD in 1880 with a thesis on chemical affinity (Figure 7.2). He then went to work at the University of Rome as assistant of Stanislao Cannizzaro and in 1887 became professor of general chemistry at the University of Padua. In 1889 he moved to the University of Bologna where he stayed for the rest of his life.

An important chapter of the scientific activity of Ciamician concerns the chemistry of pyrrole and of its derivatives (Ciamician, 1887, 1904), and in particular, the transformation of pyrrole into pyridine that he realized in collaboration with Maximilian Dennstedt (Ciamician and Dennstedt, 1881, 1882). Ciamician is, however, better known for his photochemistry research for which he was nine times nominated for the Nobel Prize (Ciamician and Silber, 1909), in particular, two times by Emil Fischer, the greatest organic chemist of the time.

The Ciamician experiments of photochemistry mostly in collaboration with his friend and coworker Paul Silber were made by exposing to sunlight tubes, flasks, glass pipes, and Erlenmeyer bottles on the terrace of the Chemistry Institute of Bologna (Ciamician and Silber, 1901). These experiments gave rise to 85 papers, published in the period 1899–1913 and listed in review articles of the same Ciamician (1908). The research led to the discovery of several new reactions (Ciamician and Silber, 1896), among them the photoreduction of aldehydes, ketones (Ciamician and Silber, 1885), quinines, and nitro compounds as well as the photodimerization and cycloaddition of olefins. Ciamician was also a pioneer of ecology, convinced that the exploitation of solar energy would set the humanity free from the use of fossil fuels (Nebbia and Kauffman, 2007). In a famous 1912 article on *Science* (Ciamician, 1912), he wrote:

civilization is the daughter of coal, for this offers to mankind the solar energy in its most concentrated form; that is, in a form in which it has been accumulated in a long series of centuries. Modern man uses it with increasing eagerness and thoughtless prodigality for the conquest of the world and, like the mythical gold of Rhine, coal is to-day the greatest source of energy and wealth. The earth still holds enormous quantities of it, but coal is not inexhaustible. The problem of the future begins to interest us … Is fossil solar energy the only one that may be used in modern life and civilization?

In 1912, Ciamician (1912) at the University of Bologna, inspired by the photosynthesis in green plants, developed perspectives for industrial solar photochemistry. Shortly before, in a lecture held at the Société Chimique de France, he presented photochemical reactions in combination with biochemical or enzyme-catalyzed transformations as elements of sustainable chemistry (Ciamician, 1908). This event may be considered the beginning of "green chemistry" (Albini and Fagnoni, 2008). He reaffirmed this idea in a subsequent famous lecture entitled *The Photochemistry of the Future*, which he delivered in 1912 at the VIII International Congress of Pure and Applied Chemistry (New York) (Ciamician, 1912), and in this instance he pronounced this prophetic sentence:

So far human civilization has made use almost exclusively of fossil solar energy. Would it not be advantageous to make a better use of radiant energy?

The final sentence of that lecture

If our black and nervous civilization, based on coal, shall be followed by a quieter civilization based on the utilization of solar energy that will not be harmful to the progress and to human happiness.

is still meaningful today, but the word "coal" would substituted with "gas and oil."

The Palermitan Emanuele Paternò (1847–1935), a teacher of chemistry in 1871 at the University of Turin, in 1982 moved to the University of Rome and in 1910, at the death of Cannizzaro, became professor of general chemistry.

In 1875, Paternò realized that under the effect of sunlight, 3-nitrocuminic acid was converted into a red amorphous product. He proved the photochemical synthesis of propyl butyrate by action of sunlight on butyric acid. In addition he studied the effect of sunlight on benzophenone and phenylacetic acid and on the photochemistry of the green substance of plants, chlorophyll.

In 1909 Paternò and his coworker Generoso Chieffi (1880–1923) discovered that a solution of benzaldehyde (or benzophenone) in 2-methyl-2-butene exposed to sunlight produces the formation of a four-membered ring, the oxetane (Paterno and Chieffi, 1909), according to the scheme

The oxetane structure was confirmed only in 1954 by a group headed by the Swiss chemist George Hermann Büchi (Buchi, Inman, and Lipinsky, 1954). For this reason the reaction is today called the Paternò–Büchi reaction. The Paternò–Büchi reaction was largely ignored by organic chemists until it was disclosed in 1963 by George S. Hammond and Nicholas J. Turro in a review of organic photochemistry (Hammond and Turro, 1963) and by John Steven Searles in 1964 in his book on heterocyclic compounds (Searles, 1964) who evidenced its utility and clarified its reaction mechanism.

Another pioneer in photocatalysis was the German chemist Alexander Schönberg (1892–1985). At the University of Cairo, over the course of twenty years, he carried out extensive studies of the chemical effect of sunlight. The results of these studies were published in important international journals. He compiled in 1958 the first book entitled *Preparative Organic Photochemistry*, revised and translated to English in 1968. Schönberg's first photochemical investigations looked at the photodimerization of thiophosgene, and in 1933, he determined the structure of the photodimeric form. During his work on sunlight-induced chemistry, Schönberg published more than 20 articles in a series entitled "Photochemical reactions." The results reported in these papers included the reaction of aldehydes with phenanthraquinoneimine to form 2-hydroxy-2,3-dihydrophenanthroxazole derivatives, photopolymerization of coumarins and related substances (in aqueous solutions), and photoaddition and photoreduction of aromatic ketones.

Serpone *et al.* (2012) have published a very accurate review on the development of photocatalysis in the first half of the twentieth century where also the contribution of the photochemistry studies performed in Russia by Alexander Terenin and coworkers is documented. Terenin, who was cited in Chapter 4 for his pioneering contribution to the spectroscopy of molecular species adsorbed on solid surfaces, is mostly known for his important contributions in the field of photophysics and photochemistry of organic molecules, dyes, and pigments and their synthetic analogues. In 1932 Terenin organized at St. Petersburg University the Laboratory of Optics of Surface Phenomena, which in 1957 became the Laboratory of Photosynthesis and since 1960 the Laboratory of Photocatalysis.

7.3 Physical Methods in Photochemistry

With the development of the modern chemical–physics instrumentation, photochemistry recovered its original kinetic vocation, and in the 1920s it was transformed into the study of the velocity of fast organic reactions, a problem closely connected to the spectroscopic identification of intermediate short-lived species, ions, or radicals, whose existence was more and more postulated in the theoretical study of reaction mechanisms.

The study of fast photochemical reaction kinetics started in England in the medical community, when in 1921 Hamilton Hartridge (1886–1976), a professor of physiology at the University of Cambridge, and his young coworker Francis John Worsley Roughton (1899–1972) invented the "stopped-flow" technique that allows two different solutions to mix very rapidly in order to study their reaction kinetics (Hartridge and Roughton, 1923). A significant leap was made at Cambridge when Ronald Norrish (1897–1978) and George Porter (1920–2002), who invented in 1949 the "flash" photolysis technique with which he succeeded in measuring, after some improvements of the initial system, times of the order of 10^{-6} s (millionth of second) and in identifying unstable intermediate species with very short lifetimes. Ronald Norrish started to be interested in photochemistry in 1915 when he was still a young student at Cambridge. His research activity under the supervision of Eric Rideal was interrupted by the World War I. In 1925 he returned to Cambridge, where he obtained his PhD degree. In 1937 he became a professor of chemical physics. At Cambridge, he started a series of works of organic photochemistry, being interested in chain reactions and in polymerization kinetics. To him we owe the study of a class of chemical reactions, called Norrish reactions, catalyzed by ultraviolet (UV) light in which aldehydes and ketones produce a great variety of compounds through the formation of intermediate radical species (Norrish and Bamford, 1936, 1938). The Norrish reactions are important also for the study of processes occurring in the upper layers of the atmosphere by action of the sunlight.

Coming back to flash photolysis in 1946, Porter had the idea of using fast light pulses to generate free radicals following their evolution in time by means of spectroscopic techniques. Starting from this idea, Porter and Norrish invented the flash photolysis technique that opened the route to research of fast kinetics at times of millionths of second (microseconds). The flash photolysis technique consists in sending on a gas or a liquid a light flash of high intensity and very short duration that gives rise to a photochemical dissociation of the molecules. Then a second light flash is sent on the sample with time delays in order to follow the reactions that take place in the system before reaching equilibrium. The first flash photolysis instrument was operative in 1947, which started a long-lasting collaboration between Norrish and Porter that led to results for which they obtained the Nobel Prize in 1967.

About the biography of George Porter, we mention that he started his research activity with Norrish studying the presence of transient species in chemical reactions, and in particular, the presence of free radicals and of excited molecular states in the gas phase (Norrish and Porter, 1949). Later he extended his research to solutions and to photosynthetic processes in green plants. In 1950 his instrument already reached the time resolution of few microseconds (10^{-6} s).

In the last 30 years, photochemistry has undergone a theoretical and experimental development that could not be even conceived until few decades before, thanks to the availability of lasers with ultrafast pulses allowing access to time pulses first of picoseconds (10^{-12} s), then of femtoseconds (10^{-15} s), and more recently of attoseconds (quintillionth of a seconds, 10^{-18} s). With the construction of solid-state lasers (titanium/sapphire and neodymium-doped YAG lasers) and with the development of modern technologies of parametric amplification, frequency tuning, and pulses time compression, a completely new era started in the study of reaction mechanisms and of molecular dynamics.

The number of physical, chemical, and biological processes investigated with the help of ultrafast lasers is very high, and new topics are added every day. Complete classes of ultrafast reactions, from isomerization to proton transfer in excited electronic states, from the identification of short-lived radical to ionic intermediates to the characterization of transition states, or of the elementary processes involved in the photosynthesis, are currently studied in the framework of the reaction mechanism theories.

In these papers, crucial findings were obtained by the Egyptian-American spectroscopist Ahmed Hassan Zewail (1946–), a professor at Caltech and 1999 Nobel laureate in chemistry, who coined first the word femtochemistry by fusion of the terms femto and chemistry.

Ahmed Zewail, born in Egypt and graduated in chemistry at Alexandria University, immigrated to the United States where he joined the University of Pennsylvania. After obtaining his PhD, he moved to California where he accepted a position at the University of California, Berkeley, as a laser spectroscopist, working on several projects, including building a picosecond laser. In 1976 he became an assistant professor at Caltech where he has remained since, becoming in 1990 Linus Pauling Professor of Chemical Physics. Zewail was awarded the Nobel Prize in Chemistry for his research on femtosecond spectroscopy. In his research activity, Ahmed Zewail had made fundamental contributions to the time evolution of chemical reactions and of biological functions and to the structural analysis, at an atomic scale resolution, of three-dimensional structures in systems ranging from small molecules to crystals and from DNA and proteins to viruses. He collected, in 1994, his scientific results in a two-volume book, entitled *Femtochemistry: Ultrafast Dynamics of the Chemical Bond* (Zewail, 1994).

7.4   Heterogeneous and Homogeneous Photocatalysis

Through heterogeneous photocatalysis, a process occurs in which light radiations having energy equal to or greater than the energy band gap of a solid semiconductor phase strike on its surface and generate electron (e−)–hole (h+) pairs.

On reaching the surface, the photogenerated electrons and holes can participate in various oxidation and reduction processes to produce the final products. However, if the charges fail to find any trapped species on the surface or if their energy band gap is too small, they recombine immediately, releasing unproductive energy as heat. Fast recombination in the bulk is the first and most dangerous obstacle against useful utilization of the reducing potentialities of generated electron and holes (Figure 7.1).

During the photocatalytic process the photo-induced charges that have escaped bulk recombination move toward the surface of the semiconductor particle. On the surface they can recombine (surface recombination) or, finally, transfer electrons and holes to the adsorbed A (acceptor) and D (donor) species following the schemes (A → A−) and (D → D+). It is evident that bulk and surface recombinations are processes that reduce the efficiency of useful photocatalytic events and that considerable work was required to overcome this problem. Classical semiconductor systems can be oxides like TiO_2 and ZnO or sulfides like ZnS and CdS. Solid semiconductors with a perovskite structure like methylammonium lead trihalide were also investigated (Snaith, 2013). Various products possessing photocatalytic functions are now commercialized. Among many photocatalyst candidates, TiO_2- and TiO_2-based systems have proved to be the most suitable materials for industrial use at present and probably also in the future. This is because TiO_2 has the most efficient photoactivity, the highest stability, and the lowest cost. Combinations of two semiconductors like TiO_2 and CdS characterized by interparticle electron transfer also were studied (Serpone, Borgarello, and Graetzel, 1984; Serpone, Borgarello, and Pelizzetti, 1988; Serpone and Emeline, 2012). The useful reactions occur on the surface, so the semiconductors usually have finely divided forms. In fact, as the size of the materials decreases, the percentage of the atoms on the surface and the surface-to-volume ratio increase dramatically. To favor the yield of the surface reaction, the addition of anchored sensitizers or noble metal particles can play a additional crucial role.

Through homogeneous photocatalyst, we intend a molecular or supramolecular complex characterized by discrete energy levels. Absorption of light can promote one electron from occupied to empty molecular levels, thus

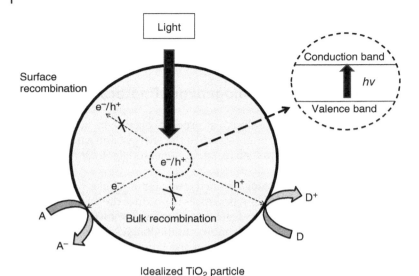

Idealized TiO_2 particle

Figure 7.1 Generation and migration of photoinduced charges in a spherical semiconductor particle generated by light irradiation. Generated electrons and holes escaping volume and surface recombination are transferred to acceptor and donor species located on the surface. *Source*: Adapted from Linsebigler, Lu, and Yates, 1995.

changing the electronic structure. In the excited state the homogeneous photoactive compound can lose ligands and become coordinatively unsaturated. Both actions can favor catalytic activity. Transition metal compounds are traditionally the commonly investigated homogeneous photocatalysts. In particular, the catalytic behavior of cobalt, iron, chromium, and tungsten carbonyls under light excitation has been widely investigated. The application of ruthenium polypyridine complexes in photoredox catalysis has also been thoroughly studied. In most instances, the commercially available $[Ru(bpy)_3]Cl_2$ (bpy=2,2'-bipyridine) was used. The structure of the cation is shown below:

Structure of $[Ru(bpy)_3]^{2+}$

Under light illumination a ligand-to-metal charge transfer transition occurs, which modifies the electronic population at the ruthenium metal site. Thus the excited $^*[Ru(bpy)_3]^{2+}$ system may induce a reductive or an oxidative transformation of impinging substrates. $[Ru(bpy)_3]^{2+}$-type complexes have been studied for several decades because of their long-lived excited state lifetimes and high chemical and thermal stability (Juris *et al.*, 1988).

Both redox processes have been applied to organic synthesis. This feature has expanded the accessibility of photochemical reactions to a broader range of synthetic organic chemists (Schultz and Yoon, 2014a, b). An exhaustive review of the very numerous homogeneous photocatalytic systems investigated so far, together with the perspectives in solar energy conversion, has been published by Balzani, Credi, and Venturi (2008), on the occasion of the 150th anniversary of Giacomo Ciamician's birthday. Balzani (1936–) has been full professor at the Dipartimento di Chimica "Giacomo Ciamician" at the University of Bologna since 1973 and is now emeritus at the same university (Figure 7.2). He can be considered the ideal continuer of the Ciamician tradition at Bologna and has greatly contributed to the development of this area (Campagna *et al.*, 2007). He is also very well known for his studies on photochemically driven molecular machines (Balzani, Credi, and Venturi, 2008) (Figure 7.2).

7.5 Natural Photosynthesis as Model of Photocatalysis

As mentioned before, the conversion of solar light into chemical energy, by splitting of H_2O with formation of molecular O_2 and H_2, could contribute to finding a solution to the greenhouse effect caused by our consumption of fossil fuels. Another important process mimicking the natural photosynthesis could be the reduction of CO_2 to produce organic compounds. However, the development of artificial photosystems able to perform these conversions is a challenging task. To show the complexity of this challenge, it is sufficient to illustrate the case of water splitting that requires the synergic combination of the several processes: (1) light harvesting, (2) charge separation, (3) electron transfer, (4) H_2O oxidation, and (5) reduction of the generated protons.

The complex nature of natural photosynthesis has inspired many scientists to produce simplified artificial versions capable of reproducing only some separate functions like water oxidation or CO_2 reduction. It can be stated with certainty that natural photosynthesis has provided the scientific community with a source of inspiration for the design of such systems.

Our comprehension of the fascinating photosynthetic processes is the result of intensive studies performed over the entire twentieth century. A very large community of researchers was directly or indirectly involved in this effort, and in the twentieth century 15 scientists were awarded the Nobel Prize for their contributions.

(a)   (b)

(c)

Figure 7.2 (a) Giacomo Ciamician (1857–1922), (b) Vincenzo Balzani (1936–), and (c) Akira Fujishima (1942–). Ciamician, inspired by photosynthesis, developed solar photochemistry and can be considered as intiator of *green chemistry*. Balzani can be considered as the ideal continuer of the Ciamician tradition at the University of Bologna. Fujishima made the important observation that water splitting could be obtained under illumination using TiO_2 (rutile) with dispersed platinum supported on the surface. *Source*: (a) The Ciamician photo is in the public domain. (b) Reproduced with permission of Vincenzo Balzani. (c) Courtesy of University of Turin.

The first was Richard Martin Willstätter (1915) for his research on plant pigments, especially chlorophyll. He was followed by Hans Fischer (1930) for his research into the constitution and synthesis of hemin (a porphyrin chelate of iron, derived from red blood cells) and chlorophyll, by Paul Karrer (1937) for his investigations of carotenoids and flavins, by Walter Haworth for his investigations of carbohydrates and vitamin C, and by Richard Kuhn (1938) for his work on carotenoids and vitamins. In the second half of the twentieth century, the research on photosynthesis substantially advanced with studies by Melvin Calvin (1960 Nobel Prize) on the carbon dioxide metabolic transformation in plants. The elucidation of the mechanism of energy transfer was the motivation of the Peter Mitchell (1978) and Rudolph Marcus (1992). In the last quarter of the twentieth century, the problem of supramolecular structural organization of the chemical components involved in the complex photosynthesis machinery started to be investigated. The 1988 Nobel Prize given to Hartmut Michel, Robert Huber, and Johann Deisenhofer for the determination of the three-dimensional structure of a photosynthetic reaction center testifies to the positive results in this area. Last elucidation of the catalytic (enzymatic) aspects of photosynthesis was the motivation of the 1997 Nobel Prize to Paul Boyer, John Walker, and Jens Skou.

This concise list of the many Nobel laureates who contributed to the understanding of photosynthesis clearly illustrates the very lively character of this branch of science throughout the twentieth century.

The interest in the structure of the photocatalytic centers is continuing, and in the last decade, the use of synchrotron radiation sources has proved especially valuable in the resolution of the complex tridimensional structures of photosynthetic centers. In this book, it is not possible to discuss in detail the history of this fascinating branch of science where catalysis is also involved.

A model of the simplified artificial photocatalytic versions may be useful in summarizing a few of the concepts on the process where by photons convert CO_2 and water into sugar and oxygen:

$$2nCO_2 + 2nH_2O + \text{photons} \rightarrow 2(CH_2O)_n + 2nO_2$$

The photosynthetic process is "catalytic" because several catalysts participate. Photosynthesis occurs in two stages (a, b). In stage a, light is captured and used for water oxidation and the reduction of protons, the basic processes present in all artificial photocatalytic processes for water splitting designed so far.

During stage b, where light is not involved, the reduction of carbon dioxide occurs.

In natural photosynthesis the first step is the absorption of a photon by chlorophyll. Because chlorophyll is green, most plants have a green color. Besides chlorophyll, plants and algae may use other pigments.

Back to plant's photosynthesis, excited chlorophylls transfer the excitation via the intervention of a chain of several molecules to generate a state where the electrons and holes are sufficiently separated and recombination is avoided. In fact only separated electrons and holes can perform the reduction and oxidation reactions. In particular, holes are used in the catalytic oxidation of water while electrons, which are shuttled through an electron transport chain (the so-called Z-scheme), are responsible for the reduction of CO_2 catalyzed by an enzyme.

In summary, the key elements of photosynthesis machinery are the light-absorbing photosensitizer (chlorophyll); a water oxidation catalyst (WOC), that is, a catalyst that uses the created holes (h+) to oxidize the water molecule; and a CO_2 reduction enzymatic catalyst using electrons (ribulose-1,5-bisphosphate carboxylase/oxygenase, commonly known by the abbreviation RuBisCO, probably the most abundant protein on Earth). In this enzyme, the catalytic center contains Mg^{2+}.

Notice that water oxidation of water to give O_2 and protons

$$4h^+ + 2H_2O \rightarrow O_2 + 4H^+$$

is a four-electron process: hence a multiple collection of holes derived from sensitizers is needed. In natural photosynthetic systems, the oxidation of water is catalyzed by a complex catalytic structure that contains four manganese ions and one Ca^{2+} connected by oxygen and hydroxyl bridges (Umena *et al.*, 2011).

The determination of the structure of this cluster and its surroundings, allowed by use of synchrotron light, can be considered a milestone in our comprehension of natural photosynthesis.

7.6 Water Splitting, CO₂ Reduction, and Pollutant Degradation: The Most Investigated Artificial Photocatalytic Processes

In the last two decades numerous photocatalytic reactions occurring on the surface of semiconductors have been investigated, and the resulting scientific literature is quite abundant (Hoffmann, 2012).

In this section only two photocatalytic reactions are discussed in some detail: water splitting to produce hydrogen and CO_2 photoreduction. The interest in hydrogen photoproduction lies in the fact that it is an environmentally friendly energy carrier because its only exothermic oxidation product is H_2O.

The interest in the photocatalytic conversion of CO_2 to either a renewable fuel or valuable chemicals using solar energy is also important because of its potential to provide an alternative clean fuel.

7.6.1 Water Splitting

In artificial water splitting systems (both homogeneous and heterogeneous), the basic strategy is substantially similar because it is based on the common presence of (1) a light-absorbing system to originate charge separation and to avoid recombination, (2) a catalyst using holes for water oxidation, and (3) a catalyst using electrons for the reaction

$$2e^- + 2H^+ \rightarrow H_2$$

This reaction is substituting the CO_2 reduction step of natural photosynthesis.

In 1972 Akira Fujishima and his coworker Kenichi Honda (Fujishima and Honda, 1972) made the pioneering experiment on water splitting by the irradiation of a TiO_2 anode coupled with a platinum dark cathode (named a PEC cell). For this discovery, Akira Fujishima, now professor emeritus at the University of Tokyo, was awarded many prizes including the Asahi Prize, the Japan Prize, and the Japan Academy Prize (Figure 7.2).

The energy gap of TiO_2 corresponds to a threshold of about 3.2 eV, which means that visible light is substantially useless and that UV light is needed to create electrons and holes.

To illustrate the basic mechanism of water splitting, it is appropriate to start by describing first aTiO_2-based heterogeneous photocatalytic water splitting system working without the intervention of an external bias (like that utilized in the Fujishima experiment) (Figure 7.3).

The TiO_2/Pt is a prototype photoactive system where the TiO_2 support, under illumination, and performing the charge separation of electron, and the holes and supported Pt particles work as co-catalysts for hydrogen and oxygen evolution reactions (Yang *et al.*, 2013).

In this solid system the overall photocatalytic water splitting reaction involves three major steps: (1) absorption of light by the TiO_2 semiconductor particles to generate electron–hole pairs, (2) charge separation and migration to the surface of particles, and (3) surface reaction for water reduction or oxidation. Photocatalytic reactions can be realized when the previous three sequential steps are completed. It is widespread opinion that the role of Pt is multiple:

i) To efficiently trap the electrons
ii) To favor the proton reduction

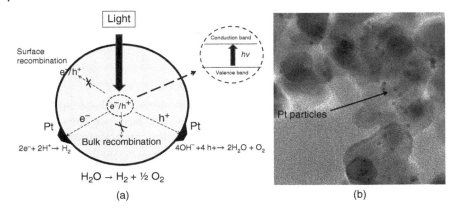

Figure 7.3 (a) Schematic representation of the function of platinum particles (black hemispheres) supported on TiO_2 that catalyze the evolution of hydrogen and oxygen. (b) Transmission electron microscopy image of a Pt/TiO_2 catalyst similar to that used by Akira Fujishima. The black dots are the Pt particles supported on the surfaces of TiO_2 microcrystals. *Source*: Adapted from Kmetykó *et al.*, 2014.

iii) To facilitate hydrogen recombination (since we know that Pt is an efficient hydrogenation catalyst on its own)
iv) To catalyze the oxygen evolution reaction (Chen *et al.*, 2012)

Recall that Pt is not the sole co-catalysts for electron trapping and proton reduction. We have seen that molecular-type co-catalysts like $[Ru(bpy)_3]^{2+}$ can possess the ability to trap electrons. Many complexes (including $Ru(bpy)_3$ derivatives) anchored on TiO_2 and other semiconductors have also been studied and tested (Eckenhoff, Brennessel, and Eisenberg, 2014) (Figure 7.4).

As we mentioned above, the 3.2 eV band gap of TiO_2 allows the utilization of only a very small fraction of the solar energy. Narrowing the band gap of TiO_2 appears to be highly desirable. To this end, a series of studies were conducted, and various metal elements were doped into the lattice of TiO_2 (Choi, Termin, and Hoffmann, 1994; Anpo, 2000). TiO_2 doped with nonmetal elements also was extensively studied. These innovative lines of research, initiated by Asahi and associated researchers at the Toyota Central R&D Laboratories (Asahi *et al.*, 2001) are universally considered a milestone in the field and are at the base of the successive explosion of the research in this area. However, we know today that doping with nonmetallic elements is not associated with narrowing the band gap as hypothesized by Asahi, but with the appearance of impurity states in the band gap (Napoli *et al.*, 2009). Despite the numerous efforts made so far to narrow the band gap and to increase the photocatalytic efficiency, pure TiO_2 still has a primary role as semiconductor.

Figure 7.4 [Ru(bpy)$_3$]$^{2+}$ grafted on an idealized TiO$_2$ particle as co-catalyst. *Source*: Adapted from Eckenhoff *et al.*, 2013.

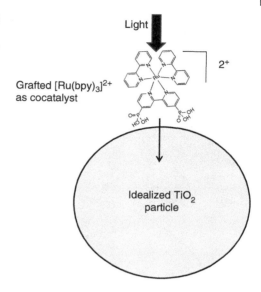

In the examples we discussed, hydrogen and oxygen were generated on different sites of the same catalyst particles. The result was a mixture of the two gases. This situation can be avoided by using photoelectrochemical cells where oxygen and hydrogen evolve at different electrodes.

A photoelectrochemical cell is substantially composed of a visible light-responsive thin film photocatalyst as a photoanode, a platinum cathode, and electrolytes separated by a proton exchange membrane that allows for the production of hydrogen at the cathode and oxygen at the anode. The TiO$_2$ film is usually deposited on a transparent conductive substrate, such as a tin-doped indium oxide (ITO) or fluorine-doped tin oxide (FTO) glass. Many types of light-responsive anodes have been experimented. Often the TiO$_2$ films are variously sensitized with activators including dyes, metal particles, metal sulfide particles, and quantum dots. Perovskite films have also been advantageously employed. Recently, several examples have appeared where a ruthenium photosensitizer was coupled to a ruthenium oxidation catalyst (RuO$_2$) (Conception *et al.*, 2009; Li *et al.*, 2010).

Following an innovative approach directly involving an enzyme/TiO$_2$ system, [NiFe] hydrogenase was deposited on nanocrystalline TiO$_2$, sensitized by a [Ru(bpy)$_3$]$^{2+}$ complex anchored to the surface with phosphonate groups. Irradiation with visible light in the presence of a sacrificial electron donor, triethanolamine (TEOA), provided an efficient (although not stable) generation of H$_2$ (Figure 7.5).

Devices mimicking the natural photosynthesis machinery based on two different photosystems were designed in 2001 by Abe *et al.* (2001). The system

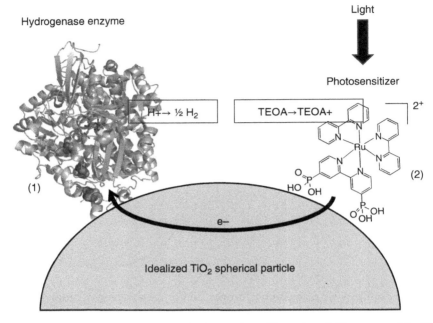

Figure 7.5 [Ni–Fe] hydrogenase attached to [Ru(bpy)$_3$]$^{2+}$-sensitized TiO$_2$ nanoparticle. (1) The hydrogenase enzyme and (2) the light-absorbing Ru photosensitizer attached to the TiO$_2$ via phosphate groups. *Source*: Adapted from Reisner *et al.*, 2009. This complex hydrogen-generating photosystem is representative of a recent direction of the scientific efforts.

was composed of a photocatalysts A (anatase/Pt) and photocatalyst B (rutile). The two photocatalysts (anatase and rutile) have slightly different band gaps, the rutile band gap being lower than that of anatase (Figure 7.6).

This device is based on two-step photoexcitation processes that mimic the photosynthesis of green plants (the so-called Z-scheme) in order to realize overall water splitting. This photocatalytic system consists of two visible light-responsive semiconducting photocatalysts specialized for respective hydrogen evolution reaction and water oxidation reaction and of a redox mediator. The photocatalyst for hydrogen evolution reaction (photocatalyst A) is excited by the irradiation of visible light, and then photo-formed electrons reduce H+ into H$_2$ and photo-formed holes oxidize the redox mediator. Simultaneously, in the photocatalyst for water oxidation reaction (photocatalyst B), photo-formed holes oxidize H$_2$O to produce O$_2$, and photo-formed electrons reduce the redox mediator under visible light irradiation. Total water splitting into H$_2$ and O$_2$ is attained.

Z-scheme constituted by two components A and B

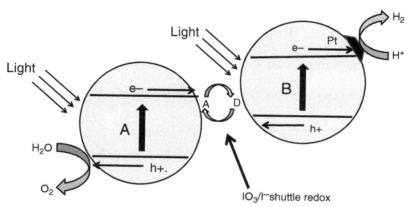

Figure 7.6 Z-scheme of a photocatalytic system constituted by two components. *Source:* Adapted from Horiuchi *et al.*, 2013.. Phocatalysts A and B have finely tuned band gaps and are connected by IO_3^-/I^- shuttle.

The Z-scheme photosystems have been successively improved by many other researchers (Horiuchi *et al.*, 2013) using different oxidic systems and activators. The recent Serpone and Emeline review (Serpone and Emeline, 2012) on the difficulties and perspectives of water photosplitting is of some utility here.

In principle, water splitting can be performed also with homogeneous catalysts, and manganese- and ruthenium-based molecules potentially capable of mediating the H_2O photooxidation under benign conditions have been synthesized and tested. However, the lesson coming from natural photosynthesis is that only a supramolecular organization of light harvesting, charge separation, charge transport, and catalytic sites where oxygen and hydrogen are formed can efficiently perform the process. A critical step toward integrated artificial photosynthetic systems is consequently to create large ordered arrays of interactive molecules. Today, while some progress has been made on each aspect of artificial photosynthesis, integration of the various components in a working system has not yet been achieved. A description of the scientific efforts and of difficulties encountered in this fascinating area is beyond the scope of this book. Detailed information can be found in the Balzani and Wasielewski contributions (Balzani, Credi, and Venturi, 2008; Wasielewski, 2009).

7.6.2 CO_2 Photoreduction

Global warming is a serious concern in the twenty-first century. The anthropogenic emissions of CO_2, mostly from combustion of fossil fuel, are the main source of global warming. Since CO_2 is green and cheap, totally abundant, and renewable feedstock, it is conceivable that its utilization as available and renewable carbon resource can be important in the development of alternative carbon-based fuels. However, the biggest obstacle in establishing industrial processes based on CO_2 chemical transformation is its stability, a fact that implies that high energy is required to transform it into useful chemicals and fuels.

The traditional approach is CO_2 catalytic reforming of methane, following the reaction

$$CH_4 + CO_2 \rightarrow 2CO + 2H_2 \Delta H = +247 \text{ kJ/mol}$$

that requires a high input of energy, which can be supplied by combustion of methane following the reaction

$$CH_4 + 2O_2 \rightarrow CO_2 + 2H_2O \ \Delta H = -800 \text{ KJ/mol}$$

The energy balance of the two reactions indicates that only a fraction of CO_2 can be usefully transformed. CO_2 reforming of CH_4 is a highly endothermic reaction; thus the metal catalyst (Pt) must be supported on high surface area solids with high thermal stability, such as SiO_2, Al_2O_3, and ZrO_2. Sintering of the active metal particles in the pores of the support and the formation of carbon residues are the major problems of this reaction. However, due to the high availability and low price of methane and the fact that CO and H_2 can react together in a well-known way to produce methanol on metal Cu particles (as described in Chapter 1), the CO_2 conversion by dry reforming with methane is still drawing the most attention (Shi *et al.*, 2013).

Yet the persisting technical difficulties (catalyst optimization and deactivation) have suggested the exploration of alternative paths for CO_2 utilization, and artificial photosynthesis processes did emerge in the final decades of the twentieth century as the most promising solution. This line of research was stimulated by the improvements obtained in photocatalytic water splitting. Indeed heterogeneous photocatalytic conversion of CO_2 was already demonstrated in 1979 using large band gap semiconductor materials like TiO_2 and $SrTiO_3$ (Inoue *et al.*, 1979). Even today, these materials are intensively studied. Similarly the earliest work on photoelectrochemical reduction of CO_2 to CH_3OH was reported by Halmann in 1978 (Halmann, 1978) using an electrochemical cell similar to that illustrated for water splitting. Since then, numerous other catalysts like WO_3, ZnO, GaP, CdS, and SiC (to mention only a few), either pure or promoted with various co-catalysts, have been tested. As we have emphasized above, water splitting research as associated with CO_2 photoreduction is considered strategic for the survival of humanity.

The CO_2 photoreduction scheme is substantially similar to that represented for water splitting, the only difference being the substitution of the reaction $2e + 2H^+ \rightarrow H_2$ with a reaction involving CO_2 reduction (for instance,

$CO_2 \rightarrow CO$) that is exploiting the electrons promoted by the absorption of photons. The net efficiency of the system is determined by the sum and synergy of these processes, in terms of both their thermodynamics and kinetics.

The CO_2 reactions that can potentially occur on a photocatalytic system are the following:

$$CO_2 + 2H^+ + 2e \rightarrow HCOOH$$
$$CO_2 + 2H^+ + 2e \rightarrow CO + H_2O$$
$$CO_2 + 4H^+ + 4e \rightarrow HCHO + H_2O$$
$$CO_2 + 6H^+ + 6e \rightarrow CH_3OH + H_2O$$
$$CO_2 + 8H^+ + 8e \rightarrow CH_4 + 2H_2O$$

From the purely thermodynamic point of view, the reduction to methanol or methane is favored. However, many electrons are involved, a fact having negative influence on the kinetics. For this reason the reduction to HCOOH and CO has been mainly considered (Handoko, Li, and Tang, 2013).

In principle, wide band gap photocatalysts like TiO_2 or ZrO_2 are not ideal because they utilize only a small fraction of the solar spectrum. The same as for water photooxidation, efforts have been made to use other oxidic systems characterized by lower band gap or containing various dopants. Anchoring on the surface of light-harvesting systems like dyes, quantum dots, and so on, has also been widely investigated.

Among the adopted strategies, we mention that reported by Woolerton *et al.* (2010) where the combination of TiO_2 promoted by the classical light-harvesting Ru-based system illustrated in a previous paragraph coupled with the CO_2-reducing enzyme already mentioned before provides a potentially innovative catalysts for CO_2 reduction. The same as for water splitting, the lesson coming from natural photosynthesis has aided the design of the new catalysts.

So far we have briefly reported about the efforts performed with heterogeneous photocatalysts. For sake of completeness, it must be mentioned that homogeneous catalysts have been tested with less success. For more information, the Windle and Perutz review (2012) should be consulted.

It is outside the scope of this book to give detailed information on this active research area, which plausibly will involve in the future the efforts of the entire scientific community. We only mention that up to now the obtained achievements did not find industrial applications (Tahir and NorAishahSaidina, 2013).

7.6.3 Photocatalysis in Environmental Protection

The industrialized nations discharge a wide variety of contaminants from residential, commercial, and industrial sources. Among them there are newly emerging pollutants like pharmaceuticals, antibiotics, and pathogens. Among the processes traditionally utilized to destroy pollutants like oxidation, UV irradiation, and ozone treatment, photocatalysis is playing an important and growing role.

The most active and stable seminconductor photocatalyst is TiO_2, and much effort has been made in the past decade to improve its performance (Fujishima, Zhang, and Tryk, 2008). The accepted opinion is that the photocatalytic activity of TiO_2 in pollutant oxidation with visible light is associated with the formation of H_2O_2 resulting from interaction of O_2 with the electrons generated during illumination.

As we documented earlier for water oxidation and CO_2 reduction, TiO_2 drawbacks are (1) the large energy band gap (3.2 eV for the anatase phase, 3.0 eV for the rutile phase), which allows only the exploitation of a very small fraction of solar light, and (2) the low quantum efficiency, a fact that is common to all semiconductor photocatalysts. For this reason anion doping with p-block elements has been largely pursued to sensitize TiO_2 toward visible light, and a large number of studies have been published on specialized journals. In particular, for pollutant photodegradation, B-N, C-, and F-doped TiO_2 have been preferentially investigated (Dozzi and Selli, 2013). It has been ascertained that the incorporation of these elements in substitutional or interstitial position can alter the TiO_2 performances in a favorable way. Doping with p-block elements is not the only investigated strategy. Surface sensitization with materials absorbing the visible light like dyes, graphene, MoS_2, and polyoxometalates (to mention only a small fraction of investigated systems) has also been tested. It is not possible to fully review this large area. The interested reader can find a complete illustration in the Park *et al.* review (2013).

References

Abe, R., Sayama, K., Domen, K., and Arakawa, H. (2001) A new type of water splitting system composed of two different TiO_2 photocatalysts (anatase, rutile) and a IO_3^-/I^- shuttle redox mediator. *Chemical Physics Letters*, **344**, 339–344.

Albini, A. and Fagnoni, M. (2008) Giacomo Ciamician and the concept of green chemistry. *ChemSusChem*, **1**, 63–66.

Anpo, M. (2000) Utilization of TiO2 photocatalysts in green chemistry. *Pure and Applied Chemistry*, **72**, 1265.

Asahi, R., Morikawa, T., Ohwaki, T. *et al.* (2001) Visible-light photo-catalysis in nitrogen-doped titanium oxides. *Science*, **293**, 269–271.

Balzani, V., Credi, A., and Venturi, M. (2008) Photochemical conversion of solar energy. *ChemSusChem*, **1**, 26–58.

Balzani, V., Ceroni, P., and Juris, A. (2014) *Photochemistry and Photophysics: Concepts, Research, Applications*. Wiley-VCH, Weinheim.

Büchi, G., Inman, C.G., and Lipinsky, E.S. (1954) Light catalyzed organic reactions. I. The reaction of carbonyl compounds with 2-methylbutene. *Journal of the American Chemical Society*, **76**, 4327–4331.

Campagna, S., Puntoriero, F., Nastasi, F. *et al.* (2007) Photochemistry and photophysics of coordination compounds: ruthenium. *Topics in Current Chemistry*, **280**, 117–214.

Chen, X., Li, C., Graetzel, M. *et al.* (2012) Nanomaterials for renewable energy production and storage. *Chemical Society Reviews*, **41**, 7909–7937.

Choi, W., Termin, A., and Hoffmann, M.R. (1994) The role of metal-ion dopants in quantum sized TiO_2 – Correlation between photoreactivity and charge carrier recombination dynamics. *Journal of Physical Chemistry*, **98**, 13669–13679.

Ciamician, G. (1887) Il pirrolo e i suoi derivati. *Atti della R. Accademia Nazionale dei Lincei. Memorie della Classe di scienze fisiche*, **4**, 274–377.

Ciamician, G. (1904) Über die Entwicklung der Chemie des Pyrrols in letzten Vierteljahrhundert. *Berichte der Bunsengesellschaft für Physikalische Chemie*, **37**, 4200–4204.

Ciamician, G. (1908) Sur les actions chimiques de la lumiere. *Bulletin de la Société Chimique de France*, **3–4**, i–xxvii.

Ciamician, G. (1912) The Photochemistry of the future. *Science*, **36**, 385–394.

Ciamician, G. and Dennstedt, M. (1881) Synthesis of pyridines from pyrroles. *Berichte der Bunsengesellschaft für Physikalische Chemie*, **14**, 1153.

Ciamician, G. and Dennstedt, G. (1882) Trasformazione del pirrolo in piridina. *Rendiconti Reale Accademia dei Lincei.*, **XII**, 1–4.

Ciamician, G. and Silber, P. (1885) Über Pyrrilendimethyldiketon. *Berichte der Bunsengesellschaft für Physikalische Chemie*, **18**, 1466–1468.

Ciamician, G. and Silber, P. (1896) Zur Kenntniss der Tropinsäure. *Berichte der Bunsengesellschaft für Physikalische Chemie*, **29**, 2975–2976.

Ciamician, G. and Silber, P. (1901) Chemische Lichtwirkungen. *Berichte der Bunsengesellschaft für Physikalische Chemie*, **34**, 1530–1536.

Ciamician, G. and Silber, P. (1909) Actions chimiques de la lumière. *Annales de Chimie et de Physique*, **16**, 474–477.

Conception, J.J., Jurss, J.W., Brennaman, M.K. *et al.* (2009) Making oxygen with ruthenium complexes. *Accounts of Chemical Research*, **42**, 1954–1965.

Dozzi, M.V. and Selli, E. (2013) Doping TiO_2 with p-block elements: effects on photocatalytic activity. *Journal of Photochemistry and Photobiology C: Photochemistry Reviews*, **14**, 13–17.

Eckenhoff, W.T., Brennessel, W.W., and Eisenberg, R. (2014) Light-driven hydrogen production from aqueous protons using molybdenum catalysts. *Inorganic Chemistry*, **53**, 9860–9869.

Eckenhoff, W., McNamara, W., Du, P., and Eisenberg, R. (2013) Cobalt complexes as artificial hydrogenases for the reductive side of water splitting. *Biochimica et Biophysica Acta*, **958**, 958–973.

Einstein, A. (1912) Thermodynamische Begründung des photochemischen Äquivalentgesetzes. *Annalen der Physik*, **342**, 832–838.

Fujishima, A. and Honda, K. (1972) Electrochemical photolysis of water at a semiconductor electrode. *Nature*, **238**, 238–242.

Fujishima, A., Zhang, X., and Tryk, D.A. (2008) TiO_2 photocatalysis and related surface phenomena. *Surface Science Reports*, **63**, 515.

Grotthuss, T. (1820) Physikalische Chemische Forschungen. *Ostwald's Klassiker der Exakten Wissenschaften*, **152**, 12–18.

Halmann, M. (1978) Photoelectrochemical reduction of aqueous carbon dioxide on p-type gallium phosphide in liquid junction solar cells. *Nature*, **275**, 115–116.

Hammond, G.S. and Turro, N.J. (1963) Organic photochemistry. *Science*, **142** (**3599**), 1541–1553.

Handoko, A.D., Li, K., and Tang, J. (2013) Recent progress in artificial photosynthesis: CO_2 photoreduction to valuable chemicals in a heterogeneous system. *Current Opinion in Chemical Engineering*, **2**, 200–223.

Hartridge, H. and Roughton, F.J.W. (1923) The velocity with which carbon monoxide displaces oxygen from, combination with hemoglobin. *Proceedings of the Royal Society of London B*, **94**, 336–367.

Hoffmann, N. (2012) Homogeneous photocatalytic reactions with organometallic and coordination compounds: perspectives for sustainable chemistry. *ChemSusChem*, **5**, 352–371.

Horiuchi, Y., Toyao, T., Takeuchi, M. *et al.* (2013) Recent advances in visible-light-responsive photocatalysts for hydrogen production and solar energy conversion from semiconducting TiO_2 to MOF/PCP photocatalysts. *Physical Chemistry Chemical Physics*, **15**, 13243–13253.

Inoue, T., Fujishima, A., Konishi, S., and Honda, K. (1979) Photoelectrocatalytic reduction of carbon dioxide in aqueous suspensions of semiconductor powders. *Nature*, **277**, 637–638.

Jablonski, A. (1935) Zur Theorie der Polarisation der Photolumineszenz von Farbstofflosungen. *Zeitschrift für Physik*, **96**, 236–239.

John, D.H.G. and Field, G.T.J. (1963) *A Textbook of Photographic Chemistry*. Chapman and Hall, London.

Juris, A., Balzani, V., Barigelletti, F. *et al.* (1988) Ru(II) polypyridine complexes: photophysics, photochemistry, electrochemistry and chemiluminescence. *Coordination Chemistry Reviews*, **84**, 85–277.

Kmetykó, A., Mogyorósi, K., Gerse, V. *et al.* (2014) Photocatalytic H_2 production using Pt-TiO_2 in the presence of oxalic acid: influence of the noble metal size and the carrier gas flow rate. *Materials*, **7**, 7022–7038.

Li, L., Duan, L., Xu, Y. *et al.* (2010) Photoelectrochemical device for visible light driven water splitting by a molecular ruthenium catalyst assembled on dye-sensitized nanostructured TiO_2. *Chemical Communications*, **46**, 7307–7309.

Linsebigler, A.L., Lu, G., and Yates, J.T. Jr., (1995) Photocatalysis on TiO_2 surfaces: principles, mechanisms, and selected results. *Chemical Reviews*, **95**, 735–758.

Napoli, F., Chiesa, M., Livraghi, S. *et al.* (2009) The nitrogen photoactive centre in N-doped titanium dioxide formed via interaction of N atoms with the solid. Nature and energy level of the species. *Chemical Physics Letters*, 135–138.

Natarajan, A., Tsai, C.K., Khan, S.I. *et al.* (2007) The photoarrangement of -Santonin is a single-crystal-to-single-crystal reaction: a long kept secret in solid-state organic chemistry revealed. *Journal of the American Chemical Society*, **129**, 9846–9847.

Nebbia, G. and Kauffman, G.B. (2007) Prophet of solar energy: a retrospective view of Giacomo Luigi Ciamician (1857–1922), the founder of green chemistry, on the 150th anniversary of his birth. *The Chemical Educator*, **12**, 362.

Norrish, R.G.W. and Bamford, C.H. (1936) Photodecomposition of aldehydes and ketones. *Journal of the Chemical Society*, 1521, 1531, 1544, 1938. Reprinted in *Nature*, **138**, 1016.

Norrish, R.G.W. and Porter, G. (1949) Chemical reactions produced by very high light intensities. *Nature*, **164**, 658.

Noyori, R. (2010) Insight: green chemistry: the key to our future. *Tetrahedron*, **66**, 1028–1030.

O'Regan, B. and Graetzel, M. (1991) Low-cost, high-efficiency solar cell based on dye-sensitized colloidal TiO2 films. *Nature*, **353**, 737–740.

Park, H., Park, Y., Kimb, W., and Choi, W. (2013) Surface modification of TiO_2 photocatalyst for environmental applications. *Journal of Photochemistry and Photobiology C: Photochemistry Reviews*, **15**, 1–20.

Paterno, E. and Chieffi, G. (1909) Synthesis in organic chemistry using light. Note II. Compounds of unsaturated hydrocarbons with aldehydes and ketones. *Gazzetta Chimica Italiana*, **39**, 341–361.

Reisner, E., Powell, D.J., Cavazza, C. *et al.* (2009) Visible light-driven H_2 production by hydrogenase attached to dye-sensitized TiO_2 nanoparticles. *Journal of the American Chemical Society*, **131**, 18457–18466.

Searles, J.S. (1964) *Heterocyclic Compounds with Three-and Four-Membered Rings*. Wiley & Sons, Ltd.

Schultz, D.M. and Yoon, T.P. (2014a) Visible light photocatalysis. *Science*, **343**, 985.

Schultz, D.M. and Yoon, T.P. (2014b) Solar synthesis: prospects in visible light photocatalysis. *Science*, **343**, 1239–1276.

Serpone, N. and Emeline, A.V. (2012) Semiconductor photocatalysis: past, present, and future outlook. *Journal of Physical Chemistry Letters*, **3**, 673–677.

Serpone, N., Borgarello, E., and Graetzel, M. (1984) Visible light induced generation of hydrogen from H2S in mixed semiconductor dispersions, improved efficiency through interparticle electron transfer. *Journal of the Chemical Society, Chemical Communications*, **342**, 342–344.

Serpone, N., Borgarello, N., and Pelizzetti, E. (1988) Utilization of the semiconductor particle as a microphotoelectrochemical cell: electrochemical evidence for interparticle electron transfer and application to photocatalysis. *Journal of the Electrochemical Society*, **135**, 2760–2766.

Serpone, N., Emeline, A., Horikoshi, S. *et al.* (2012) On the genesis of heterogeneous photocatalysis: a brief historical perspective in the period 1910 to the mid-1980s. *Photochemical and Photobiological Sciences*, **11**, 1121–1150.

Shi, L., Yang, G., Tao, K. *et al.* (2013) An introduction of CO_2 conversion by dry reforming with methane and new route of low-temperature methanol synthesis. *Accounts of Chemical Research*, **46**, 1838–1847.

Snaith, H.J. (2013) Perovskites: the emergence of a new era for low-cost, high-efficiency solar cells. *Journal of Physical Chemistry Letters*, **4**, 3623–3630.

Stark, J. (1908) Weiter Bemerkungen iiber die thermische und chemische Absorption im Banden Spektrum. *Physikalische Zeitschrift*, **9**, 889–897.

Tahir, M. and NorAishahSaidina, A. (2013) Advances in visible light responsive titanium oxide-based photocatalysts for CO_2 conversion to hydrocarbon fuels. *Energy Conversion and Management*, **76**, 194–214.

Trommsdorf, J.B. (1834) Über Santonin. *Annalen der Pharmacie*, **11**, 190–199.

Umena, Y., Kawakami, K., Shen, J.-R., and Kamiya, N. (2011) Crystal structure of oxygen evolving photosystem II at a resolution of 1.9Å. *Nature*, **473**, 55–60.

Wasielewski, M. (2009) Self-assembly strategies for integrating light harvesting and charge separation in artificial photosynthetic systems. *Accounts of Chemical Research*, **42**, 1910–1921.

Windle, C.D. and Perutz, R.N. (2012) Advances in molecular photocatalytic and electrocatalytic CO_2 reduction. *Coordination Chemistry Reviews*, **256**, 2562–2570.

Woolerton, T.W., Sheard, S., Reisner, E. *et al.* (2010) Efficient and clean photoreduction of CO_2 to CO by enzyme-modified TiO_2 nanoparticles using visible light. *Journal of the American Chemical Society*, **132**, 2132–2133.

Yang, J., Wang, D., Han, H., and Li, C. (2013) Roles of cocatalysts in photocatalysis and photoelectrocatalysis. *Accounts of Chemical Research*, **46**, 1900–1909.

Zewail, A. (1994) *Femtochemistry: Ultrafast Dynamics of the Chemical Bond*. World Scientific, Singapore.

8

Enzymatic Catalysis

8.1 Early History of Enzymes

The word "enzymes" designates a large group of biological molecules whose catalytic properties are responsible for the thousands of highly selective metabolic reactions that are at the basis of life. Most enzymes are proteins, possessing three-dimensional structures that have specific activities and may employ organic as well as inorganic helper molecules, the "cofactors," to assist in the catalytic activity.

Humans have used enzymes for centuries, either in the form of vegetables rich in enzymes or in the form of microorganisms for a variety of purposes as in the production of cheese, in brewing processes, in baking, and in the production of vines and alcohol.

The history of cheese predates recorded history, and there is no conclusive evidence indicating where cheese making originated, be it in Europe, Central Asia, the Middle East, or the Sahara. It is, however, documented by Pliny the Elder that cheese enterprises were widespread at the time of ancient Rome.

Archeological evidence suggests that brewing was already used around the sixth millennium BC in many areas, including ancient Egypt and Mesopotamia.

The brewing industry ferments the malt by means of a starch ferment such as malted barley. Similarly winemaking spans thousands of years and is closely allied with Western civilization. Grapevine and alcoholic beverages obtained from fermented juice were important beverages for many ancient civilizations including Mesopotamia, Israel, and Egypt and became essential commercial goods for Phoenicians, Greeks, and Romans. Homer quoted wine often, both in Iliad and in Odyssey. Many important Roman writers, including Cato, Horace, Pliny, and Virgil, have described the role played by wine in the Roman culture. Many of the techniques and principles developed in ancient times are found today in modern winemaking.

Although the production or consumption of wine was forbidden in Islamic countries, around AD 800 Arab alchemists such as Jābir ibn Hayyān (Latinized

The Development of Catalysis: A History of Key Processes and Personas in Catalytic Science and Technology, First Edition. Adriano Zecchina and Salvatore Califano.
© 2017 John Wiley & Sons, Inc. Published 2017 by John Wiley & Sons, Inc.

as Geber) developed the distillation of wine for obtaining medicinal and industrial products such as perfume.

In 1785 Angelo Fabroni (1732–1803), an Italian humanist, scientist, and professor at the University of Pisa, not only known as a *man of letters* but also interested in natural philosophy, was the first to provide an interpretation of the chemical composition of the ferment involved in alcoholic fermentation, which he defined as a plant–animal substance.

In a memoir presented to the Academy of Florence (1787), Angelo Fabroni compared ferments to animal substances saying that

> *The matter that decomposes sugar is a vegeto-animal substance; it resides in peculiar utricles, in grapes as well as in corn.* (Fabroni, 1873)

Subsequently, in 1837, an accomplished French physicist and engineer Charles Cagniard de la Tour (1777–1859) proved for the first time that yeast was a living organism and that it was capable of multiplying and belonged to the plant kingdom (Cagniard-Latour, 1837) (Figure 8.1). According to Cagniard, the yeast's "vital" activities were at the basis of the fermentation of sugar-containing liquids. He also studied the influence of lowering the temperature upon its life.

(a)   (b)

Figure 8.1 (a) Charles Cagniard de la Tour (1777–1859) and (b) Moritz Traube (1826–1894). These scientists actively participated in the debate on the nature of fermentation. Cagniard proved for the first time that the yeast is a living organism and that it is capable of multiplying and belongs to the plant kingdom. Traube concluded that all fermentation process caused by living organisms is due to chemical reactions. Images in the public domain.

In the *Mémoir sur la fermentation vineuse*, Cagniard argued that fermentation resulted from organic activity of yeast, with oxygen necessary to start the process that then could continue anaerobically. His observation that different yeasts produced different alcohols was a first inkling of the germ theory later developed by Louis Pasteur.

Justus von Liebig and Friedrich Wöhler mercilessly ridiculed Cagniard's views on fermentation, despite the fact that they were based on theories, later confirmed to be correct.

In 1858, Liebig's student Moritz Traube enunciated the theorem, at the time used only for alcoholic fermentation although of a more general validity, that fermentation processes caused by living organisms originate from chemical reactions rather than from the presence of a vital force (Traube, 1858, 1874) (Figure 8.1).

Around the same time, the German naturalist and physiologist Theodor Schwann (1810–1882), now acknowledged as the founder of modern physiology and histology, confirmed this theory (Figure 8.2). He named the beer yeast *zuckerpilz*, meaning sugar fungus, *Saccharomyces* in Latin. Schwann coined the term "metabolism" to describe the chemical changes observed in

(a)   (b)

Figure 8.2 (a) Theodor Schwann (1810–1882) and (b) Eduard Buchner (1860–1917). While Schwann and Anselme Payen (see Figure 1.2) are precursors of modern enzyme science, Buchner is considered the first modern enzyme scientist. We owe to Schwann the discovery of pepsin and the invention of the term "metabolism." The discovery of enzyme diastase is instead attributed to Payen. Buchner showed that a cell-free extract of yeast cells is active in fermentation. For this discovery, which put the definite end to vitalism theory, he was awarded the 1907 Nobel Prize. Images in the public domain.

living tissue by the presence of enzymes, and in 1839 he developed the cell theory that identifies cells as the basic structural units of plants and animals (Schwann, 1837, 1839). He demonstrated that heat and certain chemical products are capable of stopping alcoholic fermentation:

> *Thus, in alcoholic fermentation as in putrefaction, it is not the oxygen of the air which causes this to occur, as previously suggested by Gay-Lussac, but something in the air which is destroyed by heat.*

He also believed that pepsin was what Berzelius defined as catalysts or as "vital force" for chemical reactions in living matter. Pepsin was the first enzyme identified in an animal tissue.

Almost concurrently to Schwann (1837), Kützing (1837), a German pharmacist, botanist, and algae expert, identified yeast as a living organism that consumes the sugar it ferments, a process that they referred to the alcoholic fermentation. Kützing's (1807–1893) work on yeast confirmed the results of Cagniard and Schwann in all details (Kützing, 1837).

The principal facts derived from the publications of these scientists provided firm evidence that the metabolic activity of yeast cells is the cause of fermentation and demonstrated that sugar is used for growth and multiplication of the yeast.

In the famous investigations on wine (1866) and beer (1876), Pasteur (1866) gave credibility to the "vitalist" viewpoint of alcoholic fermentation by isolating several ferment races and species and proving that the yeast is producing secondary products such glycerol and carbon dioxide.

The idea that fermentation was an enzymatic biological process based on cells, strongly supported by Pasteur, did not convince Justus von Liebig who instead believed that fermentation was a pure chemical oxidation process. This originated a famous Liebig–Pasteur controversy that went on for several years (1857–1869).

Pasteur demonstrated that the yeast's ferment responsible for spontaneous fermentation of grape came from their surface, and to prove this, he isolated several races and species.

Diastase was the first enzyme discovered. Anselme Payen (1795–1878) and Jean-François Persoz (1805–1868), chemists at a French sugar factory (Payen and Persoz, 1833), extracted it from a malt solution in 1833. The extract, like malt itself, converts gelatinized starch into sugars, primarily into maltose. However, this extract cannot be considered a pure enzyme, as only one hundred years later scientists learned in adjusting the appropriate purification and crystallization methods.

The term enzyme, derived from the Greek word stasis, στάσις, the act of standing, was only later coined by Wilhelm Friedrich Kühne (1837–1900), and the naming of enzymes with the suffix -ase added to the chemical name of a compound was started by the French scientist Émile Duclaux (1840–1904). Duclaux intended to honor the discoverers of diastase, from the Greek διαστασις, "separation," by introducing the term in the second volume of his book *Traité de microbiologie* (Duclaux, 1899).

The German chemist Eduard Buchner (1860–1917), who had studied chemistry with Adolf von Baeyer and worked with Otto Fischer in Erlangen, uncovered the incongruence at the basis of the Liebig–Pasteur controversy over the nature of alcoholic fermentation (Figure 8.2). Buchner published his first paper in 1885 (Büchner and Curtius, 1885a, b), revealing that fermentation occurs only in the presence of oxygen, a conclusion contrary to the prevailing view, held by Louis Pasteur. For this paper, he is normally considered the person who first understood the chemistry of alcoholic fermentation, despite the fact that the Russian physician Maria Manasseina (1841–1903) claimed to have discovered cell-free alcoholic fermentation in 1872, several years earlier than Buchner.

The experiment for which Buchner won the Nobel Prize (Buchner, 1897) consisted in producing a cell-free extract of yeast cells and showing that it could ferment sugar. This dealt a blow to vitalism, showing that the presence of living yeast cells was not necessary for fermentation. Microscopic investigation confirmed that no living yeast cells were present in the extract. By 1893 Buchner had suggested that the fermentation process is caused by an enzyme, which he named zymase, and that it is the result of chemical processes both inside and outside the cells (Buchner, 1897).

Although at the Buchner's time, the structure of the proteins as natural catalysts was unknown, it is through Buchner and his experiment in fermentation that science marked one of the most important points in the history of modern chemistry.

8.2 Proteins and Their Role in Enzymatic Catalysis

Proteins often act as enzymes, that is, as catalysts that increase the rate of all the chemical reactions within cells. In absence of enzymes most biochemical reactions would not proceed at the temperature and pressure compatible with life. Instead, in presence of enzymes the rates of such reactions increase by several orders of magnitude. Cells contain many different enzymes, whose actions regulate the rates of the many possible reactions taking place within the cell.

Like all other catalysts, enzymes increase the rate of chemical reactions by reducing the activation energy without being consumed or permanently degraded and without altering the chemical equilibrium between reactants (substrates) and products.

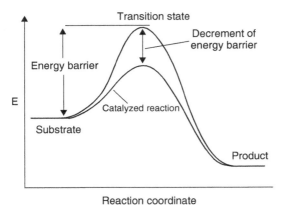

Figure 8.3 Schematic representation of the conversion of a substrate S to a product P in absence and in presence of the catalyst. The equilibrium between substrate and product is unaffected by (enzymatic) catalysis. The lowering of the energy barrier is due to the formation of the enzyme–substrate (*ES*) complex.

Figure 8.3 illustrates the simple conversion of a substrate into a final product. Because the final energy state of P (product) is lower than that of S (substrate), the reaction proceeds from left to right.

The catalytic activity of enzymes (E) involves the formation of an enzyme–substrate (ES) complex. The "active site" is the specific region of the enzyme where the substrate is bonded and where it is converted into the product, which is then released from the enzyme. The enzyme-catalyzed reaction can thus be written as follows:

$$S + E \leftrightarrow ES \leftrightarrow E + P$$

Notice that the total energy is unaltered on both sides of the equation, so indicating that the equilibrium is unaffected.

Since the evolution of discoveries concerning proteins is fundamental for the history of enzymes, we briefly summarize here the state of knowledge concerning proteins in the period ranging from the end of the nineteenth to the initial part of the twentieth centuries. This was preceded by the identification of the first amino acids, for instance, asparagine in 1806, cysteine in 1810, glycine, alanine, and leucine in 1820.

One could fix the start of protein history to 1851, when Otto Funke (1828–1879) successfully crystallized hemoglobin, a fact that must be considered a fundamental achievement in protein chemistry (Funke, 1853; Funke and Latschenberger, 1877) (Figure 8.4). Otto Funke, graduated at Heidelberg, was professor of physiological chemistry at the University of Leipzig. His work was a precursor to Felix Hoppe-Seyler's important studies of hemoglobin (Hoppe-Seyler, 1864). Hoppe-Seyler (1825–1895) was the first scientist to observe the two characteristic absorption bands in absorption spectrum in the visible of the red blood pigment. He also recognized that hemoglobin contains iron and binds oxygen reversibly to form oxyhemoglobin. Hoppe-Seyler was able to obtain hemoglobin in crystalline form (Figure 8.4). Hoppe-Seyler

(a)   (b)

Figure 8.4 (a) Otto Funke (1828–1879) and (b) Felix Hoppe-Seyler (1825–1895). In 1851 Funke succeeded in growing hemoglobin crystals. Hoppe-Seyler recognized that hemoglobin contains iron and binds oxygen reversibly to form oxyhemoglobin. The results obtained by Funke and Hoppe-Seyler were of fundamental importance for successive studies on the protein structures. Images in the public domain.

also performed investigations on the green chlorophyll pigment and isolated several different proteins (which he called proteids) including lecithin. In this regard it should be noted that lecithin, the first and characterized member of the phospholipids class, had been already identified in 1846 and in 1874 by the French chemist and pharmacist Theodore Nicolas Gobley, a recognized expert in the chemical components of brain tissues (Gobley, 1850, 1874). Lecithin was then isolated in Hoppe-Seyler's own laboratory in Strasbourg by a coworker Johannes Friedrich Miescher (1844–1895). In the Hoppe-Seyler laboratory, Miescher isolated various phosphate-rich chemicals, which he called *nuclein* (now nucleic acids), from the nuclei of white blood cells in 1869. Hoppe-Seyler preferred scientific research to medicine and held positions in applied chemistry and biochemistry in Greifswald, Tübingen, and Strasbourg.

At the end of nineteenth century, the very important early protein scientist Franz Hofmeister (1850–1922), professor of pharmacology in Prague in 1885 and later in Strasbourg in 1896, provided the chief tool for purifying proteins by precipitating them out from their solutions with the addition of salts, a method that is still in use today (Hofmeister, 1887, 1888) (Figure 8.5). Hofmeister himself may have been the first to crystallize albumin. He was one of the initiators of protein chemistry. A comprehensive review of the Hofmeister work can be found in the contribution of Baldwin (1996).

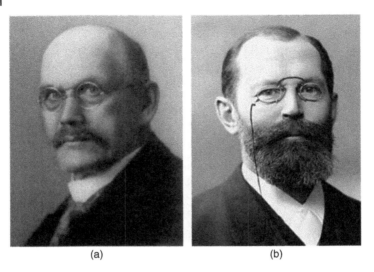

(a)   (b)

Figure 8.5 (a) Franz Hofmeister (1850–1922) and (b) Emil Fischer (1852–1919). Hofmeister was the first, with Fischer, to propose that polypeptides are amino acids linked by peptide bonds and invented a method of protein purification that is still in use today. Fischer, awarded with the 1902 Nobel Prize for his work on sugar and purine syntheses, performed the synthesis of oligopeptides. Emil Fischer was also interested in enzymes and first proposed *the key and lock model*. Images in the public domain.

Another scientist who made important contributions in the area of protein chemistry was Emil Fischer, who devoted his activity to chemistry under the influence of Adolf von Baeyer, successor of Liebig at the University of Munich (Figure 8.5). In 1888 he became professor of chemistry at the University of Würzburg and remained there until 1892, when he was asked to succeed August Wilhelm von Hofmann in the Chemistry chair at the University of Berlin where he remained until his death in 1919. His name is better known for his studies on sugars and purine. His achievement in this area was recognized in the 1902 Nobel Prize commendation *"in recognition of the extraordinary services he has rendered by his work on sugar and purine syntheses."* However, the contribution of Emil Fischer to protein chemistry is also highly important and diversified. First, he set up efficient methods for separation and identification of the individual amino acids. In fact he prepared in the laboratory several amino acids occurring in nature. In 1901 he discovered the synthesis of the dipeptide glycylglycine by reacting two glycine molecules (Fischer and Fourneau, 1901). He then performed the synthesis of the oligopeptides up to the octodecapeptide, a molecule with many properties of natural proteins (Fischer, 1905). A real fundamental step for the comprehension of the protein structure occurred in 1902, when Emil Fischer and Franz Hofmeister independently proposed that proteins are the result of the reaction between the amino group of one amino acid and the carboxyl group of another, with formation of a linear structure that Fischer named peptide (Fischer, Schulze, and Winterstein, 1903).

A simple example of peptide synthesis is shown in the succeeding text, where the formation of a peptide bond is a consequence of the water elimination between two alanine molecules:

The so formed dimer can react with another amino acid (identical or with different alkyl groups) to form a trimer, and so on. We know now that amino acids with the amine and the carboxylic acid groups attached to the first (alpha-) carbon (α-amino acids) have a particular connection to biochemistry and that the total number of these "protein-building" amino acids, characterized by different R groups, is 20. The amino acids that comprise the building blocks of proteins are characterized by their side chains, which can range from simple aliphatic groups such as that found in glycine ($R = H$) and alanine ($R = CH_3$) to more complex, functionalized side groups such as that of glutamic acid and asparagine ($R = COOH$ and/or COO^-), serine ($R = OH$), cysteine ($R = SH$), and so on. The reaction of 20 different amino acids can originate millions of combinations. This fact is at the basis of the complexity of protein structures.

While all amino acids determine the three-dimensional structure of proteins through hydrophobic and hydrophilic interactions, ionic interactions, and formation of disulfide bonds, only a smaller subset of these amino acids are implicated in enzyme activities. The important amino acids are those characterized by $R = OH$, $R = SH$, $R = NH_4^+$, $R = NH_2$, $R = COOH/COO-$, $R = $ imidazolium, and $R = $ thioether. These amino acids all have functional R groups on the side chains, performing chemical interactions with substrates or binding cofactors (*vide infra*).

We know now that the polypeptide chains can also form helical structures via hydrogen bonding. The helical and linear portions of the peptide chains can wind into globular balls.

Emil Fischer was also interested in the enzymes' chemistry. He suggested that an enzyme's specificity is due to the fact that both the enzyme and the substrate are characterized by complementary geometrical shapes that fit exactly into one another (key and lock model), an intuition that maintains a partial validity even today. A discussion of the contribution of Emil Fischer to the key and lock model can be found in the paper of Koshland (1995).

The discovery that proteins could be crystallized presented a highpoint in protein chemistry because it allowed their structure to be determined by X-ray crystallography. This fundamental achievement was at the origin of the branch of science known today as structural biology, initiated in 1958 with John Kendrew's atomic structure of myoglobin. We briefly report on the myoglobin and hemoglobin structures because they have been milestones for structural biology and for the history of enzymes. Myoglobin is an iron- and oxygen-binding protein with a globular form, present in the muscle tissue of almost all mammals. It is closely related to hemoglobin, which is the iron- and oxygen-binding protein in the red blood cells. Myoglobin was the first protein whose three-dimensional structure was revealed by X-ray crystallography. This result was published in 1958 by John Kendrew and associates (Kendrew *et al.*, 1958) (Figure 8.8). John Kendrew (1917–1997), born in Oxford, was a reader in climatology in the University of Oxford. During the World War II, he worked on radar in the Royal Air Force headquarters. He then became increasingly attracted by biochemistry and started to work on the determination of proteins structure. In 1945, he contacted Max Perutz in the Cavendish Laboratory in Cambridge where he started to work on hemoglobin. In 1947, he became a fellow of the Peterhouse at Cambridge and started to work in a research unit devoted to the study of biological systems headed by Sir Lawrence Bragg. In 1954, he became a reader at the Royal Institution in London. Although the original studies were on the structure of hemoglobin, Kendrew later embarked on the study of myoglobin from a horse's heart that is only a quarter of the hemoglobin size. An electron density map at 2 Å (0.2 nm) resolution was obtained in 1959. Myoglobin contains a porphyrin ring with an iron at its center. A histidine group is also attached directly to iron. The reversibly bonded oxygen molecule is on the opposite side (Figure 8.6).

The heme group is located into a surface pocket easily accessible to the O_2 substrate. Max Perutz's structure of hemoglobin extended this story. This protein is a tetramer of subunits similar to myoglobin but with each containing one heme group (Perutz, 1960; Perutz *et al.*, 1960) (Figure 8.7). Max Perutz (1914–2002), Jewish by ancestry but baptized Perutz in the Catholic religion, obtained his degree in chemistry at the University of Vienna in 1936

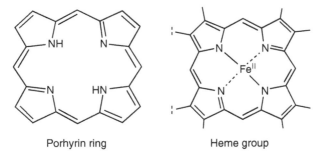

Porhyrin ring   Heme group

Figure 8.6 Structure of the porphyrin ring and of the parent porphyrin ring with an Fe at its center (heme group).

Haemoglobin

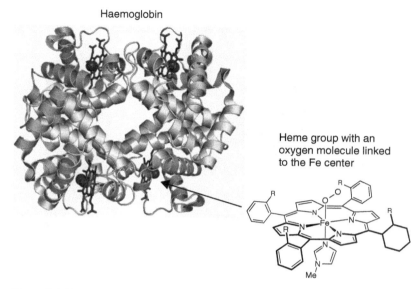

Heme group with an
oxygen molecule linked
to the Fe center

Figure 8.7 Tetrameric structure of oxygenated hemoglobin. In the inset the heme group carries an oxygen molecule and a histidine molecule. The strands are polypeptide chains. *Source*: Adapted from a figure prepared by Todd L. Mollan, CBER.

(Figure 8.8). Made aware of the advanced results in biochemistry obtained at the University of Cambridge, he succeeded to join the Cavendish Laboratory in Cambridge, where he began his doctoral thesis on the hemoglobin structure under the guidance of William Lawrence Bragg. At the start of World War II, Perutz like other persons of German or Austrian origin, was sent to Canada (on orders from Prime Minister Winston Churchill). After a several month's internment, he returned to Cambridge and in 1947, with Bragg's support, succeeded to establish the Molecular Biology Unit at the Cavendish Laboratory. Perutz's new research unit attracted talented researchers and among them were Francis Crick in 1949 and James D. Watson in 1951.

In 1959, Perutz and his colleagues started an enterprise to determine the structure of oxy- and deoxyhemoglobin at high resolution. As a result, in 1970, he was at last able to suggest its dynamical behavior as a molecular machine during the oxygen binding to the heme complex, in terms of the iron atom moving with respect to the plane of the porphyrin ring upon adsorption and desorption of the oxygen molecule.

With this interaction, the histidine moves nearer to the iron atom. This induces a similar conformational change in the other heme sites such that the binding of oxygen is facilitated. This example of dynamic cooperative effect is quite common in enzymes (Chou, 1989).

(a)   (b)

Figure 8.8 (a) John Cowdery Kendrew (1917–1997) and (b) Max Ferdinand Perutz (1914–2002). They are the founders of structural biology and shared the 1962 Nobel Prize. Kendrew determined by X-ray crystallography the structure of myoglobin, which is the first protein to have its three-dimensional structure clarified. Perutz determined the structure of oxy- and deoxyhemoglobin at high resolution. Images in the public domain.

John Kendrew and Max Ferdinand Perutz shared the Nobel Prize in 1962.

In the following decades the field grew rapidly. Already by the early 1970s there were a dozen atomic structures of proteins, and researchers were discovering that they had a gold mine of information. In very short time the number of solved protein structures became so numerous that a Protein Data Bank archive was created to solve this problem.

While in 1976 in this archive, only the structural data of 13 proteins (including enzymes) were present, in 1986 they were grown to 213, and in 1996 to about 5000. The growth was exponential, and today about 100,000 structural data are reported in the scientific literature (Figure 8.9). About 10% of them concern the enzymes. This exponential growth is indicative of a very alive and innovative branch of science that is definitely a fingerprint of the twentieth century.

Although considered as catalytically active substances, enzymes were never really isolated before 1926. The first crystallized enzyme (urease) was isolated only in 1926, thanks to the persistence of the American chemist James Batcheller Sumner (1887–1955) (Figure 8.10). Sumner in 1912 went to study biochemistry in Harvard University where he obtained the PhD degree. He then moved to Cornell University as assistant professor of biochemistry where he worked unsuccessfully for many years trying to achieve what many of his colleagues believed an impossible result: the crystallization of an enzyme (urease). Sumner (1946) was also able to demonstrate that his pure urease was a protein.

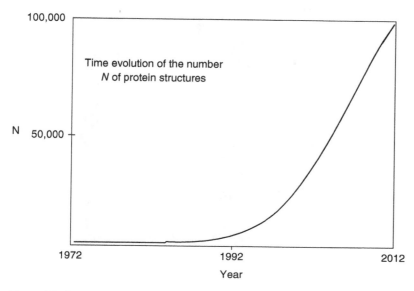

Figure 8.9 Onset and evolution of the structural biology of proteins (including enzymes). *Source*: Adapted from the Arieh Warshel Nobel lecture slides.

Sumner's result demonstrated for the first time that a pure protein can function as an enzyme, a conclusion that is at the basis of the recognition that enzymes are indeed proteins. For this result Sumner received the Nobel Prize in chemistry in 1946. The structure of urease was later solved by Paul Andrew Karplus and coworkers in 1995 (Jabri *et al.*, 1995).

In 1975, 50 years from Sumner's observation of urease crystal, Nicholas E. Dixon, professor of biological chemistry at the University of Wollongong in Australia, in collaboration with Carlo Gazzola, Burt Walters, Robert L. Blakeley, and Burt Zerner, discovered that urease is a metalloenzyme and showed that the active site contains nickel (Dixon *et al.*, 1975).

Concerning isolation and purification of pepsin, discovered in 1876 by Kühne (1877), it is worth recalling that highly active preparations were available since the end of the nineteenth century. There was uncertainty, however, about the proteic nature of pepsin. The question was not solved until 1929, when the American biochemist John Howard Northrop (1891–1987), employed at the Rockefeller Institute for Medical Research in New York City, isolated and crystallized the gastric enzyme swine pepsin (Northrop, 1921, 1929, 1946; Northrop and Kunitz, 1931) and provided convincing evidence for its identity as a protein (Figure 8.10). Northrop also isolated and crystallized pepsinogen and trypsinogen (Kunitz and Northrop, 1934; Herriott and Northrop, 1936), the inactive precursor to pepsin converted to pepsin in the presence of hydrochloric acid.

(a)

(b)

(c)

Figure 8.10 (a) James Sumner (1887–1955), (b) John Northrop (1891–1987), and (c) Otto Warburg (1883–1970). Summer and Northrop shared the 1946 Nobel Prize. Sumner demonstrated that enzymes can be crystallized and in 1926 isolated the first crystallized enzyme (urease). In 1929 Northrop isolated and crystallized pepsin. Warburg (1931 Nobel Prize) was a prominent biochemist. Among his numerous contributions, the structure and function of the respiratory enzyme must be mentioned. Images in the public domain.

Only after 1945, the amino acid sequences of many proteins, including pepsin and its precursor pepsinogen, were clarified. John Howard Northrop shared the 1946 Nobel Prize for chemistry with Wendell Meredith Stanley and James Batcheller Sumner, for his work establishing the chemical nature of enzymes and showing that bacteriophages (viruses that infect specific bacteria) contain nucleic acid. Upon approaching his ninety-sixth birthday, he viewed his future with concern and presumably felt that he was becoming an unfortunate burden to his family and friends and decided to avoid such a future and took his life on May 27, 1987.

Most of the discoveries that have enlightened the new field of enzymatic catalysis were due to the research and to the percipience in interpreting experimental facts of the German physiologist and chemist Otto Heinrich Warburg (1883–1970) (Figure 8.10). Born in Freiburg, Otto Warburg studied chemistry under the guidance of Emil Fischer and gained the degree of Doctor of Chemistry at the University of Berlin in 1906. He then obtained the degree of Doctor of Medicine at the University of Heidelberg in 1911.

Between 1908 and 1914, Warburg was affiliated with the Naples Marine Biological Station where he demonstrated that the rate of respiration of urchin eggs after fertilization increases by a factor of six.

Warburg can be considered as one of the greatest biochemists of the twentieth century. He won the 1931 Nobel Prize for medicine for its contribution to the discovery of the structure and mechanism of action of the respiratory enzyme. In 1918, Warburg was appointed professor at the Kaiser Wilhelm Institute for Biology in Dahlem, Berlin (part of the Kaiser-Wilhelm-Gesellschaft). By 1931, he was named director of the Kaiser Wilhelm Institute for Cell Physiology. Research at this institute led him to clarify that flavins and nicotinamide are the active species of the hydrogen-transferring enzymes. This result, together with the iron oxygenase discovered earlier, has formed the basis of a complete knowledge of the oxidations and reductions in living cells. For his work on nicotinamide in 1944, his name was presented again for a second Nobel Prize in physiology. He worked with Dean Burk in photosynthesis (Warburg, 1955) to discover the CO_2 splitting reaction activated by respiration (Warburg, 1952, 1955). He is undoubtedly the person who has isolated and crystallized the highest number of enzymes in his career (Warburg and Christian, 1938, 1941, 1942). Several scientists including Sir Hans Adolf Krebs (1953 Nobel Prize in medicine) worked in Warburg's laboratory. Among other discoveries, Krebs, together with Fritz Albert Lipmann, identified the citric acid cycle (or Szent-Györgyi–Krebs cycle). Indeed, there is no doubt that Warburg's combined work in physiology, cell metabolism, and oncology made him an outstanding figure in the history of biology.

8.3 Enzymes/Coenzymes Structure and Catalytic Activity

The discovery by Sumner that enzymes could be crystallized opened a specific area of structural biology devoted to enzymes. The first achievement was obtained when the structure of lysozyme was solved by David Chilton Phillips and his group (1924–1999) and published in 1965 (Blake *et al.*, 1965). This is the first enzyme containing 20 amino acids fully sequenced (Canfield, 1963) (Figure 8.11).

Lysozyme (McKenzie and White, 1991) is able to hydrolyze peptidoglycan, a polymer consisting of sugars and amino acids that forms a layer outside the membrane of bacteria. For this reason lysozyme is part of the innate immune system. The chemical groups responsible for the hydrolysis reaction are carboxylic and carboxylate groups of glutamic acid and aspartate amino acids located in a pocket. The peptidoglycan/lysozyme pocket obeys the "lock and key model" hypothesized by Fischer, even if all the structure is responding due to cooperative dynamic effect.

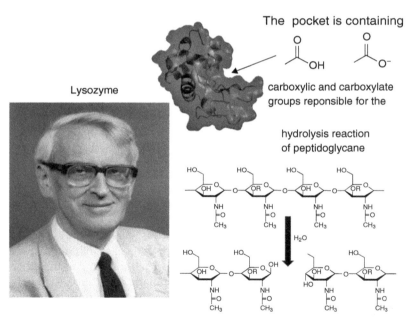

Figure 8.11 David Chilton Phillips (1924–1999), the first, in 1965, to determine in atomic detail the structure of an enzyme (lysozyme) by X-ray crystallography. Lysozyme is classified as a hydrolase and is part of the innate immune system. Chilton Phillips image is in the public domain (author: Tomos Antigua Tomos).

Starting in 1965, several enzymes structures have been solved, but it is beyond the scope of this book to tell the whole story. Only a few enzymes and the catalytic mechanisms, in which they play a role, will be illustrated. Some enzymes do not need any additional components to exhibit full activity. We have already illustrated the case of lysozyme. Other enzymes of this type are pepsin and trypsin. However, many others require nonprotein molecules (*cofactors*) to be bound to their proteic structure for efficient activity. Their primary function is to assist in enzyme activity in performing certain reactions that the proteic part of the enzyme cannot perform alone. The discovery of the structure and function of cofactors has represented an important chapter of enzyme science. Cofactors can be either partially inorganic (containing metal ions and iron–sulfur clusters) or fully organic compounds, which are also known as *coenzymes*. In some enzymes both inorganic and organic cofactors can be present.

Most cofactors are not covalently bound to an enzyme but are closely connected to the proteic part via electrostatic forces, hydrogen bonding, and van der Waals interactions. However, some cofactors, known as *prosthetic groups*, are tightly bound to the enzyme through covalent bonds. Most cofactors serve as carriers during the reaction to transfer electrons, atoms, or functional groups from an enzyme to a substrate.

Because of the complexity of the enzyme world constituted by several thousand components, it is appropriate to spend a few sentences on enzyme classification and nomenclature, based on the fact that enzymes are known for their specificity, that is, they often interact with only one substrate to catalyze a particular reaction. Thus enzymes have often been named by adding the suffix *-ase* to the name of the substrate. Enzymes are broadly organized into six categories based on the types of reactions they catalyze:

- *Oxidoreductases* catalyze oxidation–reduction reactions, which involve electron transfer.
- *Transferases* transfer a chemical group (such as a methyl or phosphate group) from one molecule to another.
- *Hydrolases* catalyze the cleavage of chemical bonds through the addition of a water molecule hydrolysis.
- *Lyases* cleave various bonds by means of mechanisms different from hydrolysis and oxidation.
- *Isomerases* transfer a group within a single molecule to form an isomer.
- *Ligases* join two molecules with covalent bonds.

Examples of coenzymes present in oxidoreductases are nicotine adenine dinucleotide (NAD), nicotine adenine dinucleotide phosphate (NADP), and flavin adenine dinucleotide (FAD). As for coenzymes in transferases, we mention coenzyme A, thiamine pyrophosphate (Vit B1), and biotin. The development of knowledge as to the structure and role of coenzymes has been central in the understanding of catalytic activity of the enzymes. For this reason in the following discussion, the contributions of the main researchers in this area will be briefly described.

The first organic cofactor to be discovered was NAD$^+$, which was identified by the British Arthur Harden (1865–1940) in 1906 (Figure 8.12). Arthur Harden entered the University of Manchester in 1882, where he studied under Sir Henry Enfield Roscoe, a pupil of Bunsen. In 1887–1888, he worked with Otto Fischer at Erlangen. In 1907, he was appointed head of the Biochemical Department of the University of Manchester, a position that he held until his retirement. He studied the breakdown products of glucose and the chemistry of the yeast cell. His classic study on the fermentation of sugar by yeast juice lasted for many years and advanced the knowledge of the metabolism processes in all living forms. In 1929, he was awarded, together with the Swedish von Euler-Chelpin, the Nobel Prize in chemistry. Hans von Euler-Chelpin (1873–1964), who studied at the University of Berlin under Emil Fischer, Jay A. Rosenheim, Emil Warburg, and Max Planck, took in 1895 his doctorate at the University of Berlin. Through a long and difficult purification from yeast extracts, he identified this heat-stable factor as a nucleotide sugar phosphate.

In 1930, von Euler-Chelpin published the structure of nicotinamide adenosine dinucleotide (NAD$^+$) and its phosphate derivative (NADP$^+$), a coenzyme working in anabolic reactions and requiring NADPH as reducing agent (Figure 8.12). In his research on yeast cells, von Euler-Chelpin showed that in enzyme chemistry also a coenzyme is needed to do the work. The structure of NAD$^+$ is shown in the succeeding text.

Structure of NAD$^+$

An important part of von Euler-Chelpin's work was done in collaboration with Karl Olof Josephson on the enzymes saccharase and catalase.

Throughout the early twentieth century, numerous cofactors were progressively identified. In particular, ATP was isolated in 1929 by Karl Lohmann (1898–1978), and the important coenzyme A was discovered in 1945 by Fritz

(a)

(b)

(c)

Figure 8.12 (a) Arthur Harden (1865–1940), (b) Hans von Euler-Chelpin (1863–1964), and (c) Fritz Lipmann (1899–1986). These scientists were all involved in the understanding of the role of coenzymes. Harden and von Euler-Chelpin obtained the 1929 Nobel Prize for their investigations on the fermentation of sugar and fermentative enzymes. Fritz Lipmann, 1953 Nobel Prize, discovered the coenzyme A. *Source*: (a, c) Images in the public domain. (b) Reproduced with permission of Pieter Kuiper.

Albert Lipmann (1899–1986), who received the 1953 Nobel Prize for this discovery (Figure 8.12). Lipmann was born a German, from a Jewish family. In 1926, he was at the Kaiser Wilhelm Institute and then moved to Heidelberg. From 1939 on, he lived and worked in the United States at Harvard Medical School and Rockefeller University.

As for inorganic cofactors, we mention Zn^{2+} in carbonic anhydrase and alcohol dehydrogenase, Fe^{2+}/Fe^{3+} in cytochromes, methane monooxygenase (MMO), hemoglobin, and ferredoxin, and Cu^{2+}/Cu^{+} in cytochrome oxidase. In many enzymes, inorganic and organic cofactors are simultaneously present. We will briefly describe the structure and activity of metalloenzymes in the following text.

8.4 Mechanism of Enzyme Catalysis

In principle, the mechanism of enzyme catalysis is similar to other types of chemical catalysis in that it provides an alternative reaction route for reducing the activation energy.

Enzymes are very specific in their action, and already in 1894, the Nobel Laureate organic chemist Emil Fischer hypothesized that this was due to the complementary geometric shapes of enzymes and substrates that fit exactly into one another (*lock and key* model) (Fischer, 1894). This was the first attempt to explain the extraordinary selectivity of enzymes.

About 70 years later (1965), Jacques Monod, Jeffries Wyman, and Jean-Pierre Changeux proposed a more sophisticated concerted model, known as the MWC model (Monod, Wyman, and Changeux, 1965). The model was based on the postulate that proteins (enzymes) exist in different interconvertible states whose ratio is determined by thermal equilibrium.

The MWC model, although widely utilized in enzymology, has been shown inappropriate in many cases. In 1966 Daniel Koshland suggested another modification of the Fischer lock and key model, proposing that distortion of the enzyme and the substrate is important to account for the substrate and enzyme interaction (Koshland, 1958). With his coworkers George Nemethy and David Filmer, Koshland developed the subsequent model of enzyme–substrate cooperativity (Koshland, Nemethy, and Filmer, 1966), which was an improvement over the MWC model (Monod, Wyman, and Changeux, 1965). The basic idea of the new model is that the active site is not rigid because during the interaction with substrate, the amino acid side chains that make up the active site adapt their tridimensional structure to the geometry of the substrate. In other words, the active site gradually changes until the substrate is completely bound. In some cases, also the substrate molecule slightly changes shape as it enters the active site (induced fit model). Induced fit may enhance the molecular recognition in the presence of competition of different substrates. The ability to change shape upon interaction with substrates differentiates enzymes from inorganic selective catalysts like zeolites or other microporous materials.

It is evident that Daniel Koshland's research has dramatically changed the knowledge of enzymes and protein chemistry and that the "induced fit" theory is occupying the topmost position. Koshland, with a PhD in 1949 from the University of Chicago and a successive postdoctoral at Harvard University, moved in 1951 to Brookhaven National Laboratory and in 1965 to Berkeley as head of the Biochemistry Department and as a member of the Chancellor's Advisory Council on Biology. The induced fit theory belongs to the Berkeley period.

Due to the extremely rapid development of the area, it is impossible to report about the structures of all enzymes discovered in period comprised between the end of twentieth century and the initial part of the twenty-first. For this reason we only mention the enzymes that are able to catalyze reactions such as hydrocarbon partial oxidation, hydrogen, and nitrogen activation that, as documented in the previous chapters, have stimulated along the twentieth century the efforts of scientists involved in homogeneous and heterogeneous catalysis.

In particular, among the enzymes that have most inspired the homogeneous and heterogeneous catalysis community and the synthesis of bioinorganic synthetic analogues, MMO that belongs to the class of oxidoreductases is occupying a preeminent position. MMO is an enzyme able to perform the partial oxidation of methane to methanol under very mild conditions. This reaction, which is involving the very stable methane molecule, has been attempted by many researchers without success and can be considered as a catalysis still open problem. The detailed structure of MMO was solved only in 2005 by the group of Amy C. Rosenzweig (Lieberman and Rosenzweig, 2005), professor of chemistry, biochemistry, molecular biology, and cell biology at Northwestern University.

There are two well-studied MMO forms: the soluble methane monooxygenase (sMMO) and the particulate methane monooxygenase (pMMO) form. The active site in sMMO contains two iron centers bridged by oxygen atoms (Fe–O–Fe), whereas the active site in pMMO utilizes a copper dimer. The probable catalytic mechanism of sMMO is illustrated in Figure 8.13. The presence of two iron ions connected by two oxygen bridges in the active center has stimulated the synthesis of solids and homogeneous compounds mimicking the enzyme structural situation, that is, containing iron dimers or small clusters. The synthesis and properties of heterogeneous catalysts like Fe–ZSM-5, containing dispersed iron centers, was described in Chapter 6. Enzymatic and ordinary homogeneous and heterogeneous catalysis are inextricably connected in modern chemistry. For the same reason the group of metalloenzymes including the cytochrome P450 family (with a heme iron center in the active site) (Rittle and Green, 2010), the nitrogenases family (enzymes used by some organisms to fix atmospheric nitrogen gas and constituted by proteins containing Fe and Mo and iron–sulfur clusters as

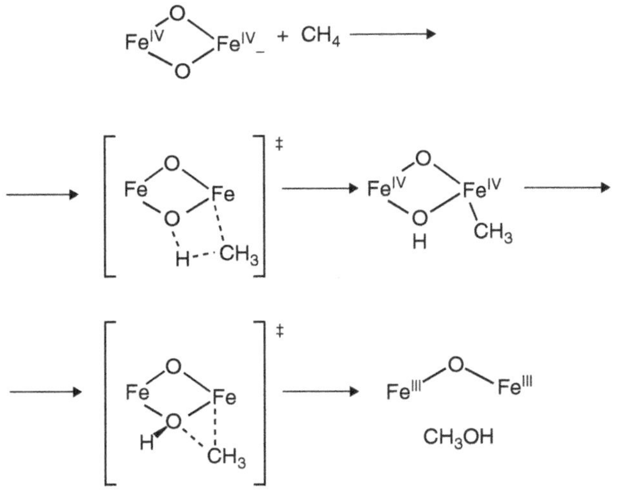

Figure 8.13 Catalytic center of methane monooxygenase (sMMO) that contains an Fe dimer. The mechanism of methane to methanol reaction is likely involving the coordination of methane to an Fe–O group, followed by C–H bond dissociation and methanol elimination. The missing oxygen is then replaced in a successive oxidation step. *Source:* Adapted from Huang, Shiota, and Yoshizawa, 2012.

cofactors) (Raymond *et al.*, 2003), and the hydrogenases family (enzymes that catalyze the reversible oxidation of molecular hydrogen and contain Ni–Fe, Fe–Fe, and Fe centers) (Fontecilla-Camps *et al.*, 2007) build a fundamental section of modern enzymatic chemistry. They have also attracted the attention and inspired the investigation of bioinspired partial oxidation, nitrogen and hydrogen activation, and homogeneous and heterogeneous catalysts.

Notice that all the previously mentioned enzymes have a common feature, namely that the metal ion or a cluster of metal ions are bound to the protein and have one labile coordination site. The most involved metals (under divalent or high valence form) are Fe, Cu, Ni, Zn, Mg, Co, Mn, Mo, and V. Although in many cases the metal ions are under monoatomic form, often associated with sulfur (Zn^{2+} in carbonic anhydrase, Fe–S in the iron–sulfur protein rubredoxin, Cu–S in the blue copper protein plastocyanin, etc.), dimers and clusters are present (Fe and Cu dimers in MMO, Ni–Fe–S clusters in hydrogenase, Fe_4S_4 and $MoFe_7S_3$ clusters in nitrogenase, etc.). As for all enzymes, the structure and shape of the active sites are playing a fundamental role for the catalytic properties. The metal ion or cluster of ions are usually located in a pocket whose shape fits the substrate following roughly the key and lock model. Notice that the same metals in form of oxides, mixed oxides, or sulfides form the basis of many heterogeneous catalysts. It is clear that metalloenzymes are a permanent

Figure 8.14 Shneior Lifson (1914–2001). Lifson used statistical mechanics to investigate the structural changes of macromolecules in solution. He collaborated to the formulation of the theory of the helix–coil transition in biological macromolecules and of protein folding processes. .*Source*: Reproduced with permission of Aoriliron.

source of inspiration for the scientists involved in homogenous and heterogeneous catalyst design.

Besides the rich activity surrounding the isolation and crystallization of enzymes and the metalloenzymes and resolution of their structure, fundamental research has continued on the dynamical properties of enzymatic systems. Major contributions are due to the work of the Austrian Martin Karplus who has shared the Nobel Prize for chemistry with Michael Levitt and Arieh Warshel. Before describing the contributions of these scientists, it is appropriate to describe Shneior Lifson (1914–2001) because both Levitt and Warshel were his students (Figure 8.14). Shneior Lifson, born in Israel in 1914, to Russian immigrants, was one of the most brilliant representatives of the highest level of European–Jewish culture. Lifson played a fundamental role in the development of protein dynamics and he has created an entire new field, Computational Structural Biology. In 1949, he pursued doctoral studies at the Hebrew University of Jerusalem on polyelectrolyte solutions. The research he conducted on this topic is representing a cornerstone for understanding the behavior of macromolecules in water solutions. To investigate the structural changes of these molecules in solution, Lifson used the methods of statistical mechanics. Lifson contributed to the formulation of the theory of the helix–coil transition in biological macromolecules (Lifson, 1958, 1959) and of protein folding (Lifson, 1964). Warshel developed a method, known as the consistent force field method, by which phase transitions and folding processes can be predicted and calculated (Lifson and Warshel, 1968).

Even today, numerous studies of protein folding originate from predictions obtained through such computations, which enable the study of chemical reactions, of the interactions between different proteins, and of the binding of ions to biomolecules. With this brief illustration of the Lifson contributions, we can return to the 2013 Nobel laureates Martin Karplus (1930–), Michael Levitt (1947–), and Arieh Warshel (Figure 8.15).

Karplus was a child when, in 1938, his family escaped from the Nazi occupation in Austria and then immigrated to the United States. After graduation

(a)

(b)

(c)

Figure 8.15 (a) Martin Karplus (1930–), (b) Michael Levitt (1947–), and (c) Arieh Warshel (1940–). They are known for having developed computer modeling methods of proteins and enzymes. They shared the 2013 Nobel Prize. Karplus developed the CHARMM program for molecular dynamics simulations. Levitt conducted molecular dynamics simulations of DNA and proteins. Warshel developed methods known today as computational enzymology. Images in the public domain.

at the California Institute of Technology, in 1953 he completed his PhD under the guidance of Linus Pauling (1954 Nobel Prize). In the 1953–1955 period, he was a postdoctoral fellow at Oxford University with Charles Coulson. In the 1960–1967 period, Karplus was at the University of Illinois, at Columbia University, and finally (1967) at Harvard. His group developed the famous CHARMM program for molecular dynamics simulations. During his career he studied several enzymatic systems, among these ribozyme and liver alcohol dehydrogenase. His attention was also focused on how the dynamic properties of proteins are linked to enzyme catalysis (Henzler-Wildman and Kern, 2007).

The second 1913 Nobel laureate in chemistry, Michael Levitt, was born in South Africa. In 1980 he became an Israeli citizen, and in the 1980–1987 period, he covered the position chemical physics professor at Weizmann Institute, and from 1987, he became professor of structural biology at Stanford University. Being interested in catalytic reactions of enzymes, he constructed a methodology based on the use of both classical and quantum mechanics on different parts of the enzyme. The interaction of both parts with the surrounding medium was also taken into consideration. In 1976, in collaboration with Gaurav Chopra and Nir Kalisman, he applied the methodology to construct the first computer model of an enzymatic reaction. Their scheme proved to be of general applicability and could be applied to model any enzyme (Chopra, Kalisman, and Levitt, 2010).

The third 2013 Nobel Laureate Arieh Warshel was born in 1940 in Palestine (now Israel). He was at Harvard University until 1972 with a postdoctoral position and then at the Weizmann Institute and at the Laboratory of Molecular Biology, Cambridge, England. In 1976, he joined the department of Chemistry of the University of Southern California.

Warshel is known for the studies of computational biochemistry and biophysics. He has greatly contributed to the development of what is today defined as computational enzymology (Warshel and Levitt, 1976; Warshel, 2002).

The concise picture of the enzyme area given previously is intended to offer a full idea of how vital, creative, interdisciplinary, and rapidly evolving is this area of science. It is a matter of fact that, in order to provide geometrical and electronic metal site structures and global protein architecture, enzyme science has stimulated the development and use of powerful experimental methodologies, like a variety of spectroscopies and associated magnetism investigations, of computational methods, and of new and expanding crystallographic methods based on synchrotron radiation.

In addition, contributions in the novel fields of metalloimaging, medicinal inorganic chemistry, metallosensors, and synthetic analogues must be mentioned. Especially in the synthetic analogues area (biomimetic inorganic chemistry), the gained knowledge at the atomic level of the catalytic sites structure in metalloenzymes, together with the more detailed information on the catalytic mechanisms, has been a powerful stimulus for the rational design of the ligand sphere and in the implementation of a synthetic analogues structure containing metal centers. The recent advances in this field are fully documented in the editorial of the Volume 114, Issue 8, 2014 of *Chemical Reviews* (Holm and Solomon, 2014) devoted to bioinorganic enzymology, where the structure and action mechanism of the nitrogen fixation enzyme (nitrogenase), of hydrogenase, of photosynthetic enzymes, and of cytochromes (to mention only a few) are reviewed and discussed.

8.5 Biocatalysis

Humans have empirically utilized enzymes for thousands of years in the form of fermentation as a means to produce and preserve foodstuffs such as cheese, beer, vinegar, and wine. This utilization can be considered an unconscious application of biocatalysis. Since biocatalysis deals with enzymes and microorganisms, it is historically considered a branch of catalysis separate from "homogeneous catalysis" and "heterogeneous catalysis." However, from a strict mechanistic point of view, biocatalysis is simply a special application of heterogeneous catalysis.

Biocatalytic conversion offers many potential advantages in the context of *green chemistry* because it is performed in water and under very mild conditions (ambient temperature and pressure). Furthermore it often involves fewer steps than conventional chemical procedures. Despite these outstanding potentialities, the use of enzymes as catalysts in synthetic organic chemistry and biotechnology suffers from some limitations. For instance, many useful enzymes cannot be mass produced. Other major limitations include the often-observed poor stereo- and/or regioselectivity, a narrow substrate scope, and sometimes insufficient stability under operating conditions. These challenges had limited the development of the entire field. However, this situation changed drastically in the late 1970s. In fact bioreactor system engineering, bioprocess design, enzyme immobilization, and the more ecologically and economically viable processes paved the way to many industrial applications of biocatalysis, a trend likely to continue in the future (DiCosimo *et al.*, 2013).

In particular, metabolic engineering has supported the production of complex natural products and even of simple organic molecules, such as biofuels (Collins *et al.*, 2014; Cameron, Bashor, and Collins, 2014; Way *et al.*, 2014). Furthermore, as the simultaneous development of recombinant DNA technology leads to the advent of so-called directed evolution in protein engineering, the problems associated with poor enzyme stability may be more effectively addressed and generally solved. This improvement has been indicated not only by numerous academic studies but also by the rapidly increasing number of industrial applications (Reetz, 2013). It is thus expected that special synthetic importance will be acquired in the near future by those chemical transformations that are difficult or impossible to implement using nonbiological catalysts.

In conclusion, biocatalysis is a rapidly and impressively evolving branch of catalysis and will likely dominate industrial enzymes applications in the near future.

References

Baldwin, R.L. (1996) How Hofmeister ion interactions affect protein stability. *Biophysical Journal*, **71**, 2056–2063.

Blake, C.C., Koenig, D.F., Mair, G.A. *et al.* (1965) Structure of hen egg-white lysozyme: a three-dimensional Fourier synthesis at 2 Angstrom resolution. *Nature*, **206**, 757–761.

Buchner, E. (1897) Alkoholische Gärung ohne Hefezellen (Vorläufige Mitteilung). *Berichte der Deutschen Chemischen Gesellschaft*, **30**, 668–678.

Buchner, E. and Curtius, T. (1885a) Synthese von Ketonsäureäthern aus Aldehyden und Diazoessigäther. *Berichte der Deutschen Chemischen Gesellschaft*, **18**, 2371–2377.

Buchner, E. and Curtius, T. (1885b) Über die Einwirkung von Diazoessigester auf aromatische Kohlenwasserstoffe. *Berichte der Deutschen Chemischen Gesellschaft*, **18**, 2377–2379.

Cagniard-Latour, C. (1837) Recherches Microscopiques sur les Substances qui produisent la Fermentation Spiritueuse. *Annales de Chimie et de Physique*, **68**, 206–210.

Cameron, D.E., Bashor, C.J., and Collins, J.J. (2014) A brief history of synthetic biology. *Nature Reviews Microbiology*, **12**, 381–390.

Canfield, R.E. (1963) The amino acid sequence of egg white lysozyme. *Journal of Biological Chemistry*, **238**, 2698–2707.

Chopra, G., Kalisman, N., and Levitt, M. (2010) Consistent refinement of submitted models at CASP using a knowledge-based potential. *Proteins: Structure, Function, and Bioinformatics*, **78**, 2668–2678.

Chou, K.C. (1989) Low-frequency resonance and cooperativity of haemoglobin. *Trends in Biochemical Sciences*, **14**, 212.

Collins, J.J., Maxon, M., Ellington, A. *et al.* (2014) Synthetic biology: how best to build a cell. *Nature*, **509**, 155–157.

DiCosimo, R., McAuliffe, J., Poulose, A.J., and Bohlmann, G. (2013) Industrial use of immobilized enzymes. *Chemical Society Reviews*, **42**, 6437–6474.

Dixon, G.H., Candido, E.P.M., and Honda, B.M. (1975) Louie A. J., The Structure and Function of Chromatin, **28**, 229–258.

Duclaux, E. (1899) *Traité de Microbiologie*, vol. **2**. Masson and Co., Paris.

Fabroni, A. (1873) in *Laurentii Medicis Magnifici vita, Auctore Angelo Fabronio Academiae Pisanae curatore*, vol. 1, Pisa, 1784, (eds. G. Ripley and C.A. Dana). American Cyclopaedia, D. Appleton and Co., New York.

Fischer, E. (1894) Einfluss der Configuration auf die Wirkung der Enzyme. *Berichte der Deutschen Chemischen Gesellschaft*, **27**, 2985–2993.

Fischer, E. (1905) Synthese von Polypeptiden, IX: Chloride der Aminosauren und ihrer Acyl derivate. *Berichte der Deutschen Chemischen Gesellschaft*, **38**, 605–619.

Fischer, E. and Fourneau, E. (1901) Über einige Derivate des Glykokolls. *Berichte der Deutschen Chemischen Gesellschaft*, **34**, 2868–2877.

Fischer, E., Schulze, E., and Winterstein, E. (1903) Für die quantitative Bestimmung der aus Eiweiss darstellbaren Aminosäuren. *Zeitschrift für Analytische Chemie*, **42**, 674–684.

Fontecilla-Camps, J.C., Volbeda, A., Cavazza, C., and Nicolet, Y. (2007) Structure/function relationships of [NiFe]- and [FeFe]-hydrogenases. *Chemical Reviews*, **107**, 4273–4303.

Funke, O. (1853) *Atlas der physiologischen Chemie*. Engelman, Leipzig.

Funke, O. and Latschenberger, J. (1877) Über die Ursachen der respiratorischen Blutdruckschwankungen im Aortensystem. *Pflüger, Archiv für die Gesammte Physiologie des Menschen und der Thiere*, **15**, 405–429.

Gobley, T.N. (1850) Recherches chimiques sur les oeufs de carpe. *Journal de Pharmacie et de Chimie*, (a) **18,** 401; (b) **18**, 107.

Gobley, T.N. (1874) Sur la lécithine et la cérébrine. *Journal de Pharmacie et de Chimie*, (a) **19**, 346–354; (b) **20**, 98–104.

Henzler-Wildman, K.A. and Kern, D. (2007) Dynamic personalities of proteins. *Nature*, **450**, 964–972.

Henzler-Wildman, K.A., Thai, V., and Lei, M. (2007) Intrinsic motions along an enzymatic reaction trajectory. *Nature*, **450**, 838–844.

Herriott, R.M. and Northrop, J.H. (1936) Isolation of crystalline pepsinogen from swine gastric mucosae and its autocatalytic conversion into pepsin. *Science*, **83**, 469–470.

Hofmeister, F. (1887) Über Resorption und Assimilation der Nährstoffe – Dritte Mittheilung. *Archiv für Experimentelle Pathologie und Pharmakologie*, **22**, 306–324.

Hofmeister, F. (1888) Zur Lehre von der Wirkung der Salze – Zweite Mittheilung. *Archiv für Experimentelle Pathologie und Pharmakologie*, **24**, 247.

Holm, R.H. and Solomon, E.I. (2014) Introduction: bioinorganic enzymology II. *Chemical Reviews*, **114**, 4039–4040.

Hoppe-Seyler, F. (1864) Uber die chemischen und optishen eigenshaften de blutforbstoff. *Archiv fur Patologische Anatomis und Physiologie*, **29**, 233–235.

Huang, S., Shiota, Y., and Yoshizawa, K. (2012) DFT study of the mechanism for methane hydroxylation by soluble methane monooxygenase (sMMO): effects of oxidation state, spin state, and coordination number. *Dalton Transactions*, **42**, 1011–1023.

Jabri, E., Carr, M.B., Hausinger, R.P., and Karplus, P.A. (1995) The crystal structure of urease from *Klebsiella aerogenes*. *Science*, **268**, 998.

Kendrew, J.C., Bodo, G., Dintzis, H.M. *et al.* (1958) A three-dimensional model of the myoglobin molecule obtained by X-ray analysis. *Nature*, **181**, 662–666.

Koshland, D.E. Jr., (1958) Application of a theory of enzyme specificity to protein synthesis. *Proceedings of the National Academy of Sciences of the United States of America*, **44**, 98–104.

Koshland, D.E. Jr., (1995) The key-lock theory and the induced fit theory. *Angewandte Chemie International Edition*, **33**, 2375–2378.

Koshland, D.E. Jr.,, Nemethy, G., and Filmer, P. (1966) Comparison of experimental binding data and theoretical models in protein containing subunits. *Biochemistry*, **5**, 365–385.

Kühne, W. (1877) *Zur Photochemie der Netzhaut*. Carl Winter's Universitätsbuchhandlung.

Kunitz, M. and Northrop, J.H. (1934) The isolation of crystalline trypsinogen and its conversion into crystalline trypsin. *Science*, **80**, 505–506.

Kutzning, F. (1837) Microscopische Untersuchungen über die Hefe und Essigmutter nebst mehreren andern dazu gehörigen *Vegetabilischen*. *Journal fur praktische Chemie*, **11**, 385–409.

Lieberman, R.L. and Rosenzweig, A.C. (2005) Crystal structure of a membrane-bound metalloenzyme that catalyses the biological oxidation of methane. *Nature*, **434**, 177–182.

Lifson, S. (1958) Neighbor interactions and internal rotations in polymer molecules. I. Stereospecific structure and average dimensions. *Journal of Chemical Physics*, **29**, 80–88.

Lifson, S. (1959) Neighbor interactions and internal rotations in polymer molecules III, statistics of interdependent rotations and their application to the polyethylene molecule. *Journal of Chemical Physics*, **30**, 964–967.

Lifson, S. (1964) Partition functions of linear chain molecules. *Journal of Chemical Physics*, **40**, 3705–3710.

Lifson, S. and Warshel, A. (1968) Consistent force field calculations, vibrational spectra, and enthalpies of cycloalkane and *n*-alkane molecules. *Journal of Chemical Physics*, **49**, 5116–5129.

McKenzie, H.A. and White, F.H. (1991) Lysozyme and alpha-lactalbumin: structure, function, and interrelationships. *Advances in Protein Chemistry*, **41**, 173–315.

Monod, J., Wyman, J., and Changeux, J.P. (1965) On the nature of allosteric transitions: a plausible model. *Journal of Molecular Biology*, **12**, 88–118.

Northrop, J.H. (1921) The mechanism of an enzyme reaction: as exemplified by pepsin digestion. *Science*, **53**, 391–393.

Northrop, J.H. (1929) Crystalline pepsin. *Science*, **69** (**1796**), 580.

Northrop, J.H. (1946) Preparing pure proteins. Nobel Lecture.

Northrop, J.H. and Kunitz, M. (1931) Isolation of protein crystals possessing tryptic activity. *Science*, **73**, 262–263.

Pasteur, L. (1866) *Études sur le vin, ses maladies, causes qui les provoquent, procédés nouveaux pour le conserver et pour le vieillir*. Imprimerie impériale, Paris.

Payen, A. and Persoz, J.-F. (1833) Mémoire sur la diastase, les principaux produits de ses réactions et leurs applications aux arts industriels. *Annales de Chimie et de Physique*, **53**, 73–92.

Perutz, M.F. (1960) Structure of haemoglobin. *Brookhaven Symposia in Biology*, **13**, 165.

Perutz, M.F., Rossmann, M.G., Cullis, A.F. *et al.* (1960) Structure of haemoglobin. *Nature*, **185**, 416–422.

Raymond, J., Siefert, J.L., Staples, C.R., and Blankenship, R.E. (2003) The natural history of nitrogen fixation. *Molecular Biology and Evolution*, **21**, 541–544.

Reetz, M.T. (2013) Biocatalysis in organic chemistry and biotechnology: past, present, and future. *Journal of the American Chemical Society*, **135**, 12480–12496.

Rittle, J. and Green, M.T. (2010) Cytochrome P450 compound I: capture, characterization, and C-H bond activation kinetics. *Science*, **330**, 933–937.

Schwann, T. (1837) Vorläufige Mittheilung betreffend Versuche über die Weingährung und Fäulniss. *Annalen der Physik Chemie*, **41**, 184–193.

Schwann, T. (1839) *Mikroskopische Untersuchungen über die Übereinstimmung in der Struktur und dem Wachstum der Thiere und Pflanzen*. Sander, Berlin.

Sumner J.B. (1946) The chemical nature of enzymes. Nobel Lecture.

Traube, M. (1858) *Theorie der Fermentwirkungen*. Verlag Ferd, Dümmler, Berlin.

Traube, M. (1874) Zur Theorie der Gährungs- und Verwesungs-Erscheinungen, wie der Fermentwirkungen überhaupt. *Zeitschrift für Analytische Chemie*, **13**, 349–356.

Warburg, O. (1952) Energetik der Photosynthese. *Die Naturwissenschaften*, **39**, 337–341.

Warburg, O. (1955) Photodissoziation und induzierte atmung, die fundamental reaktionen der photosynthese. *Naturwissenschaften*, **42**, 449–450.

Warburg, O. and Christian, W. (1938) Koferment der d-alanin-dehydrase. *Naturwisssenschaften*, **26**, 235.

Warburg, O. and Christian, W. (1941) Isolierung und Kristallisation des Gärungsferments Enolase. *Naturwissenschaften*, **29**, 589–590.

Warburg, O. and Christian, W. (1942) Isolierung und Kristallisation des Gärungsferments Zymohexase. *Naturwissenschaften*, **30**, 48.

Warshel, A. (2002) Molecular dynamics simulations of biological reactions. *Accounts of Chemical Research*, **35**, 385–395.

Warshel, A. and Levitt, M. (1976) Theoretical studies of enzymatic reactions: dielectric electrostatic and steric stabilization of the carbonium ion in the reaction of lysozyme. *Journal of Molecular Biology*, **103**, 227–249.

Warshel, A. and Lifson, S. (1970) Consistent force field calculations. II. Crystal structure, sublimation energies, molecular and lattice vibrations, molecular conformations and enthalpies of alkanes. *Journal of Chemical Physics*, **53**, 582–594.

Way, J.C., Collins, J.J., Keasling, J.D., and Silver, P.A. (2014) Integrating biological redesign: where synthetic biology came from and where it needs to go. *Cell*, **157**, 151–161.

9

Miscellanea

9.1 Heterogeneous and Homogeneous Catalysis in Prebiotic Chemistry

The origin of life is a fascinating problem. Over time scientists have proposed many different theories for the origin of life, but we are still far from a full understanding. Among the science branches involved in this interdisciplinary endeavor (biology, chemistry, physics, Earth and planetary sciences), it is clear that chemistry and more specifically catalysis, both the heterogeneous and homogeneous versions, had role in the generation of living organisms. In the second half of the twentieth century, increased funding for research enabled scientists to verify the contribution in synthetic organic chemistry of organometallic catalysis, which pushed the boundaries of materials to recreate a type of biogenesis resembling our biological evolution. The knowledge of the nature and evolution of these compounds also helped bridge the catalytic processes and the existing theories on the origin of life on our planet and even on the origin of the universe.

In the last 15 years, interest of the scientific community in the chemical problems associated with biogenesis has soared as a result, and about 2000 papers have been published in this short period of time (ISI Web of Science).

In physics and planetary science, spectroscopic evidence is being accumulated on the existence of huge clouds occurring in interstellar space, made of atoms, molecules, and even solid particles. The components of these clouds are mainly hydrogen (90%) and helium (10%) but there also are small amounts of oxygen, carbon, nitrogen, sulfur, calcium, aluminum, and iron. The interstellar clouds reach to very high temperature regions with extremely low densities up to few molecules/cm^3 and down to very cold regions with temperatures around 3K and densities on the order of 10^4–10^{10} molecules/cm^3. Interstellar clouds containing solid particles are normally called obscure because the solid particles of cosmic dust block the visible and ultraviolet (UV) part of the radiation emitted by the stars located behind them.

The Development of Catalysis: A History of Key Processes and Personas in Catalytic Science and Technology,
First Edition. Adriano Zecchina and Salvatore Califano.
© 2017 John Wiley & Sons, Inc. Published 2017 by John Wiley & Sons, Inc.

Thanks to the development of radio-frequency spectroscopy and to the infrared telescopes mounted on satellites rotating in outer space, some poly-atomic molecules, including ammonia NH_3, discovered in space as early as in 1968, ions, and even radicals have been detected (Solomon, 1973; Shu, 1982). Some others include formaldehyde (HCHO), ethyl alcohol (C_2H_5—OH), the radical (—$(C\equiv C)_3$–CN), cyanotriacetylene (H—$(C\equiv C)_3$—CN), acetone (($CH_3)_2CO$), methane (CH_4), acetic acid (CH_3—COOH), benzene (C_6H_6), and even molecules suggesting glycine (H_2N—H_2C—COOH), dihydroxypropanone ($CO(CH_2OH)_2$), methyl-ethyl-ether (CH_3—O—C_2H_5), acetamide (CH_3—CO—NH_2), propionaldehyde (CH_3—CH_2—CHO), the radical (H—$(C\equiv C)_4$–), and propylene (CH_3—$CH\equiv CH_2$).

In 1993 the cation $H_3{}^+$, the most abundant ion in the universe, was discovered in space. There is further some evidence of the existence in space of more complex molecules like aromatic hydrocarbons, amino acids, polyacetylenes, among several other complex structures. Recently, large carbonaceous species like the fullerenes C_{60} and C_{70} was unambiguously detected (Cami *et al.*, 2010).

In quiescent dark clouds, grains provide a surface on which species can be adsorbed, meet, and react and to which they can donate excess energy. Surface chemistry and heterogeneous catalysis on grains and particles can so be at the basis of many elementary processes whereby the reactants are initially adsorbed onto the surface and in thermal equilibrium with the solid. Three major mechanisms operate for surface reactions on a regular grain surface: Langmuir–Hinshelwood, Eley–Rideal, and hot atom mechanism. In the simplest type of reaction, a Langmuir–Hinshelwood mechanism (described in Chapter 4) operates, with the result that one or more reactants diffuse on the grain surface and then give rise to the formation of a new product reaction. In an Eley–Rideal reaction, a particle coming from the gas phase reacts more or less directly with a surface-adsorbed species. In the hot atom mechanism (which is a version of the Eley–Rideal mechanism), a gas-phase species lands on the surface and moves around before thermalization forces the species to collide and react with an adsorbate.

Understanding how the simple molecules present in the interstellar space could have reached the Earth's ground, giving rise to the complex systems and processes, is widely regarded as one of chemistry's greatest unsolved questions. The problem of biogenesis is extensively treated by Horst Rauchfuss of the University of Dortmund and now in Varberg, Sweden, in his book *Chemical Evolution and the Origin of Life* (Rauchfuss, 2008), first published in German as *Chemische Evolution und der Ursprung des Lebens* in 2012. This problem is challenging because there is still not one fully accepted theory of the emergence of life. The existing hypotheses and theories all try to explain how small "building block" molecules such as amino acids and sugar molecules were formed and how these molecules underwent polycondensation to give origin to macromolecules (for instance, peptides) in later stages of development. Rauchfuss (2008) also addresses the basic theoretical questions, including the so-called chirality feature in some molecules. Among the scientific fields covered in the book, chemistry and catalysis are the most prominent. In particular, as will be described below, heterogeneous and organic catalysis are active in the efficient polycondensation of amino acids and nucleotides in heterogeneous aqueous solutions. Consequently Rauchfuss suggests that the first functional biopolymers (e.g., polypeptides and polynucleotides) emerged without the aid of biological catalysts as likely did the appearance of homochirality.

Much earlier at the beginning of the twentieth century, the German chemist Walther Löb in 1906 had achieved the chemical syntheses of some aldehydes and simple amino acids such as glycine by exposing wet formamide to a silent electrical discharge and to UV light (Löb, 1906). He found that an electric discharge acting on HCHO in the presence of water vapor gives rise to a mixture of CO, CO_2, H_2, and CH_4, which is essentially the reverse of the well-known reaction demonstrated later by Miller in 1953. Then, in 1924, the Soviet biochemist Aleksandr Oparin suggested that life on Earth developed through a gradual chemical evolution of carbon-based molecules (Oparin, 1924). At almost the same time, the British-born, later Indian naturalized, John Burdon Sanderson Haldane (1892–1964), recipient of the Darwin medal and of the Feltrinelli Prize in 1952 and 1961, hypothesized that in presence of a reducing atmosphere and under the action of lightening or UV light, a wide variety of organic compounds might have been synthesized.

Oparin had suggested that in aqueous solution, the initially formed organic molecules could have formed colloid aggregates (named "coacervates"), able to absorb and exchange organic compounds from the medium in a way reminiscent of metabolism and eventually triggering the first life forms.

The same as to Oparin, Haldane suggested that under the action of light, the primordial sea behaved as a chemical laboratory (named "prebiotic soup") in which organic monomeric and polymeric species were formed and acquired lipid membranes, eventually behaving as precursors of the first living cells. Today the term "prebiotic soup" is representative of the Oparin–Haldane theories on the origin of life.

The classic experiment describing the formation of amino acids when an electric discharge is passed through mixtures of methane, ammonia, water, and hydrogen, now widely known as the Miller–Urey experiment, was reported by Stanley Miller in *Science* in 1953 (Miller, 1953). Two scientists conducted this experiment: Harold C. Urey (1893–1981) and Stanley Miller (1930–2007) (Figure 9.1).

(a)   (b)

Figure 9.1 (a) Harold Clayton Urey (1893–1981) and (b) Stanley Lloyd Miller (1930–2007) with his apparatus (images of public domain). Urey (1934 Nobel Prize for the discovery of deuterium) was also interested in the chemistry at the birth of planets and Earth and in the chemical aspects of the origin of life. Miller, in collaboration with Urey, designed and realized the experiment that propelled him to instant fame.

Harold C. Urey was a descendant of pioneers who had settled in Indiana. In 1914, he entered the University of Montana where he received his Bachelor of Science in 1917. After a short period in industry, he entered the University of California, Berkeley, where he received the PhD degree in 1923. From 1929, he was a professor of chemistry at Columbia University, where during the 1940–1945 period he was also director of the Atomic Bomb Project. In 1958, he moved to the University of California, San Diego. Among the numerous Urey researches, the discovery of a method for the concentration of heavy hydrogen isotopes by fractional distillation of liquid hydrogen led to the discovery of deuterium. For this discovery, he was awarded the 1934 Nobel Prize in chemistry.

In the 1950s, he developed an interest in the measurement of paleo-temperatures, in the chemistry at the basis of the birth of the planets and Earth, and in the chemical aspects of the origin of life.

This late scientific interest might have been due to his collaboration with Stanley Miller, a descendant of Jewish immigrants from Belarus and Latvia. Miller studied chemistry at Berkeley and in 1951 became a teaching assistant at the University of Chicago on the synthesis of elements in the stars. However, after attending the chemistry seminar of the Nobel laureate Harold Urey in 1952 on the origin of the solar system and how a synthesis of organic molecules could have caused a reduction in the atmosphere of the primitive Earth, Miller approached Urey and persuaded him to accept his collaboration on this topic.

This collaboration is at the basis of the famous Urey–Miller experiment. After the Chicago period, Miller moved to the California Institute of Technology and then to the Department of Biochemistry at Columbia University, New York. In 1962, he finally moved to the newly established University of California at San Diego where he became associate professor and then full professor in 1968.

Urey and Miller decided to test their prebiotic soup theory by exposing to electrical discharge a mixture of steam, methane (CH_4), ammonia (NH_3), and hydrogen (H_2) in order to simulate the oceanic-atmospheric condition of the primitive Earth. After a week of continuous discharge, Miller was able to detect, by chromatography, the formation of amino acids such as glycine and alanine. Informed by Miller, Urey suggested immediate publication of these results. The manuscript with Miller as sole author was submitted to *Science*. The 1953 publication on *Science* propelled Miller to instant fame as the "father of prebiotic chemistry." The results, however, are usually ascribed to both men as the Urey–Miller experiment.

Miller continued and improved his research until his death in 2007. In this period, he refined the details and methods, succeeded in the synthesis of more varieties of amino acids, and widened the variety of organic molecules essential for life metabolism. With all the continual revelations that under strongly reducing conditions the primitive atmosphere could contain other gases in different proportions, new experiments became necessary. Miller's last work, posthumously published, still succeeded in obtaining arrays of all such organic molecules arrays of organic molecules (Cleaves *et al.*, 2008).

These experiments showed that the question of the origin of life is a scientific problem could be addressed and likely solved by using a scientific methodology. But the Miller–Urey experiment is particularly remarkable for the simplicity of the experimental apparatus as shown in Figure 9.1.

Subsequent experiments building on the Miller discovery have suggested many more prebiotic organic reactions. Reactions leading to the formation of amino acids seem to be limitless. For example, it is now understood that the amino acids were not necessarily formed only in the gas phase; they can be synthesized also in the liquid phase by reacting HCN, NH_3, and H_2O.

Today, we also know that by using spark discharge, UV irradiation, or shock waves in various gaseous mixtures of methane, carbon oxides (CO and CO_2), ammonia, nitrogen, and water, a large variety of molecules can be obtained. Many contain two or three carbon atoms, such as hydrogen cyanide, cyanate, cyanogen, formaldehyde, formamide, formic acid, ammonium formate, ammonium cyanide, urea, acetaldehyde, cyanoacetylene, and cyanoacetaldehyde. From this pool of simple molecules, myriad more complex compounds could have formed, via heterogeneous catalysis and organocatalysis. It is a

(a)   (b)

Figure 9.2 (a) Joan Orò (1923–2004) and (b) Fred Hoyle (1915–2001). The synthesis of adenine by Orò is, together with the Miller–Urey experiment, one of the fundamental pillars of prebiotic chemistry. Hoyle was an expert in nucleosynthesis and promoted the idea that the first life on Earth came from outer space. *Source:* (a) Courtesy of Astrofels. (b) The Hoyle image is in the public domain.

fact that new catalytic effects induced by small organic molecules are being discovered every year in the field of organocatalysis that could explain how the condensation of initially formed simple molecules into more complex systems could have happened.

Only a few years after the publication of the Miller–Urey experiment, Joan Orò was able to synthesize one of the most important biomolecules, adenine (Figure 9.2). This purine derivative is not only a component of the nucleic acids but, like adenosine triphosphate (ATP) (in combination with ribose and three phosphate residues), plays a key role in the metabolism of all living creatures (Orò and Kamat, 1961).

This result, which was obtained in the period 1959–1962, is, together with the Miller–Urey experiment, one of the fundamental pillars of prebiotic chemistry. In fact it opened up a research direction potentially leading to the complete synthesis of other components of nucleic acids. Joan Orò (1923–2004), a Catalan biochemist graduated at the University of Barcelona, moved to the United States in 1952, where he obtained at the University of Houston his PhD in biochemistry. He became full professor in 1955 and founded the Department of Biochemistry and Biophysics at the University of Houston. From the 1960s, he worked with NASA on the Viking mission, which explored the planet Mars. In 1961, he advanced the idea that comets were the key carriers of organic molecules to the Earth's primitive biosphere, a hypothesis widely accepted today. Although the idea was not entirely new, it was verified by the results of space exploration and prebiotic chemistry experiments. Comets proved to be rich in carbon and water, and capable of bearing precursor molecules based on the carbon chemistry of amino acids, for example. Orò's idea that life on the Earth started from primitive molecules falling on its soil from the outer space followed, however, a previous theory developed by Sir Fred Hoyle (1915–2001). Hoyle was an English astronomer mainly known for his contribution to the theory of stellar nucleosynthesis and for his opposition to the theory of "Big Bang," which was an expression originally coined by him in the *Third Programme* transmission, broadcasted by the BBC radio on 28 March 1949 (Figure 9.2).

In the 1950s, Hoyle coordinated a group of very talented physicists, including William Alfred Fowler (awarded the 1983 Nobel Prize). This group realized the basic theories of the genesis of all the chemical elements in the universe, a field now called nucleosynthesis. Famously, in 1957, this group produced the cornerstone B^{4}FH paper (from the initials of the four authors) in which the field of nucleosynthesis is defined. Hoyle and Fowler proposed that all the elements from which our world is made, from carbon atoms to uranium atoms, were cooked during an immeasurably long period of time, from a basic fuel made of hydrogen.

Hoyle resigned his institute directorship in 1973 and started writing novels often with his son Geoffrey Hoyle, the most popular being the "black cloud" as well as books of science fiction, for instance, the well-known A for Andromeda. In his later years, Hoyle in collaboration with Nalin Chandra Wickramasinghe advanced the hypothesis that the first life began in space via the contribution of comets (panspermia). Notice that Hoyle's view that comets were containing organic molecules was well ahead of his time, as in 1970s the common opinion was that comets were essentially constituted of ice and that the presence of organic matter was negligible.

The panspermia theory suggesting that life seeds landed on Earth from outer space has had in time a large number of supporters among the astronomers influenced by Hoyle's reputation. Yet the idea that life came from outer space was a common belief of Berzelius (1834), Lord Kelvin (1871), and Arrhenius (1903, 2009).

In this chapter we will describe first the processes where the intervention of solid surfaces has likely played a critical role. In particular, many researchers have discussed the potential role of the surfaces of oxidic systems as catalysts for the condensation reaction, inducing the formation of peptide bonds between amino acids, which is the starting point of the protein formation and is one of the most fundamental reactions in biology. The discussion of this possible synthetic application has implications for the origin of life and also for the role of the mineral world in the emergence of biochemistry. It can further be hypothesized that biochemical complexity is due to a *transduction* of inorganic chemical information driving from the surface structure of solids (Cairns-Smith, 2008). Following this hypothesis, the first step of the biogenesis process would be the adsorption of certain molecules on the surface of common inorganic crystals (silicates) that catalyze their condensation. The last step would occur when these complex molecules leave the surface and become independent loci of replication via, for instance, homogeneous catalysis. A typical reaction illustrating a condensation reaction via peptide bond formation is the reaction scheme

Peptide bond

where the reaction is associated with a dehydration process. The formation of this bond is thus expected to not be thermodynamically favored in diluted aqueous solutions. However, the situation would be reversed as the amino acids are adsorbed on the surface of oxides or oxidic materials where the concentration of amino acids is greatly increased by the adsorption process (Martra *et al.*, 2014). Then the presence of surface species like hydroxyl groups (Brønsted sites) and coordinatively unsaturated cations (Lewis sites) would have had an extra catalytic function in water elimination (Rimola, Sodupe, and Ugliengo, 2007). In the case of a realistic feldspar surface containing neighboring Brønsted and Lewis sites, the proper pre-reactant complex of the amino acid with the surface would be highly stabilized by a simultaneous interaction with the Lewis and Brønsted sites. This reaction would occur in such a way that the Lewis site strongly attaches the amino acid (glycine) to the surface, whereas the Brønsted site efficiently catalyzes the condensation reaction.

It has been proved that a condensation reaction can occur on silica surface (Lambert *et al.*, 2013). It is likely that the surface condensation reaction would also be favored by high temperature conditions and high pressures present in hydrothermal sources in deep sea.

Other oxides like iron oxide and iron oxo-hydroxides, quite abundant in nature, have shown some activity in amino acid dimerization and oligomerization. It has been also hypothesized that the chemistry of amino acids in the confined spaces of small pores and open cavities, where the concentration of amino acids could be greatly enriched with respect to the solution, was favorable for amino acid meeting and reaction. Such porousness can be seen today in silicate materials like clays, mesoporous silica, and pumice – all solids whose presence is well documented in the prebiotic conditions. This brings in porous materials as also having relevance in the origins of life. Additionally, it has been suggested that pyrite, which is an abundant sulfide mineral in the Earth's crust, could have played a role during the prebiotic surface synthesis of peptides, affirming what is called the *iron–sulfur world* hypothesis. Interestingly pyrite displays high reactivity because of the presence of sulfur vacancies. Schreiner *et al.* (2011) present a comprehensive study of the mechanisms of polymerization of glycine and peptide hydrolysis at the water–pyrite interface under extreme thermodynamic conditions.

Yet, besides amino acids and their condensation products, prebiotic chemistry includes the formation of sugars from simple building blocks. For many decades, the formaldehyde oligomerization in the presence of alkaline earth hydroxides (formose reaction) has been cited as the prebiotic source of sugars. Conceptually, the formose reaction is the simplest synthesis of sugars like pentoses via iterative homologation, starting from formaldehyde and followed by the glycolaldehyde dimer and then the glyceraldehyde trimer, as illustrated in the scheme below. Indeed, when formaldehyde is treated with calcium hydroxide, conversion of the formaldehyde into glycolaldehyde and glyceraldehyde and even larger sugars is observed.

To explain this behavior, it was proposed by Breslow many years ago that an autocatalytic cycle was operating (Breslow, 1959) (Figure 9.3).

Formose sequence

Figure 9.3 Ronald Breslow (1931–). Breslow is known for his contribution on the autocatalytic mechanism of the formose reaction and on the origin of homochirality. *Source*: Reproduced with permission of Lockard.

The basic idea is that the first molecule of glycolaldehyde, formed by a slow process catalyzed by the mild solid base, reacts with formaldehyde to form the dimer (glyceraldehyde in equilibrium with dihydroxyacetone), and then it reacts with another formaldehyde molecule, and so on, via an autocatalytic process.

This early contribution of Ronald Breslow (1931–) was initiated at Harvard University with Professor R. B. Woodward and was followed by relevant contributions in the field of organic chemistry. He is now professor of chemistry at Columbia University. His recent scientific interest is centered on the study of molecules whose transformation imitates enzymatic reactions and artificial enzymes. During his long career, he was awarded many prizes. For the prebiotic chemistry history, another relevant contribution of Breslow (2011) concerns the origin of homochirality, which will be discussed next.

The formose scheme has been criticized by Ritson and Sutherland (2012). They argue that the sugars can be formed from hydrogen cyanide and formaldehyde by UV irradiation via a homogeneous photocatalytic process. The formation of the first HCN-promoted homologation reaction of formaldehyde is supposed to occur following a Kiliani–Fischer synthesis approach (Fischer and Hirschberger, 1889).

The formation of HCN by participation of the copper compound as the catalyst and UV light may be one of the first examples of photochemistry involving transition metal complexes. Ritson and Sutherland propose instead that a pentose fragment reacting with a pyrimidine fragment produces an intermediate form already containing the C−N that acts as a catalyst for successive reactions. Notice that in all these cases, heterogeneous and homogeneous catalysis (including photocatalysis) are simultaneously invoked.

Currently the discussion is still open concerning the successive polycondensation of peptides and sugars to give macromolecules in later stages of development. It has been recently reported that the presence of Zn^{2+} ions can have a beneficial effect (Rufo *et al.*, 2014).

An unanswered question in all the studies of amino acids and small sugar molecules relates to chirality. In the absence of some form of a homochiral template, these reactions necessarily produce a racemic product. The homochirality occurring in such reactions is a key point. Consequently the origin of the homochirality in biological compounds, including *l*-amino acids and *d*-sugars, is a long-standing puzzle and of broad interest in many scientific fields.

Homochirality is indeed one of the essential features of life, and clarification of the origin of chirality in key biological compounds will contribute to our understanding the chemical origin of life. Several chiral factors have been suggested as the origin of homochirality, including circularly polarized light, spontaneous chiral crystallization of achiral compounds, asymmetric catalysis by chiral inorganic crystals such as quartz, calcite, and cinnabar, and statistical fluctuations in enantiomeric excess. However, as the enantiomeric excess caused by all these factors is usually low, catalytic amplification processes that enhance the small enantio-imbalances induced by each chiral factor mentioned earlier have been invoked (Morowitz, 1969).

The emerging consensus is that chirality is a natural auto-amplification process associated with the second law of thermodynamics (Jaakkola, Sharma, and Annila, 2008). As far as auto-amplification is concerned, it can be useful at this point to recall the definition of autocatalysis, the most probably involved factor. Indeed, since the last decade of the twentieth century, it has been hypothesized that autocatalysis has played a prominent role in the life processes (Kauffman, 1993). A single chemical reaction is said to undergo autocatalysis, or to be autocatalytic, if the reaction product is itself the catalyst for that reaction. The aforementioned hypothesized catalytic role of glycolaldehyde and of pyrimidine-containing molecules to give further reaction products is an example of autocatalytic processes. A branch of autocatalysis is asymmetric autocatalysis occurring when the reaction product is chiral and thus acts as a chiral catalyst for its own production. Reactions of this type, such as the Soai reaction described in detail below), can amplify a very small chiral excess into a large one. This autocatalytic process has been considered in the origin of biological homochirality (*vide infra*).

Many researchers have suggested that the surfaces of solid catalyst have a role in chiral synthesis. An example of such a catalyst is quartz, a widespread mineral. This is because mineral crystals like quartz that do not have a center of symmetry in their bulk structures can expose chiral surfaces. Quartz is the most abundant chiral material in nature, and its chirality arises from the helical arrangement of SiO_4 tetrahedra in the bulk structure. Its surfaces are also chiral and can potentially exhibit enantiospecific interaction with the two enantiomers of chiral molecules (Horvath and Gellman, 2003) (Figure 9.4). In 1986 a hypothesis was advanced that quartz could have been the source of chirality, at least for the simplest organic molecules (Cairns-Smith and Hartman, 1986). A very small imbalance of chirality had earlier been observed by Bonner *et al.* in an asymmetric adsorption of chiral molecules by quartz (Bonner *et al.*, 1974). The Bonner contribution is in fact the first one reported in the literature. In 2009 Kahr *et al.* (2009) reported on an overlooked observation made in 1919 at the Institute of Physics of the University of Turin by Eligio Perucca (1890–1965) (Perucca, 1919). Perucca's observation concerned the anomalous optical rotatory dispersion of chiral $NaClO_3$ crystals, colored from having been grown from a solution containing an equilibrium racemic mixture of a triarylmethane dye, extra China blue. The ability of sodium chlorate to selectively adsorb enantiomers was confirmed in 2000 (Sato, Kadowaki, and Soai, 2000).

In 2001 Hazen, Filley, and Goodfriend (2001) advanced another plausible scenario for chiral separation by mineral surfaces. Robert Hazen is a professor at George Mason University and he had been awarded the Ipatieff Prize (1986) and the Educational Award (1992). Hazen and coworkers found that crystals of the common mineral calcite ($CaCO_3$), when immersed in a racemic aspartic acid solution, show significant adsorption and chiral selectivity of D- and L-enantiomers (Figure 9.4). It is evident that the displayed selectivity on this widespread solid, although still low, may be affirming the idea that solid surfaces did play a role in homochirality appearance. Since the enantiomers form bidimensional arrays on the surfaces, a sterically oriented polymerization path has been suggested, which preserves homochiral features.

Quartz *d*- and *l*-crystals

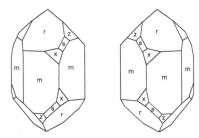

d- and *l*-quartz promote
asymmetric addition reactions

Calcite

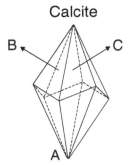

B and C faces adsorb
preferentially *l*- and *d*-aspartic
acid

Figure 9.4 Quarz and calcite crystals. *Source*: Adapted from Soai, Sato, and Shibata, 2001, and Hazen, Filley, and Goodfriend, 2001.

An important breakthrough came in 1999 when Soai *et al.* (1999) reported that *d*- and *l*-quartz promote the asymmetric addition of diisopropylzinc (i-Pr$_2$Zn) to 2-alkynylpyrimidine-5-carbaldehyde 1, giving rise, in combination with asymmetric autocatalysis, to the formation of chiral pyrimidyl alkanol possessing S (from *d*-quartz) and R (from *l*-quartz) configurations, respectively, with a significantly high (93–97%) enantiomeric excess.

For this discovery, Soai has received several awards, including the Medal with Purple Ribbon from the Japanese government. He graduated from the University of Tokyo where he got the PhD degree in 1979. He was then appointed associate and full professor at the Tokyo University of Science.

The large enantiomeric excess obtained with this process, although initially promoted by the chiral properties of the surface, can be justified only by the presence of the asymmetric autocatalysis that amplifies the small initial enantiomeric excess induced by the surface chirality (Sato *et al.*, 2004). The Soai opinion (Soai, Sato, and Shibata, 2001) is that the initially formed chiral alkanols (with only less than 1% initial enantiomeric excess) act as highly enantioselective asymmetric autocatalysts in the enantioselective addition of diisopropylzinc to the corresponding aldehydes. This explains the 99% final enantiomeric excess, since clearly the enantiomeric excess of the resulting alkanol is much higher than that of the chiral initiator. Interestingly, leucine with very low enantiomeric excess induced by circularly polarized light has also been observed to serve as a chiral initiator and produce alkanol with high enantiomeric excess. We can gather from these facts that enantiomeric autocatalytic processes have some important and very general role.

The Soai reaction is certainly not typical of what can be imaged about prebiotic synthesis reactions. However, it is the first example of asymmetric synthesis of an organic compound with a high enantiomeric excess where a chiral inorganic crystal was used as the initial catalyst.

Cinnabar is another known chiral mineral naturally occurring in nature. Due to the spiral arrangement of atoms in a trigonal crystal structure, the natural HgS crystal, cinnabar, is chiral and similar to quartz. The crystal is comprised of long helices of alternating mercury and sulfur atoms that spiral around the *c*-axis of the hexagonal unit cell.

A recent synthesis by Ben-Moshe, Govorov, and Markovich (2013) shows HgS nanocrystals to have a large enantiomeric excess. The synthesis was made in the presence of chiral surfactant molecules (penicillamine). The generation of truly chiral nanoparticles of tailored size may open the way toward enantioselective catalytic synthesis of nanostructures of other chiral inorganic crystals.

One such experimental demonstration is from around the same time by Shindo *et al.* (2013). They show that when pyrimidine-5-carbaldehyde is treated with iPr$_2$Zn in the presence of cinnabar with right-handed helicity,

(R)-pyrimidyl alkanol is obtained in 88% enantiomeric excess and 87% yield. In contrast, in the presence of the (M)-HgS, the opposite enantiomer of (S) remains isolated in 91% yield and 92% enantiomeric excess. Hitosugi *et al.* (2014) have more recently found that single-wall carbon nanotubes with helical chirality efficiently catalyze the same reaction.

So far it is evident that enantiomeric surfaces in combination with autocatalysis can be one of the key points for enantiomeric amplification, which may explain why a chiral compound with an extremely low initial enantiomeric excess can act as an asymmetric autocatalyst and amplify its own enantiomeric excess significantly during a process of auto-multiplication. As this process is operative on several chiral inorganic crystals like quartz (Soai, Sato, and Shibata, 2001), HgS (Shindo *et al.*, , 2013), carbon nanotubes (Hitosugi *et al.*, 2014), and sodium chlorate (Sato, Kadowaki, and Soai, 2000), it appears to suggest a general characteristic of chiral compounds.

Other examples that illustrate how prebiotic chemistry results from catalytic amplification of enantiomeric excesses of amino acids and sugars were initially also aided by physical processes, including crystallization and circularly polarized light (Hein and Blackmond, 2012; Weissbuch *et al.*, 2009; Weber and Pizzarello, 2006).

As for enantiomeric sugar synthesis, the aforementioned Breslow contribution (Breslow, 2011) shows the effect of L-amino acids on the formose reaction. Breslow found that L-amino acids catalyze the reaction of glycolaldehyde with formaldehyde with the formation of a small excess of the D-enantiomer. Although the excess is not high, this result is more evidence that organocatalysis is involved in prebiotic asymmetric chemistry.

In summary, although the prebiotic branch of science is relatively young and rapidly developing, it can be concluded that most scientists agree that adsorption on surfaces, heterogeneous catalysis, and the mineral world in conjunction with autocatalytic amplification of the organocatalytic type are essential to understanding prebiotic processes. For a comprehensive view of the contributions of heterogeneous and organocatalysis to theories on the origin of life, the interested reader can consult the review by Ruiz-Mirazo, Briones, and De La Escosura (2014).

9.2 Opportunities for Catalysis in the Twenty-First Century and the Green Chemistry

In previous chapters we documented the role of catalysis in the production of fertilizers, fuels, polymers, chemicals, drugs, paints, and cosmetics. Today over 90% of all industrial chemicals are produced with the aid of catalysts (Armor, 2011) with a ratio of applications of heterogeneous to homogeneous catalysis of approximately 70–60/30–40 (Cornils and Herrmann, 2003). In the previous chapters we also mentioned briefly the many processes waiting for new catalysts. Among these challenging processes, the most relevant today is the partial oxidation of methane to methanol, which is only one of the "dream" reactions involving the transformation of natural gas into useful basic materials like methanol, ethane, and ethylene. These reactions in the past led to the production of great variety of useful chemicals (Fechete, Wang, and Vedrine, 2012; Schlogl, 2015). Other challenging processes are the synthesis of H_2O_2 from hydrogen and oxygen, the partial oxidation of benzene to phenol, and the epoxidation of ethylene (to mention only a few).

However, "dream" processes represent only a minor part of the grand challenge of catalysis. Currently the most urgent challenge is the threat to humanity primarily associated with global warming and excessive release of toxic substances. This calls for the creation of alternative fuels and the reduction of by-products in industrial synthesis by increasing selectivity, preventing pollution, and cleaning up the environment. Catalysts of unprecedented selectivity are thus needed to meet these challenges, but their complexity demands a revolution in catalyst design. This revolution can be realized only through the application of entirely new synthesis and characterization methods. Simply stated, the grand challenge for catalysis science in the twenty-first century is to learn how to design catalyst structures that can control catalytic activity and selectivity.

In addition, a common language is needed between the branches of catalysis science. At the level of the atom and of the chemical bond there is no distinction between heterogeneous, homogeneous, and enzymatic catalysts. We have attempted in this book to meet this objective by sorting out the fundamental properties of each catalyst based on its ability to activate chemical bonds.

This awareness motivated also the structure of the book in which we have treated the evolution and advances of heterogeneous, homogeneous, and enzymatic catalysis as the essential constituent parts of catalysis science.

With regard to the alternative fuel production problem, it must be underlined that this cannot be procrastinated any longer. As is becoming increasingly evident, the usage of fossil fuels is contributing to huge increments of CO_2 emissions and dangerous environmental effects.

It is a general opinion that as far as the future energy production is concerned, the renewable energies, in particular wind and solar, can compete favorably with fossil resources, especially if environmental benefits are also taken into account.

Hydrogen gas has been proposed as an excellent and clean fuel that produces, besides energy, only water when combusted. The "hydrogen economy" based on electrochemical and photocatalytic water splitting into hydrogen and oxygen has been greatly popularized by Jeremy Rifkin (1945–), an American economic and social theorist and writer.

Rifkin is the author of 20 books on the impact of scientific and technological changes on the economy, society, and the environment. Among them *The Biotech Century* (1998), *The Hydrogen Economy* (2002), and *The Third Industrial Revolution* (2011) can be cited. Already in 1989, Rifkin organized the first meeting of the Global Greenhouse Network in Washington, DC.

In his book devoted to the hydrogen economy (Rifkin, 2002), he defended the political cause of the use of renewably generated hydrogen as the global energy fuel able to solve the environmental problem associated with the production of energy from fossil fuels.

Currently, however, due to its physical and chemical properties, hydrogen gas has a number of serious drawbacks. This is mainly because the infrastructure requires secure transport, and to store, and dispense hydrogen gas safely would be very expensive. Efficient hydrogen storage materials are still waiting for substantial innovations.

A liquid is preferable to a gaseous fuel in most applications. In the transportation sector in particular, a transition from traditional liquid fossil fuels (gasoline, diesel fuel, kerosene, etc.) to a renewable and sustainable liquid fuel would be highly desirable. This would enable the existing infrastructure to be used with only minor modifications. Among the possible candidates fulfilling these requirements, methanol, the simplest liquid compound containing only one carbon, has been proposed. This proposal form the basis of the famous "methanol economy," proposed and popularized by the Nobel laureate Olah (2013a; Olah, Goeppert, and Prakash, 2009). About the catalytic methods adopted so far for the synthesis of methanol from CO_2, we have already concisely reported them in Chapter 5. This issue is extensively explored by Olah in a paper entitled "The Role of Catalysis in Replacing Oil by Renewable Methanol Using Carbon Dioxide Capture and Recycling (CCR)" (Olah, 2013b). We only add that both hydrogen and methanol economies require highly innovative homogeneous and heterogeneous catalysts and photocatalysts and that a unified approach to research that combines advances in bioscience, molecular science, surface science, computational science, and nanoscience is required to address all the problems associated with the transition from fossil fuels to hydrogen and methanol (Goeppert *et al.*, 2014). Coming to the reduction of harmful by-products in manufacturing by cleaning up the environment and preventing future pollution, it can be stated that due to unprecedented expansion of the transportation industry and chemical processes, characteristic of the modern twentieth century, we are obligated to seek technological solutions that can provide society with energy and materials in an environmentally responsible way.

In 1987 the Brundtland Commission (Brundtland, 1987) defined sustainable development as

> Meeting the needs of the present generation without compromising the ability of future generations to meet their own needs.

Few years later, Anastas and Warner (1998) coined the concept of green chemistry in terms of

> the utilisation of a set of principles that reduces or eliminates the use or generation of hazardous substances in the design, manufacture and application of chemical products.

Paul Anastas (1962–) is the director of the Center for Green Chemistry and Green Engineering and professor at the School of Forestry & Environmental Studies of Yale University. John Warner (1964–), the second founder of field of green chemistry and founder in 2007 of the Warner Babcock Institute for Green Chemistry (a research organization developing green chemistry technologies), is recipient of the 2014 Perkin Medal.

The titles of the famous Anastas–Warner 12 principles of green chemistry are as follows:

1. Prevention
2. Atom economy
3. Less hazardous chemical syntheses
4. Designing safer chemicals
5. Safer solvents and auxiliaries
6. Design for energy efficiency
7. Use of renewable feedstocks
8. Reduce derivatives
9. Catalysis
10. Design for degradation
11. Real-time analysis for pollution prevention
12. Inherently safer chemistry for accident prevention

These principles, of course, derive from the twentieth-century history of industrial chemistry and catalysis, but they are a compelling exhortation for the future, and involve selective catalysis as the central pillar, because the rational utilization of renewable feedstocks, derivative reduction, atom economy, and pollution prevention cannot be performed without highly selective heterogeneous and homogeneous catalysts and photocatalysts.

Some principles of green chemistry have already been applied in the past. We have shown in the previous chapters that chemical and petrochemical industries have always attempted to improve process efficiency and reduce waste by applying novel heterogeneous and homogeneous catalysts. A good example is the replacement of hazardous catalysts like H_2SO_4, HF, and $AlCl_3$ with acidic zeolites that revolutionized the petrochemical industry in the second half of the twentieth century.

Another example has been the discovery and use of enantioselective catalysts that revolutionized the drug production and eliminated the need of costly enantiomer separation.

Yet another example has been the tendency to replace the costly noble metal catalysts with more abundant and less costly materials.

Although these examples represent only a fraction of the improvements of catalysis science contributing to safer production systems made during the twentieth century, the writings of Paul Anastas and John Warner (1998) provide the stimulus toward adopting a compelling ethic of goods production and consumption.

One of the principles of green chemistry is that rather than using every possible chemical process, only those that are environmental benign should be used. Green chemistry promotes the use of renewable biomass instead of nonrenewable fossil feedstocks. Biomass is the ensemble of all products from plants (crops, algae, agricultural wastes), and the energy obtained from it is defined as renewable because it is not potentially contributing to the increment of CO_2 emissions.

In recent years the energy produced from biomass in the United States and Europe has surpassed hydroelectric energy, even if it still representing a small fraction (few %) of the total energy produced from other conventional sources.

On this basis, it can be fully understood why biomass conversion in a sustainable manner has become an important area of research (Sheldon, 2014). The full exploitation of biomass requires the development of new catalytic technologies and chemical pathways, called "biorefinery".

To illustrate the complexity of the biorefinery processes associated with biomass conversion, it is sufficient to recall that biomass valorization requires gasification to syngas, pyrolysis, or other attack strategies, all processes needing improvements in catalytic technologies. The transformation of cellulose and lignin into small molecules is still quite challenging.

Following Fechete, Wang, and Vedrine (2012), some analogies can be drawn between the petroleum refinery processes developed in the twentieth century and the modern biorefinery, a branch of catalysis science at its infancy.

It is necessary nevertheless to add that the exploitation of biomass is subjected to severe limitations, the most important being represented by the competition with the use of natural resources for food production. For this reason it is expected that in the future the energy obtained from biomass will remain a modest fraction of all consumed energy.

References

Anastas, P.T. and Warner, J.C. (1998) *Green Chemistry: Theory and Practice.* Oxford University Press, New York, p. 30.

Armor, J.N. (2011) A history of industrial catalysis. *Catalysis Today*, **163**, 3–9.

Arrhenius, S. (1903) The propagation of life in space. *Die Umschau*, **7**, 32–33.

Arrhenius, S. (2009) The spreading of life throughout the universe. Worlds in the Making, 1908, Reprinted from *Journal of Cosmology*, **1**, 91–99.

Ben-Moshe, A., Govorov, A.O., and Markovich, G. (2013) Enantioselective synthesis of intrinsically chiral mercury sulfide nanocrystals. *Angewandte Chemie International Edition*, **52**, 1275–1979.

Berzelius, J.J. (1834) Über Meteorsteine, Meteorstein von Alais. *Annual Review of Physical Chemistry*, **33**, 113–123.

Bonner, W.A., Kavasmaneck, P.R., Martin, F.S., and Flores, J.J. (1974) Asymmetric adsorption of alanine by quartz. *Science*, **186**, 143–144.

Breslow, R. (1959) On the mechanism of the formose reaction. *Tetrahedron Letters*, **1**, 22–26.

Breslow, R. (2011) The origin of homochirality in amino acids and sugars on prebiotic Earth. *Tetrahedron Letters*, **52**, 4228–4236.

Brundtland, C.G. (ed.) (1987) *Our Common Future, The World Commission on Environmental Development.* Oxford University Press, Oxford, p. 400.

Cairns-Smith, A.G. (2008) Chemistry and the missing era of evolution. *Chemistry – A European Journal*, **14**, 3830–3839.

Cairns-Smith, A.G. and Hartman, H. (1986) *Clay Minerals and the Origin of Life.* Cambridge University Press, Cambridge, UK.

Cami, J., Bernard-Salas, J., Peeters, E., and Malek, S.E. (2010) Detection of C_{60} and C_{70} in a young planetary nebula. *Science*, **329**, 1180–1182.

Cleaves, H.J., Chalmers, J.H., Lazcano, A. *et al.* (2008) A reassessment of prebiotic organic synthesis in neutral planetary atmospheres. *Origins of Life and Evolution of Biospheres*, **38**, 105–115.

Cornils, B. and Herrmann, W.A. (2003) Concepts in homogeneous catalysis: the industrial view. *Journal of Catalysis*, **216**, 23–31.

Fechete, J., Wang, Y., and Vedrine, J.C. (2012) The past, present and future of heterogeneous catalysis. *Catalysis Today*, **189**, 2–27.

Fischer, E. and Hirschberger, J. (1889) Über Mannose II. *Berichte der deutschen chemischen Gesellschaft*, **22**, 365–376.

Goeppert, A., Czaun, M., Jones, J.P. *et al.* (2014) Recycling of carbon dioxide to methanol and derived products – closing the loop. *Chemical Society Reviews*, **43**, 7995–8048.

Hazen, R.M., Filley, T.R., and Goodfriend, G.A. (2001) Selective adsorption of L- and D-amino acids on calcite: implications for biochemical homochirality. *Proceedings of the National Academy of Sciences of the United States of America*, **98**, 5487–5490.

Hein, J.E. and Blackmond, D.G. (2012) On the origin of single chirality of amino acids and sugars in biogenesis. *Accounts of Chemical Research*, **45**, 2045–2054.

Hitosugi, S., Matsumoto, A., Kaimori, Y. *et al.* (2014) Asymmetric autocatalysis initiated by finite single-wall carbon nanotube molecules with helical chirality. *Organic Letters*, **16**, 645–650.

Horvath, J.D. and Gellman, A.J. (2003) Naturally chiral surfaces. *Topics in Catalysis*, **25**, 9–15.

Jaakkola, S., Sharma, V., and Annila, A. (2008) Cause of chirality consensus. *Current Opinion in Chemical Biology*, **2**, 53–58.

Kahr, B., Bing, Y., Kaminsky, W., and Viterbo, D. (2009) Turinese Stereochemistry: Eligio Perucca's enantioselectivity and Primo Levi's asymmetry. *Angewandte Chemie International Edition*, **48**, 3744–3748.

Kauffman, S.A. (1993) *The Origins of Order*. Oxford University Press, New York.

Kelvin, W.T. (1871) On geological time. *Transactions of the Geological Society of Glasgow*, **3**, 1–28.

Lambert, J.F., Jaber, M., Georgelin, T., and Stievano, L. (2013) A comparative study of the catalysis of peptide bond formation by oxide surfaces. *Physical Chemistry Chemical Physics*, **15**, 24292–24306.

Löb, W. (1906) Studien über die chemische Wirkung der stillen elektrischen Entladung. *Zeitschrift für Elektrochemie*, **15**, 282–312.

Martra, G., Deiana, C., Sakhno, Y. *et al.* (2014) Self-assembly of long prebiotic oligomers produced by the condensation of unactivated amino acids on oxide surfaces. *Angewandte Chemie International Edition*, **53**, 4671–4674.

Miller, S.L. (1953) Production of amino acids under possible primitive earth conditions. *Science*, **117**, 528–529.

Morowitz, H.J. (1969) A mechanism for the amplification of fluctuations in racemic mixtures. *Journal of Theoretical Biology*, **25**, 491–494.

Olah, G. (2013a) Towards oil independence through renewable methanol chemistry. *Angewandte Chemie International Edition*, **52**, 104–107.

Olah, G. (2013b) The role of catalysis in replacing oil by renewable methanol using carbon dioxide capture and recycling (CCR). *Catalysis Letters*, **143**, 983.

Olah, G., Goeppert, A., and Prakash, G.K.S. (2009 (1st ed.)) *Beyond Oil and Gas: The Methanol Economy*, 2nd ed. Wiley-VCH, Weinheim.

Oparin, A.I. (1924) *The Origin of Life* (in Russian). Moscow Worker publisher, Moscow. English translation: Oparin, A.I. (1968) *The Origin and Development of Life* (NASA TTF-488). Washington, DC.

Orò, J. and Kamat, S.S. (1961, 2008) Amino-acid synthesis from hydrogen cyanide under possible primitive earth conditions. *Nature*, **190**, 442–443.

Perucca, E. (1919) Anomalous optical rotatory dispersion of NaClO3 crystals. *Nuovo Cimento*, **18**, 112–154.

Rauchfuss, H. (2008) *Chemical Evolution and the Origin of Life*. Springer-Verlag, Berlin.

Rifkin, J. (2002) *The Hydrogen Economy*. Tarcher/Putnam, New York.

Rimola, A., Sodupe, M., and Ugliengo, P. (2007) Aluminosilicate surfaces as promoters for peptide bond formation: an assessment of Bernal's hypothesis by ab initio methods. *Journal of the American Chemical Society*, **129**, 8334.

Ritson, D. and Sutherland, J.D. (2012) Prebiotic synthesis of simple sugars by photoredox systems chemistry. *Nature Chemistry*, **4**, 895.

Rufo, C.M., Moroz, Y.S., Moroz, O.V. *et al.* (2014) Short peptides self-assemble to produce catalytic amyloids. *Nature Chemistry*, advance on line publication, **6**, 303–309.

Ruiz-Mirazo, K., Briones, C., and De La Escosura, A. (2014) Prebiotic systems chemistry: new perspectives for the origins of life. *Chemical Reviews*, **114**, 285.

Sato, I., Kadowaki, K., and Soai, K. (2000) Asymmetric synthesis of an organic compound with high enantiomeric excess induced by inorganic ionic sodium chlorate. *Angewandte Chemie International Edition*, **39**, 1510.

Sato, I., Kadowaki, K., Ohgo, Y., and Soai, K. (2004) Highly enantioselective asymmetric autocatalysis induced by chiral ionic crystals of sodium chlorate and sodium bromate. *Journal of Molecular Catalysis A: Chemical*, **216**, 209.

Schlogl, R. (2015) Heterogeneous catalysis. *Angewandte Chemie International Edition*, **54**, 3465.

Schreiner, E., Nair, N.N., Wittekindt, C., and Marx, D. (2011) Peptide synthesis in aqueous environments: the role of extreme conditions and pyrite mineral surfaces on formation and hydrolysis of peptides. *Journal of the American Chemical Society*, **133**, 8216.

Sheldon, R.A. (2014) Green and sustainable manufacture of chemicals from biomass: state of the art. *Green Chemistry*, **16**, 950.

Shindo, H., Shirota, Y., Niki, K. *et al.* (2013) Asymmetric autocatalysis induced by cinnabar: observation of the enantio-selective adsorption of a 5-pyrimidyl alkanol on the crystal surface. *Angewandte Chemie International Edition*, **52**, 9135.

Shu, F.H. (1982) *The Physical Universe: An Introduction to Astronomy*. University Science Books, Sausalito, CA.

Soai, K., Osanai, S., Kadowaki, K. *et al.* (1999) d- and l-quartz-promoted highly enantioselective synthesis of a chiral organic compound. *Journal of the American Chemical Society*, **121**, 11235.

Soai, K., Sato, I., and Shibata, T. (2001) Asymmetric autocatalysis and the origin of chiral homogeneity in organic compounds. *Chemical Record*, **1**, 321.

Solomon, P.M. (1973) Interstellar molecules. *Physics Today*, **26**, 32–40.

Weber, A.L. and Pizzarello, S. (2006) The peptide-catalyzed stereospecific synthesis of tetroses: a possible model for prebiotic molecular evolution. *Proceedings of the National Academy of Sciences of the United States of America*, **103**, 12713.

Weissbuch, I., Illos, R.A., Bolbach, G., and Lahav, M. (2009) Racemic β-sheets as templates of relevance to the origin of homochirality of peptides: lessons from crystal chemistry. *Accounts of Chemical Research*, **42**, 1128.

Index

The Development of Catalysis: A History of Key Processes and Personas in Catalytic Science and Technology,
First Edition. Adriano Zecchina and Salvatore Califano.
© 2017 John Wiley & Sons, Inc. Published 2017 by John Wiley & Sons, Inc.